Bacteriophages: Practical Applications for Nature's Biocontrol

Sabah A.A. Jassim · Richard G. Limoges

Bacteriophages: Practical Applications for Nature's Biocontrol

Springer

Sabah A.A. Jassim
Applied Bio Research Inc.
Windsor, ON
Canada

Richard G. Limoges
Applied Bio Research Inc.
Windsor, ON
Canada

ISBN 978-3-319-54050-4 ISBN 978-3-319-54051-1 (eBook)
DOI 10.1007/978-3-319-54051-1

Library of Congress Control Number: 2017932425

Printed on acid-free paper

This Springer imprint is published by Springer Nature
The registered company is Springer International Publishing AG
The registered company address is: Gewerbestrasse 11, 6330 Cham, Switzerland

First and foremost, I give thanks to the Almighty God for giving me the knowledge, the strength and the direction to complete this summary of my life's work on bacteriophages. I dedicate this book to my wife Ghazal, and my three daughters: Maryam, Sarah and Amna in thanks for their unwavering support and understanding. In recognition for his dedication and assistance in articulating this work, I have named my friend Richard Limoges as my co-author. Last but not least, I share this information as guidance to researchers and students who seek a greater understanding of natural biological processes and wish to serve mankind with their knowledge and talents.

Sabah A.A. Jassim

Preface

We find ourselves in the twenty-first century with a world of disenchantment, a self-imposed return to the dark ages of medicine! Most antibiotics are unable to treat multidrug-resistant bacteria, which are causing serious diseases. Prior to the discovery of penicillin, there were fewer bacteria that caused diseases, fewer bacterial mutations, less food poisoning, less water contamination. Our own interventions have caused bacterial mutations resulting in more lethal bacteria with fewer remedies. Throughout much of the twentieth century, antibiotics have been our primary defense against bacterial diseases. The excessive and inappropriate use of antibiotics particularly in animal husbandry is at the root of this problem and threatening their efficacy. The pharmaceutical industry appears unlikely to offer the necessary countermeasures because of the objective difficulties with synthesis of new antibiotics. The inexorable rise in the incidence of antibiotic resistance in bacterial pathogens, coupled with the low rate of emergence of new, clinically useful antibiotics, have encouraged researchers to revisit the bacteriophage and the potential utility of bacteriophages in biocontrol and for preventing or treating human and animal bacterial diseases.

The proper use of lytic 'virulent' bacteriophages through dietary and environmental application shows promise in livestock and poultry in particular. Bacteriophages may also be used to enhance or rekindle the effectiveness of antibiotics in numerous applications. Bacteriophages are known to have some advantages associated with human therapy over the use of antibiotics. However, we urge caution since the mechanism that caused the spread of antibiotic resistance genes between bacteria occurs most often through lysogenic bacteriophage-mediated transduction. Inappropriate use of bacteriophages could similarly lead to bacterial development of bacteriophage resistance. Furthermore, bacteriophage proteins including those that are genetically modified for commercial purposes, may also integrate into human and animal society with unknown effect. Therefore, it would be wise to approach such methodologies with caution in order to avoid repeating mistakes that were made with the improper use of antibiotics.

We suggest the use of properly developed and highly virulent lytic bacteriophages for environmental biocontrol to selectively reduce or eliminate problematic bacteria from sensitive environments. Bacteriophages can be effective in decontamination and sanitation of both natural and manmade environments, including farms, factories, in workplaces, crowded places, and healthcare settings or in the laboratory. When strategically applied, they can be used without harmful effect on and around people and animals to eliminate harmful bacteria while supporting beneficial microflora. The ability of bacteriophages to recognize precisely their target hosts, renders them as favorable antibacterial agents compared to broad-spectrum antibiotics which kill target bacteria along with other beneficial bacteria. In this book we discuss the safe use of bacteriophages as antidotes or as a biocontrol from farm to fork and as a biodefence or to prevent biothreats while recognizing the obstacles associated with their use.

Windsor, ON, Canada
December 2016

Sabah A.A. Jassim
Richard G. Limoges

Contents

About the Authors

Professor Sabah A.A. Jassim Adjunct Professor, Civil and Environmental Engineering, University of Windsor, is CEO of Applied Bio Research Inc., Canada. His research and academic contributions span 29 years. Sabah was awarded his M. Phil and Ph.D. degrees from Nottingham University and Loughborough University, respectively, both in the UK. He has worked as a faculty research fellow and an adjunct professor at Nottingham University, UK and University of Guelph in Canada, respectively, focussing in phage biotechnologies. He was also a visiting professor at Universiti Putra Malaysia supervising a research postdoctoral team working on phage design technology. Sabah has also worked extensively in the private sector focussing on practical applications for his scientific research, especially relating to biota as well as bacteriophages and related topics.

Sabah was also once listed 13th of Power 500 the World's Most Influential Arabs/Middle East by Arabian Business Journal. Winner of several best research awards, he made trend-setting achievements to the state of the art in bacteriophage breeding and design technology to produce large-scale highly lytic phages for biocontrol systems. These include using phages for rapid bacterial detection, rapid drug susceptibility testing, biocontrol, alternative therapy, molecular detection and characterization of bacterial pathogens, control of pathogens in environmental industries, microbial bioluminescence, deletion of bacterial biofilm, bacterial stress response, controlling harmful algal blooms and novel methods in wastewater treatment.

His more recent innovative phage programming technology represents a model for smart phages to gain a high-speed infection against their counterpart bacterial pathogens which can play a significant role in decreasing bacterial pathogenic risk, preventing loss of life and reducing the use of antibiotics in animal agriculture industries. Dr. Jassim holds 18 international patents in several biological sciences including phage biotechnologies. Three of these technologies have been transferred to industrial practice. He has published extensively in prestigious journals and conferences including peer-reviewed research and review articles as well as book chapters and is a consistent leader in R&D to enhance bacteriophages infectious activity to their target bacteria. Sabah has devoted much of his research to using phages as a novel, environmentally friendly biocontrol, particularly in agricultural applications from farm to fork.

Mr. Richard G. Limoges currently a businessman, is operating two successful small businesses in Windsor, Ontario where he met Sabah several years ago. Mr. Limoges now acts as Chief Administrative Officer of Applied Bio Research, Inc., a company dedicated to the commercialization of Dr. Jassim's various innovative technologies. Rick has a long history of community service having served for 14 years (5 terms) as a Member of Windsor City Council and Chair of numerous Local Boards and Committees. He was next elected Member of Parliament, Windsor-St. Clair in Canada's 36th Parliament. Prior to his election as an M.P., Mr. Limoges worked as a Senior Manager in one of Canada's largest banks. Rick is a graduate of the University of Windsor with Honours in Business Administration, and has applied his communication skills to assist Dr. Jassim in disseminating his research and life's work into several peer-reviewed publications in scientific journals and now this book. In recognition of Rick's efforts and dedication to assisting Dr. Jassim with this work, he is honored to be named as co-author in several of Sabah's publications.

Acronyms and Abbreviations

AAP	American Academy of Pediatrics
ABHRs	Alcohol-Based Hand Rubs
AGPs	Antibiotic Growth-Promoters
AMR	Antimicrobial Resistance
ATP	Adenosine Triphosphate
BGA	Blue-Green Algae
BoNT	*Clostridium botulinum* Neurotoxin
BSL-2	Biosafety Level-2
BTA	Biothreat Alarm System
BZ	Burst Size
CA	Community-Associated
CCN	Conical Cyanophage Net
CDC	Centers for Diseases Control and Prevention
CFR	Code of Federal Regulations
CFU	Colony-Forming Units
$CHAP_K$	Cysteine- and Histidine-Dependent Amidohydrolase/Peptidase
CoNS	Coagulase-Negative Staphylococci
CoPS	Coagulase-Positive Staphylococci
CRE	Carbapenem-Resistant Enterobacteriaceae
CyanoHABs	Cyanobacterial Harmful Algal Blooms
DFPS	Dry Fog Phage System
DNA	Deoxyribose Nucleic Acid
ECDC	European Centre for Disease Prevention and Control
EFSA	European Food Safety Authority
EHEC	Enterohemorrhagic *E. coli*
ELISA	Enzyme-Linked Immunosorbent Assay
EMRSA	Epidemic MRSA
EPA	Environmental Protection Agency
ESBL	Multidrug-Resistant Extended-Spectrum β-Lactamase
ESR	Institute for Environmental Science and Research

EU	European Union
FAO	Food and Agriculture Organization
FDA	Food and Drug Administration
FIGE	Field Inversion Gel Electrophoresis
FnBPA	Fibronectin-Binding Protein A
FSEP	Food Safety Enhancement Program
FSIS	Food Safety Inspection Services
GFP	Green Fluorescent Protein
GHG	Greenhouse Gas
GMP	Good Manufacturing Practices
GRAS	Generally Recognized as Safe
HA	Hospital-Associated or Acquired
HABs	Harmful Algal Blooms
HACCP	Hazard Analysis Critical Control Point
HCW	Healthcare Worker
HGT	Horizontal Gene Transfer
HICPAC	Healthcare Infection Control Practices Advisory Committee
ICMSF	International Commission for Microbiological Safety of Foods
IDSA	Infectious Diseases Society of America
IR	Infective Ratio
LA	Luria Agar
LA-MRSA	Livestock-Associated MRSA
LB	Luria Broth
LEAD	Livestock, Environment and Development
LODs	Limits of detections
	LuxAB-PASA LuxAB-Phage Anthracis Spore Alarm
MDRB	Multidrug-Resistant Bacteria
MRSA	Methicillin-Resistant *Staphylococcus aureus*
MRSP	Methicillin-Resistant *Staphylococcus pseudintermedius*
MSSA	Methicillin-Susceptible *Staphylococcus aureus*
NDM	New Delhi Metallo-β-Lactamase
OIE	World Organisation for Animal Health
PAD	Phage Alarm and Detector
PBHR	Phage-Based Hand Rubs
PBS	Phosphate Buffered Saline
PCR	Polymerase Chain Reaction
PDA	Phage Alarm and Detector
PFGE	Pulsed-Field Gel Electrophoresis
PFU	Plaque Forming Units
PIA	Polysaccharide Intercellular Adhesion
PLPs	Phage-Like Particles
PRE	Pomegranate Rind Extract
PVL	Panton-Valentine Leukocidin
Q&Q	Qualitative and Quantitative

QRA	Quantitative Risk Assessment
RNA	Ribonucleic Acid
SE	*Salmonella enterica* serovar Enteritidis
SEs	Staphylococcal Enterotoxins
SHEA	Society for Healthcare Epidemiology of America
STEC	Shiga-Toxin producing *E. coli*
SUR	Solar Ultraviolet Radiations
UK	United Kingdom
US FDA	United States Food and Drug Administration
USA	United States of America
USDA	United States Department of Agriculture
USDA-FSIS	United States Department of Agriculture- Food Safety and Inspection Service
US-FSIS	US-Food Safety and Inspection Service
UTIs	Urinary Tract Infections
UV	Ultraviolet
VFA	Volatile Fatty Acids
WHO	World Health Organization

Chapter 1
Enhanced Bacteriophages

Abstract The bacterial cell wall is the most important part of the bacterial structure for bacteriophage attachment, which is required to initiate infection. The rapid and precise attachment of the phage onto a susceptible host cell is the first step of infection. In this chapter, methods are described to control phage-host interactions and to produce highly lytic phages with no or far less phage-resistant mutants, along with broad host targeting capabilities. These methods do not employ genetic modification to breed 're-tailored' wild phages with auxiliary mechanisms for phage adherence, adsorption, binding and uptake which are critical for plaque formation. The purpose of these tactics is to gain new sub-strains of phages that are able to infect previously resistant bacteria and to play an important role in future applications.

Keywords Bacteriophage · Phage design · Phage breeding · Phage reprogramming technology

Contents

S.A.A. Jassim and R.G. Limoges, *Bacteriophages: Practical Applications for Nature's Biocontrol*, DOI 10.1007/978-3-319-54051-1_1

1.1 Introduction

Bacteriophage (phage) interactions with bacteria can be divided into two broad life cycles based on the biological outcome of infection. The lysogenic cycle, also called 'temperate' or 'dormant' typically integrates phage DNA within the host bacteria. The prophage "endogenous phage", or latent form of phage, also called lysogens (this is discussed in greater detail in Chap. 2), has a circular form of the phage's genome, crossing over between the prophage genome and the circular bacterial genome (Jassim and Limoges 2014). This allows prophage genome to replicate along with the bacterial genome, also protecting the bacterium from further infection. The phage's genome in this state is called a prophage and during this time, the bacteria may appear to be dividing in phage-free conditions. The phage and bacterial genomes will remain integrated until the prophage is induced or activated by adverse conditions to then proceed to cause lysis (Jassim and Limoges 2014). Typical plaques made by lysogenic phage lambda are pinpoint turbid plaques, due to lysogenization of bacteria within the plaque.

The phage lytic cycle involves a host-specific parasitic relationship in which the phage uses the energy and the metabolic machinery of a bacterium for reproduction. Unlike the lysogenic cycle, they do not integrate their genetic material into the host cell chromosome and usually the host cell undergoes lysis and dies, simultaneously liberating a large number of progeny phages, which are each then ready to start another cycle by infecting the surrounding bacteria (Jassim and Limoges 2014). The lytic phage fits in the class of 'natural living antimicrobial agents' and are arguably the most abundant biological entities on the planet (Jassim and Limoges 2014). Most lytic cycles take between 30 and 40 min depending on factors such as bacterial growth conditions, target bacterial type, phage host-interactions, phage infective ratio and phage biokinetic values, such as phage binding or adsorption time, phage amplification, burst time and burst size (Jassim and Limoges 2013, 2014). Understanding these phage-host interactions is very important for the programming of successful bacterial pathogen biological controls. If lysis occurs too quickly, it will not produce enough new phages to infect neighbouring bacterial cells. If it is too slow, the phage loses an opportunity to infect new host cells for further replication, perhaps giving the bacterial cell time to adapt and become resistant to the phages. This bacterial adaptation, aided by the influence of environmental forces (Jassim and Limoges 2013), may also give rise to the lysogenic cycle, allowing phage DNA to integrate into the host DNA (see above). If extracellular phage concentrations are too high in this regard some lytic phages undergo a phenomenon known as lysis inhibition, where completed phage progeny will not immediately lyse out of the cell. This mechanism is not identical to that of temperate phage going dormant and is usually temporary.

Phages and bacteria are continually and rapidly evolving in the ecosystem, with bacteria becoming phage-resistant and phages evolving to maintain or improve infectivity of host bacteria (Jassim and Limoges 2014; Maxwell 2016). It may therefore be better to view phages as not merely parasitic on the host cells but as opportunistic biological, living agents where survival depends not only on the host cell for its propagation but is also influenced by several environmental biophysical factors including virophages, nutrients (macronutrients and micronutrients), temperature, light, solar UV radiation, pH, cation (+ve ion) concentrations, salinity, seasonal variations, microbial community and other environmental forces (Jassim and Limoges 2013, 2014). These factors can either impede or expedite phage–host interactions and the resultant viral infection and killing of the bacterial cell, which is a necessary outcome of successful phage therapies and biocontrols using lytic phages (Jassim et al. 2012; Abdulamir et al. 2014; Jassim and Limoges 2014).

Phages that are best equipped to survive and reproduce perpetuate the highest frequency of their dominant genes to descendant populations. This is the principle known colloquially as 'survival of the fittest,' where fitness denotes the overall ability to pass copies of phage genes on to successive generations within their bacterial communities. The evolutionary survival of phages is attributed to five factors: genetic variability, variety in means of transmission, efficient replication within host cells, ability to remain dormant within the host (lysogeny), and environmental or external forces (Jassim and Limoges 2014).

Based on the above concepts, phage vertical breeding protocol was developed by Jassim et al. (1995) to select and breed the best phage to target host cells. This is a method of selecting a virulent phage from a population and breeding the selected phage to create more assertive progeny. Steps include:

(i) selecting a target cell sample of the test phage;
(ii) incubating a mixture of this phage population with the target cell sample so as to effect attachment to and infection of target cells by test phage particles;
(iii) incubating the target cells to complete infection by the test phage and to cause phage progeny to be released, then recovering the phage progeny;
(iv) an additional step performed between steps (ii) and (iii) comprises neutralizing the extracellular phage with an antiviral agent in order to control the conditions of the exercise and to select only the most virulent phages.

Horizontal phage design technology (Jassim et al. 2010) was developed to address phage-host interactions and to produce highly lytic phages with no or far less phage-resistant mutants, along with broad host targeting capabilities. Both methods, phage vertical breeding (Jassim et al. 1995) and horizontal phage design (Jassim et al. 2010), do not employ genetic modification to breed "re-tailored" wild phages on the host cells in order to gain newly bred sub-strains of phages which are able to overcome the host defence mechanisms in order to infect previously resistant bacteria and to play an important role in future applications (Jassim et al. 1995, 2012; Hibma et al. 1997; Abdulamir et al. 2014; Jassim and Limoges 2014).

Recently we have developed newer methodologies to reprogram phages to possess (Fig. 1.1) (Jassim and Limoges 2014) auxiliary mechanisms for phage adherence/adsorption/binding and uptake that are critical for plaque formation, in order to gain new sub-strains of phages able to infect previously resistant host cells, namely prophage. This non-genetic approach of the technology is

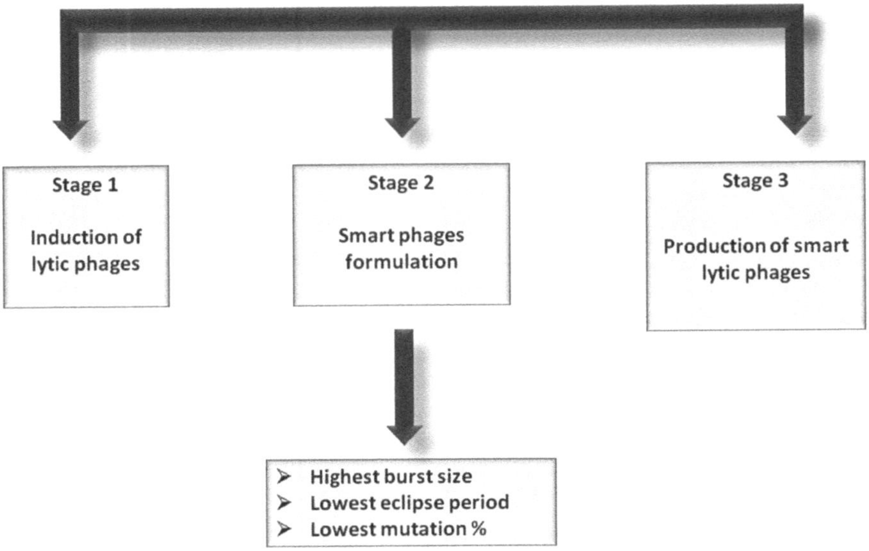

Fig. 1.1 Non-genetic phage programming technology to produce smart lytic phage. *Source* Applied Bio Research Inc.; Jassim and Limoges (2014)

environmentally-driven and so mimics natural selection or evolution of the phage by reproducing vast numbers of mixed populations of the most robust wild-type phages. Phage programming technology, illustrated in Fig. 1.1, remains a proprietary technology which permits a better selection and adaptation of robust lytic phages for each potential application. This technology is capable of converting naturally occurring wild phages including prophages to smart phages with a broader range of host specificity that can overcome a bacterium's resistive defense mechanisms. Finally, several suggestions are shared in order to give direction for overcoming common obstacles in applied phage technology.

The key to successful use of phages in modern scientific, farm, food processing and clinical applications is to understand the common obstacles as well as best practices and to develop answers that work in harmony with nature. Here, we describe various methodologies for adapting phages with enhanced virulent properties; such approaches have been extensively modified from our earlier published work with vertically bred phages (Jassim et al. 1995).

1.2 Materials

All media, solutions and buffers are prepared using water distilled from drinking water and analytical grade reagents. Prepare and store all media, solutions and buffers at room temperature (unless indicated otherwise). Diligently follow all waste disposal regulations when disposing waste materials. Work with moderately hazardous bacteria should be done in a biosafety level-2 (BSL-2) facility. A laboratory coat or gown and gloves should be worn. There should be no mouth pipetting, eating, or drinking in the laboratory and materials used in laboratory (pipettes, slides, culture vessels, etc.) as well as work surfaces should be decontaminated. In working with hospital and other environmental samples, it should be remembered that these can carry other dangerous pathogenic organisms and therefore, must be conducted in a biological safety cabinet for level 2 organisms. Laboratory infections can occur as a result of splashes and failure to wash hands after handling potentially infectious materials. It is recommended that laboratory personnel be offered the vaccine for cretin bacterial pathogens. Vaccine will not decrease the risk of infection but can reduce the potential health risk of laboratory-acquired disease.

1.2.1 Media

1. Luria broth (LB): tryptone 10 g l^{-1} (HiMedia, Mumbai, India), yeast extract 5 g l^{-1} (HiMedia, Mumbai, India), and sodium chloride 10 g l^{-1} (HiMedia, Mumbai, India) adjust at pH 7.2.
2. Luria-agar (LA) consisted of the above with the addition of 14 g l^{-1} agar (HiMedia, Mumbai, India) use for culture maintenance.
3. Bacterial dilutions from 18 h LB cultures grown at 37 °C to be carried out in phosphate buffered saline (PBS, Oxoid, UK).
4. Top layer agar: Prepare LB with Lambda-buffer supplemented with 4 g l^{-1} agar bacteriology No. 1 (No. 1 Oxoid).

1.2.2 Buffers

1. PBS: Dissolve one tablet in 100 ml of a 1X solution. 1X PBS solution contains 10 mM phosphate buffer, 137 mM sodium chloride, and 2.7 mM potassium chloride. Each tablet prepares a 1X PBS solution when dissolved in 100 ml of distilled water H_2O. Adjust the pH to 7.2 with HCl. Dispense the solution into aliquots and sterilize them by autoclaving for 20 min at 15 psi (1.05 kg/cm^2) on liquid cycle or by sterile filter membrane 0.45 μm units. Store PBS at room temperature.
2. Lambda-buffer: Prepare from 6 mmol l^{-1} Tris pH 7.2, 10 mmol l^{-1} Mg $(SO_4)_2.7H_2O$, 50 μg ml^{-1} gelatin (Oxoid, UK). Adjust the pH to 7.2 with HCl.

Dispense the buffer into aliquots and sterilize them by autoclaving for 20 min at 15 psi (1.05 kg/cm^2) on liquid cycle or by Sterile filter membrane 0.45 μm units. Store PBS at room temperature.

1.2.3 Bacteria

Phage cultures require host cells (bacteria) in which the phages multiply. The standard reference bacterial culture can be obtained from microbial culture collections. The target bacteria can also be isolated from animal farms, hospital, soil, mud, sewage disposals, etc.

1.2.4 Preparation of $FeSO_4.7H_2O$ Solution

1. First freshly prepare solution (0.01%) of $FeSO_4.7H_2O$ in Lambda-buffer, pH 6.3.
2. Sterilized by membrane filtration (0.45 μm, Whatman).

1.2.5 Preparation of 13% Pomegranate Rind Extract (PRE)

1. Blend pomegranate rind in distilled water (25% w/v) and boil for 10 min.
2. Centrifuge (20,000×*g*, 4 °C, 30 min) (see **Note 1.4.1**).
3. Autoclave supernatant (121 °C, 15 min) and cool.
4. Store at −20 °C.
5. Prior to use, mix 1.3 ml of stock solution of PRE (25% w/v) with 8.7 ml of Lambda buffer. The final PRE concentration is 13% (see **Note 1.4.2**).

1.2.6 Antiviral (Anti-phage) Agent (Jassim et al. 1995)

Immediately prior to use, transfer 3.3 ml of 13% PRE added to 7 ml of freshly prepared ferrous sulphate solution (0.01%). After about 30 s the colour of the mixture changes to greenish then to black. These mixtures of ferrous sulphate and PRE should be protected from light. The mixture is active for 45 min after preparation.

1.3 Methods

Carry out all procedures at room temperature unless otherwise stated.

1.3.1 Bacterial Stock Culture

1. Prior to procedure, a stock culture of the phage target bacteria should be maintained on LA plate (see **Note 1.4.3**).
2. Inoculate 1 loopful of the bacteria into a 100 ml Erlenmeyer flask containing 10 ml of LB (supplemented with 10 mmol l^{-1} $CaCl_2.2H_2O$) at pH 7.2 (see **Note 1.4.4**).
3. Incubate for 18 h at 37 °C in an orbital shaker [200 revolutions per minute (rev min^{-1})].
4. One ml serial dilutions prepare in LB (supplemented with 10 mmol l^{-1} $CaCl_2.2H_2O$).
5. The cell concentrations used in this study are estimated as 10^1, 10^2, 10^3, 10^4, 10^5, 10^6 and 10^7 colony-forming units per millilitre (cfu ml^{-1}).

1.3.2 Phage Hunting/Isolation Techniques

1. Collect crude specimens of approximately 50 g of environmental dirt, sewage disposal, and poultry or cattle waste in a sterile sample collection tube (100 ml).
2. Mix each specimen thoroughly and transfer 5 g of each specimen into 90 ml of LB in 250 ml Erlenmeyer flask with cotton-plug and mix using a vortex mixer for 30 s.
3. Then add 5 ml of 10 h LB cultures of the target bacterial isolate or reference strains and mix by a vortex mixer for 30 s then incubate at 37 °C in an orbital shaker (50 rev min^{-1}).
4. After 18 h, withdraw 20 ml of the mixture into a sterile 50 ml test tube, cell debris is pelleted by centrifugation at $5000\times g$ at room temperature for 5 min.
5. Divide the supernatant into two portions (2×10 ml). Transfer each 10 ml portion into 2 sterile 15 ml test tubes:

   (i) Add to one portion of 10 ml supernatant, 1 ml of chloroform (Sigma, USA) with gentle shaking of tubes for 5 min then incubate on crushed ice for 5 min. A milky solution appears due to bacterial protein digestion by chloroform. Bacterial cell debris and bacterial ghosts are discarded by centrifugation at $5000\times g$ for 5 min at room temperature. Collect the top aqueous supernatant into 15 ml sterile tube and store at 4 °C as a possible phage solution and add 5 μl of chloroform to prevent microbial contamination.

(ii) Filter the second portion of 10 ml supernatant via sterile membrane filters 25 mm diameter 0.45 nm pore size (Millipore) then by 25 mm diameter 0.25 nm pore size (Millipore). Collect the filtrate solution into a 15 ml sterile tube and store at 4 °C as a possible phage solution and add 5 μl of chloroform to prevent microbial contamination.

1.3.3 Testing for the Presence of Crude Wild Phages (Phage Spot Lysis Test)

1. Prepare bacterial lawns of target bacterial isolates or reference bacterial strains by adding 500 μl of LB 18 h cultures on LA plates, allowing the liquid bacterial culture to soak into the LA, with a half lid cover on the plate, at room temperature or incubator at 37 °C for 20 min (see **Note 1.4.5**).
2. Transfer 10 μl of the possible phage solution on the bacterial lawns and then incubate at 37 °C.
3. Check for plaques or lysis spots which are observed after 6–18 h. The detection of phage presence is based on visual appearance of a lysis zone at the site where the 10 μl solution was added onto the surface of the target bacterial lawn. Positive results are expressed by either clear or semi-clear (turbid) lysis zones, while negative results are expressed by the absence of such lysis zones.

1.3.4 Production of the Transient Phage Stock (Aldoori et al. 2015)

1. Cut 5–10 plaques with a sterile Pasteur pipette and place them in 1.5 ml sterile Eppendorf micro-centrifuge tubes (polypropylene; 1.5 ml; Sarstedt).
2. Add into 300 μl of Lambda buffer and incubate for 30 min at room temperature with intermittent gentle shaking every 10 min.
3. Add 1:10 chloroform:lysate ratio with gentle shaking for 10 min at room temperature in order to elute the phages from the agar to lyse the bacterial cells.
4. Incubate the mixture for 3 min in crushed ice.
5. Cell debris is removed by centrifugation at 5000×*g* for 15 min at room temperature and the supernatant is transferred to a 1.5 ml sterile Eppendorf micro-centrifuge tube.

1.3.5 *Optimization of the Phages Lytic Characteristics (Aldoori et al. 2015)*

1.3.5.1 Plaque-Based Optimization

The isolates of wild lytic phages from the transient stocks are propagated with the corresponding host bacterial isolates and the representative standard reference bacterial strains using the plate method as follows:

1. Prepare 10-folds serial dilutions (10^{-1}–10^{-7}) with Lambda buffer for the phage stock solutions by taking 100 μl of the phage solution into 900 μl of lambda buffer.
2. Transfer of 100 μl of each dilution for each phage stock solution into 15 ml volume sterile plastic containers containing 100 μl of 10^9 cfu ml^{-1} of 18 h bacterial host LB culture and incubate at 37 °C.
3. After 10 min incubation add 2.5 ml of top layer agar cooled to 45 °C and poured over LA plates and incubate at 37 °C for 18–24 h.
4. Plaques morphology and growth characteristics are recorded according to the following parameters:

   (i) plaque visible time,
   (ii) clarity or turbidity of the plaque,
   (iii) diameter (mm) of the plaque,
   (iv) shape of the plaque,
   (v) depth of the plaque,
   (vi) margin cut.

1.3.5.2 Phage Enhancement of Lytic Activity

1. Select the best plaques according to the above categories in Sect. 1.3.5.
2. Repeat step in Sect. 1.3.5.1 several times in order to magnify the outcome of the biased selection of the larger and clearer plaques each time, until obtaining the largest and clearest plaques, which represent the upper limits of plaque-based optimization, reflecting the best possible enhancement of the lytic characteristics of the phages.

1.3.6 *Vertical Optimization for Phage-Host Interaction*

Use the phage breeding protocol, modified from our earlier published data (Jassim et al. 1995) to increase the vertical biokinetic interaction of the phages resulting from the steps in Sect. 1.3.5.2, for their host cell. In brief, this process will increase

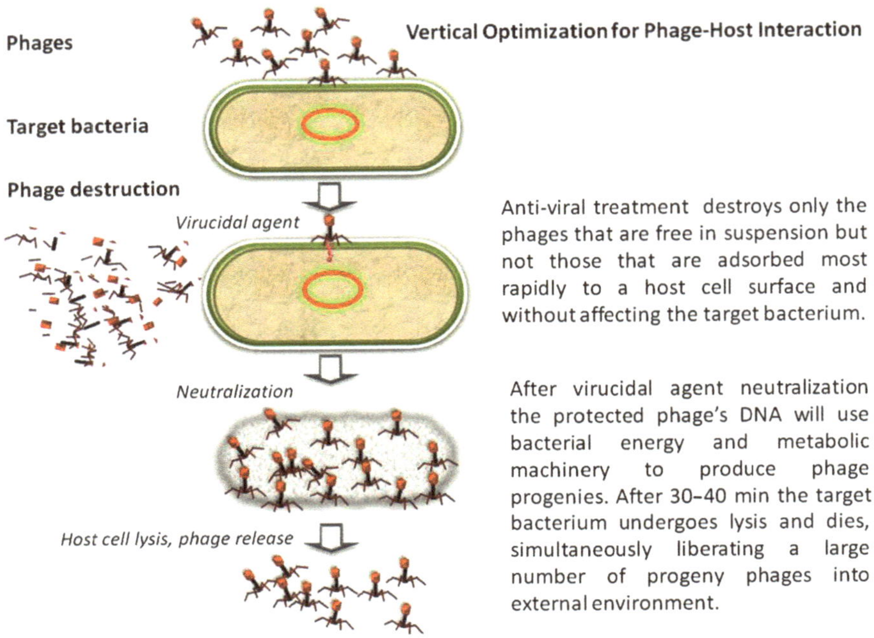

Fig. 1.2 Principle of vertical optimization for phage-host interaction

the virulence of the isolated phages within their bordering host population (Fig. 1.2).

This process uses an antiviral compound derived from pomegranate extract mixed with ferrous sulphate, which destroys phages that are free in suspension but not those that are bound to a host cell surface. When added to broth containing both phage and host bacteria, the process will select those phages that adsorb most rapidly, as the unbound or slower phages will be destroyed. The progeny from the above infections will be passed through repeated rounds of this breeding process to produce the ultimate surviving progeny phage that has demonstrated enhanced virulence.

1.3.6.1 Phage Mass-Action Biokinetics to Select the Elite Phages

1. Place 200 μl of an appropriate Lambda dilution of bacterial host cell 10^5 cfu ml^{-1} in a sterile Eppendorf micro-centrifuge tube.
2. Add phage (see Sect. 1.3.5.2) 200 μl of the highest phage titer (10^9–10^{12} pfu ml^{-1}) (where pfu represents plaque forming units per ml) in Lambda buffer. Mix gently to avoid causing any bubbles.

3. At appropriate contact time 2, 5, 7, 10, and 15 min transfer 20 μl from phage-bacteria mixture into a sterile Eppendorf micro-centrifuge tube.
4. Add 150 μl of the PRE/$FeSO_4$ solution (Antiviral Agent see Sect. 1.2.6) and mix gently to avoid causing any bubbles.
5. Incubate for 5 min at room temperature.
6. The activity of the virucidal agent is neutralized by adding an equal volume (150 μl) of 2% (v/v) Tween-80 (Sigma Chemical Co., St Louis, MO, USA) in Lambda buffer.
7. After 30 s transfer the mixture into 15 ml sterile tube and add 680 μl of 10^9 cfu ml^{-1} of 18 h bacterial host LB culture and mix.
8. Immediately add 2.5 ml of top layer agar cooled to 45 °C and pour over LA plates and incubate at 37 °C for 18–24 h.
9. Plaques morphology and growth characteristics are recorded according to the following parameters:

   (i) plaque visible time,
   (ii) clarity of the plaque,
   (iii) diameter (mm) of the plaque,
   (iv) shape of the plaque,
   (v) depth of the plaque,
   (vi) margin cut.

1.3.6.2 Propagation of the Elite Phages

1. Select the best plaques according to the morphology and growth characteristics above in Sect. 3.6.1.
2. Cut the plaques using a sterile Pasteur pipette and place them in 1.5 ml sterile Eppendorf micro-centrifuge tube. Each plaque will contain approximately 10^7 pfu ml^1.
3. Add into 300 μl of Lambda buffer and incubate for 30 min at room temperature with intermittent gentle shaking every 10 min.
4. Add 1:10 chloroform:lysate ratio with gentle shaking for 10 min at room temperature in order to elute the phages from the agar and to lyse the bacterial cells.
5. Incubate the mixture for 3 min in crushed ice.
6. Centrifuge at 5000×*g* for 15 min at room temperature and transfer the supernatant into a 1.5 ml sterile Eppendorf micro-centrifuge tube and store at 4 °C as a phage stock solution.

1.3.6.3 Phage Titer

1. Make 10-fold serial dilutions of the selected phage in lambda buffer.

10^{-1} dilution is 10 μl of selected phage particles from the phage stock solution into 990 μl of lambda buffer; mix by pipetting up and down,
10^{-2} dilution is 10 μl of 10^{-1} dilution into 990 μl of lambda buffer; mix as above.
10^{-3} dilution is 10 μl of 10^{-2} dilution into 990 μl lambda buffer; mix as above.
10^{-4} dilution is 10 μl of 10^{-3} dilution into 990 μl lambda buffer; mix as above.
10^{-5} dilution is 10 μl of 10^{-4} dilution into 990 μl lambda buffer; mix as above.
10^{-6} dilution is 10 μl of 10^{-5} dilution into 990 μl lambda buffer; mix as above.
10^{-7} dilution is 10 μl of 10^{-6} dilution into 990 μl lambda buffer; mix as above.

2. Transfer 100 μl of each dilution for each phage stock solution into a 15 ml volume sterile plastic container contain 100 μl of 10^9 cfu ml^{-1} of 18 h bacterial host LB culture and incubate for 15 min at 37 °C.
3. Add 2.5 ml of melted supplemented (10 mmol l^{-1} $MgSO_4$) LB top agar (cooled to ~48 °C). Vortex gently and pour onto pre-warmed (37 °C) LA plates. Allow the top agar to solidify and then incubate overnight at 37 °C for 18 h.
4. Count the plaques and determine the titer (pfu ml^{-1}) and packaging efficiency (See sample calculations).

1.3.6.4 Phage Titer Calculations

If there were 35 plaques on a 10^{-7} dilution plate, then the phage titer, pfu ml^{-1}, of this reaction would be:

$$\frac{(\#\text{ of plaques})\,(\text{dilution factor})\,(1000\,\mu l/ml)}{(\text{volume of phage plated }(\mu l))}$$

$$\text{For this example}: \frac{(35\text{ pfu})\,(10^7)\,(1000\,\mu l/ml)}{(100)} = 3.5\times 10^9\,\text{pfu ml}^{-1}$$

1.3.6.5 Improve Phage Adsorption 'Attachment' Time

To optimize the time for the encounter between bacterial hosts and their specific phages giving the highest number of phage particles at the pre-burst era or yielding the highest infective ratio.

1. Place 200 μl of an appropriate Lambda dilution of bacterial host cell 10^5 cfu ml^{-1} in a sterile Eppendorf micro-centrifuge tube.
2. Add 200 μl of the selected elite (see Sect. 1.3.6.2) from the highest phage titer in Lambda buffer. Mix gently to avoid causing any bubbles.
3. At phage-host contact time intervals 1, 2, 3, 4, and 5 min transfer 20 μl from phage-bacteria mixture into a sterile Eppendorf micro-centrifuge tube.

4. Add 150 µl of the PRE/$FeSO_4$ solution (Antiviral Agent see Sect. 1.2.6) and mix gently to avoid causing any bubbles then incubate at room temperature for 5 min.
5. The activity of the virucidal agent is neutralized by adding an equal volume (150 µl) of 2% (v/v) Tween-80 (Sigma Chemical Co., St Louis, MO, USA) in Lambda buffer.
6. After 30 s transfer the mixture into 15 ml sterile tube and add 680 µl of 10^9 cfu ml^{-1} of 18 h bacterial host LB culture and mix.
7. Immediately add 2.5 ml of top layer agar cooled to 45 °C and pour over LA plates then incubate at 37 °C for 18 h.
8. Plaques morphology, growth characteristics are recorded as above.
9. Repeat Sects. 1.3.6.2, 1.3.6.3 and 1.3.6.4.

1.3.6.6 Infective Ratio (IR)

The ratio between the number of phage particles at the pre-burst era and the number of the bacterial hosts used in the assay is the infective ratio (IR).

1. Add 200 µl of an appropriate Lambda dilution of bacterial host cells 10^5 cfu ml^{-1} in sterile Eppendorf micro-centrifuge tubes.
2. Add 200 µl of the selected elite phage (according to Sects. 1.3.6.2 and 1.3.6.5) at titer 10^5, 10^6, or 10^7 pfu ml^{-1} in Lambda buffer to each tube containing the host cells. Mix gently to avoid causing any bubbles.
3. After the contact time (see Sect. 1.3.6.5), for example 3 min, transfer 20 µl from each phage-bacteria mixture into a sterile Eppendorf micro-centrifuge tube.
4. Add 150 µl of the PRE/$FeSO_4$ solution (Antiviral Agent see Sect. 1.2.6) and mix gently to avoid causing any bubbles then incubate at room temperature for 5 min.
5. The activity of the virucidal agent is neutralized by adding an equal volume (150 µl) of 2% (v/v) Tween-80 (Sigma Chemical Co., St Louis, MO, USA) in Lambda buffer.
6. After 30 s transfer the mixture into 15 ml sterile tube and add 680 µl of 10^9 cfu ml^{-1} of 18 h bacterial host LB culture and mix.
7. Immediately add 2.5 ml of top layer agar cooled to 45 °C and pour over LA plates and incubate at 37 °C for 18 h.
8. Phage titer calculations (Sect. 1.3.6.4).
9. Calculate the IR

$$= \frac{\#\text{ of phage particles in the pre} - \text{burst era at a given dilution}}{\#\text{ of the bacterial hosts used in the assay at the same given dilution}}$$

$$\text{or}: \frac{6\text{ pfu }10^6\text{ ml}^{-1}}{\text{cfu }10^5\text{ ml}^{-1}} = 60$$

From this arithmetic: The closer the number of plaques in the pre-burst era to the bacterial titer used, the higher the IR.

1.3.6.7 Determination of Phage Burst Time and Phage Burst Size (BZ)

1. Add 50 μl of an appropriate Lambda dilution of bacterial host cells 10^9 cfu ml^{-1} in a sterile Eppendorf micro-centrifuge tube.
2. Add 50 μl of the selected elite phage (according to Sects. 1.3.6.2, 1.3.6.5 and 1.3.6.6) at titer 10^{11} pfu ml^{-1} in Lambda buffer to the tube containing the host cells. Mix gently by pipetting up and down to avoid causing any bubbles.
3. Allow 5 min contact time at room temperature or at an appropriate temperature depending on bacterial strain and species in use.
4. Add 200 μl of the PRE/$FeSO_4$ solution (Antiviral Agent see Sect. 1.2.6) and mix gently to avoid causing any bubbles then incubate at room temperature for 5 min. This process will select those phages that adsorb most rapidly, as the unbound phages will be destroyed by the solution.
5. The activity of the virucidal agent is neutralized by adding an equal volume to antiviral agent (200 μl) of 2% (v/v) Tween-80 (Sigma Chemical Co., St Louis, MO, USA) in Lambda buffer.
6. Immediately add 500 μl of LB supplemented with 10 mmol l^{-1} $CaCl_2.2H_2O$ and 10 mmol l^{-1} $MgSO_4$ at pH 7.2 to make it up to 1 ml (see **Note 1.4.6**).
7. At the following timely intervals of 5 min (this is the actual time the phage and bacteria take to pass the neutralization step), 10, 15, 20, 25, 30, 35, 40, 45, 50, 55, 60, 65, 70 min transfer 10 μl of the above phage-bacteria mixtures in sterile Eppendorf micro-centrifuge tubes containing 990 μl Lambda buffer and conduct 10-fold serial dilution as the above (Sect. 1.3.6.3) (see **Note 1.4.7**).
8. Spot 10 μl on of each dilution onto the surface of the appropriate bacterial lawn of LA plates and incubate the plates at 37 °C for 18–48 h depending on bacterial strain species used in the experiment.
9. Count the plaques of the new phage progenies.
10. Calculate the BZ

$$= \frac{\text{Maximum yield pfu at a given time past the neutralization step}}{\#\,\text{bacterial hosts (assuming this also represents the initial phage}\,\#\,\text{that infected the bacterial host}\,\#)}$$

The highest value of the phage yield per single host cell at a given time past the neutralization step will determine the burst size and the lysis time, respectively (see **Note 1.4.8**).

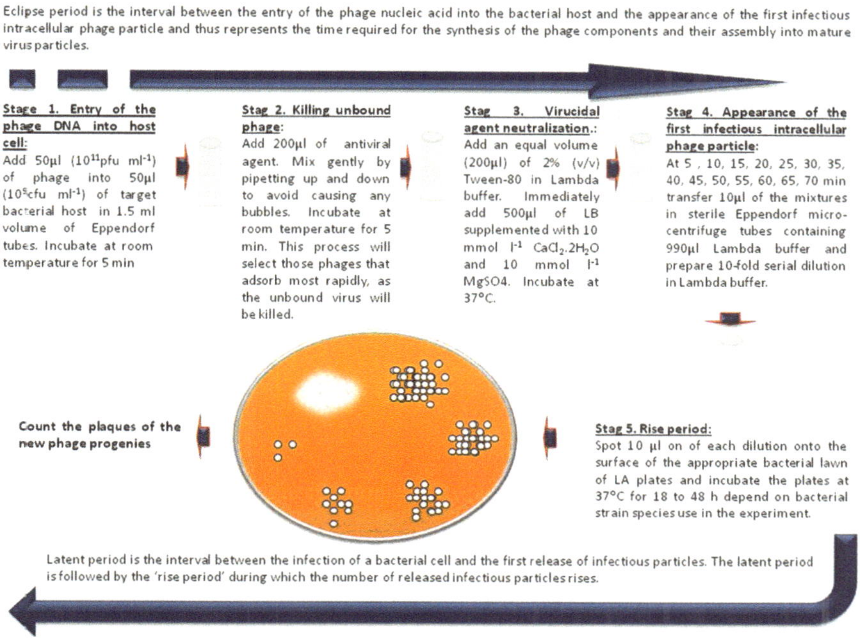

Fig. 1.3 Production of smart phages via vertical optimization (see **Note 1.4.9**)

1.3.6.8 Formation of the Optimized Definitive Vertical Phage-Bred Stock

The elite phages from the vertical-breeding assay above are propagated on their appropriate host strains by a plate lysis procedure, essentially equivalent to growing phage Lambda-derived vectors (Ausubel et al. 1991). Briefly, preparation of a large volume of the optimized phages is conducted by using the soft layer plaque technique as follows:

1. An aliquot (100 µl) of the phage sample (10-fold serially diluted with lambda-buffer) is mixed with 100 µl of an overnight LB culture of target bacteria in a sterile Eppendorf micro-centrifuge tube (polypropylene; 1.5 ml; Sarstedt) and incubate for 10 min at 37 °C to facilitate attachment of the phage to the host cells.
2. Add 2.3 ml of molten top layer agar approximately 45 °C. The contents of each bottle are then well mixed by gentle rolling and pour onto plates of LA and allow setting for 15 min at room temperature.
3. Incubate inverted plates at 37 °C for 18 h.
4. Plates showing almost confluent plaques are used to prepare a concentrated phage suspension by overlaying with 5 ml of lambda-buffer (titer 10^{10-12} ml^{-1}) (see **Note 1.4.10**).

5. Transfer the lysate into sterile 50 ml centrifuge tubes (polypropylene; Sarstedt).
6. Add 1:10 chloroform to lysate ratio with gentle shaking for 10 min at room temperature to lyse the bacterial cells and separate the phage from the bacterial cells and agar (see **Note 1.4.11**).
6. Incubate the mixture for 10 min in crushed ice.
7. Cell debris is removed by centrifugation at 5000×*g* for 15 min at room temperature.
8. Collect the top aqueous supernatant into 15 ml sterile tube (polypropylene; Sarstedt) and add 10 µl of chloroform to prevent microbial contamination.
7. The phage stocks are then maintained in lambda-buffer at 4 °C.

1.4 Notes

1.4.1 Optional: A further purification of the pomegranate extract can be achieved by membrane filtration through a CentriconTM-10 (Amicon, Danvers, MA, USA) and the filtrate stored as above (2.4 Preparation of Antiviral Agent).

1.4.2 Other plant extracts which can be prepared in the same way as PRE and used successfully in the method described in 2.6 are extracts of *Viburnum plicatum* leaves or flowers, maple leaves and commercial tea leaves.

1.4.3 The cell wall composition is one of the most important factors providing attachment sites for phages. For example, phages-teichoic acids attach to the outer surface of the wall which is like a landing pad for viruses that infect bacteria. The selection of media for each bacterial species should be considered carefully before starting the protocol of training phages for enhanced properties, as these play a major roll by providing optimum nutrient and growth environments for the target host to create a suitable phage-host interaction.

1.4.4 The basic requirement of all phage culture media is that these should include the source of energy, the source of carbon, and the source of trace and major elements. The pH, growth temperature and oxygen tensions are further requirements that should be considered to suit the needs and limits of the desired phage amplification. For many phages the optimum pH range is 6.5–7.5 and optimum growth temperature is about 25–37 °C.

1.4.5 Culture cells are healthy and grow rapidly therefore, to prevent bacterial growth and too much thickening of the bacterial lawn, the plate should be used within 1 h at room temperature.

1.4.6 To permit the desired phage amplifications to provide rapid phage lytic activity, consider supplementing the general phage media for lyset with additional phage amplification factors such as Ca and Mg.

1.4.7 If the doubling generation time of the target bacterium is longer, such as in *Mycobacterium smegmatis*, the time intervals can be extended by up to 240 min.

1.4.8 The interpretation of the optimizations for vertical phage-host interaction classifies into two eras; the pre-burst era and post-burst-era. In the pre-burst era, the number of the pfu or plaques that infect host cells is equal to the number of the bacteria used in the test for the given dilution. In this era, each plaque was formed by lysis of one bacterial cell releasing a high number of phage progenies in situ, leading to formation of a plaque. Each bacterial cell shelters or harbours a certain number of replicating phages which will then form a plaque. The period after the burst is considered as the post-burst era. In this era, each plaque represents a new phage progeny that released after an appropriate time of incubation in the tube before spotting onto the lawn. Hence in this assay, plaques represent the following according to the pre- or post-era of the assay:

1. We address the challenges in 3 steps: (1) estimating the equal number of phage that have infected bacterial cells, (2) estimating the lysis time, and (3) estimating the burst size.
2. Phage binding time (PBT) or adherence/adsorption time: The time for the encounter between bacterial hosts and their specific phages that gives the highest number of phage particles in the pre-burst era or yields the highest infective ratio.
3. Infective ratio (IR): Is the ratio between the number of phage particles in the pre-burst era and the number of bacterial hosts used in the assay. IR = No. of phage particles in the pre-burst era at a given dilution/No. of bacterial hosts used in the assay at the same given dilution. The closer number of plaques in the pre-burst era to the bacterial titer used, the higher the IR.
4. Burst 'lysis' time (BT): Is the time measured before a sharp increase is observed in the number of the progeny phage particles, compared to the number of bacteria used for the given dilution. In other words, it is the time when the new phage progenies became responsible for the formation of plaques rather than their infected host cells.
5. Burst size (BZ): Is the average yield of progeny phage per infected bacterial cell.
6. Eclipse period: The time between infection of the host cell by a phage and the appearance of mature virus within the cell; an interval of time during which viral infectivity cannot be recovered.
7. Latent period: Is the period, phase, or elapsed time between the infection of a bacterium and the release of newly resulting phage upon lysis of the bacterial cell. The latent period encompasses the active phage infection of an individual bacterium. At higher smart phage density e.g., $\geq 10^9$ pfu ml^{-1} versus 10^7 cfu ml^{-1} the phage populations can grow more rapidly simply because phage can find bacteria faster. Higher smart phage densities, therefore, should rapidly infect their target host cell displaying shorter latent periods, while un-bred phage are less effective and thus even with higher phage densities and higher bacterial densities, will display longer latent periods. This suggests and

we have consistently found that higher smart phage densities ($\geq 10^9$ pfu ml^{-1}) display shorter latent periods with greater burst sizes.

1.4.9 Figure 1.3 illustrates vertical phage breeding for optimization and biokinetic interaction of the phage values. This protocol can illustrate that the increments in the IR, BS, and plaque size values after the optimization can be correlated positively with each other. This technique can demonstrate the consistency of phage optimization with three parameters of the phage lytic cycle optimized similarly and can be correlated in a significant linear pattern.

1.4.10 To increase phage titer, it is recommended to recycle the 5 ml of lambda-buffer washing on many plates showing almost confluent plaques.

1.4.11 Some phages are sensitive to chloroform; in this case it is recommended to Filter the supernatant using sterile membrane filters 25 mm diameter 0.45 nm pore size (Millipore) then by 25 mm diameter 0.25 nm pore size (Millipore). Collect the filtrate solution into 15 ml sterile tubes (polypropylene; Sarstedt) and add 10 μl of chloroform to prevent microbial contamination.

References

Abdulamir AS, Jassim SAA, Abu Bakar F (2014) Novel approach of using a cocktail of designed bacteriophages against gut pathogenic *E. coli* for bacterial load biocontrol. Ann Clin Microbiol Antimicrob 13:39. http://www.ann-clinmicrob.com/content/pdf/s12941-014-0039-z.pdf

Aldoori AA, Mahdii EF, Abbas AK, Jassim SAA (2015) Bacteriophage biocontrol rescues mice bacteremic of clinically isolated mastitis from dairy cows associated with methicillin-resistant *Staphyloccocus aureus*. Adv Microbiol 5:383–403. http://www.scirp.org/journal/PaperInformation.aspx?PaperID=56931

Ausubel FM, Brent R, Kingstone RE et al (1991) Growing lambda-derived vectors. In: Asdf A (ed) Current protocols in molecular biology, vol 1. Wiley Interscience, New York, pp 12.1–12.3

Hibma AM, Jassim SAA, Griffiths MW (1997) Infection and removal of L-forms of *Listeria monocytogenes* with bred bacteriophage. Int J Food Microbiol 34:197–207

Jassim SAA, Limoges RG (2013) Impact of external forces on cyanophage—host interactions in aquatic ecosystems. World J Microbiol Biotechnol 29:1751–1762

Jassim SAA, Limoges RG (2014) Natural solution to antibiotic resistance: bacteriophages 'The Living Drugs'. World J Microbiol Biotechnol 30:2153–2170

Jassim SAA, Denyer SP, Stewart GSAB (1995) Selective virus culture. WO/1995/023848. http://patentscope.wipo.int/search/en/WO1995023848

Jassim SAA, Abdulamir AS, Abu Bakar F (2010) Methods for bacteriophage design. WO/2010/064044. http://www.wipo.int/pctdb/en/wo.jsp?WO=2010064044

Jassim SAA, Abdulamir AS, Abu Bakar F (2012) Novel phage-based bio-processing of pathogenic *Escherichia coli* and its biofilms. World J Microbiol Biotechnol 28:47–60

Maxwell KL (2016) Phages fight back: inactivation of the CRISPR-Cas bacterial immune system by anti-CRISPR proteins. PLoS Pathog 12(1):e1005282. doi:10.1371/journal.ppat.1005282

Chapter 2
Bacteriophage and Antimicrobial Resistance

Abstract The antibiotic resistance crisis is considered one of the world's major health threats. The use of antibiotic growth-promoters (AGPs) in animal agriculture over the last 70 years has been implicated in the development of antimicrobial resistance (AMR) in bacteria that cause common infections. Many studies and researchers have proposed bacteriophages as alternatives for AGPs to help maintain current animal production. The mechanisms that caused the spread of antibiotic resistance genes between bacteria occur through the gene transfer process of plasmid mediated conjugation and by phage-mediated transduction. Many studies have confirmed that phages have contributed to horizontal gene transfer (HGT) of AMR genes and virulence factors from other bacteria and would also integrate into human and animal biomes with unknown effect. This chapter will explore the complicated relationship between animal environmental factors and describe the significant role bacteriophages have played in development of AMR via HGT. The purpose of this chapter is to provide an updated overview on the use of antimicrobial agents in livestock, the pitfalls of using AGPs, and the role of bacteriophages in horizontal gene transfer between animals and humans. We wish to further expand current knowledge on the effects to both human health and animal production of using bacteriophages in animal therapy or environmental biocontrol to reduce AMR in the livestock reservoir. We also describe factors that will need consideration should we seek to overcome these obstacles in order to employ bacteriophages as an alternative or supplement to antibiotics in various applications.

Keywords Antibiotics · Antibiotic growth-promoters · Antimicrobial resistance · Animal farms · Bacteriophage · Horizontal gene transfer · Livestock · Mobile genetic elements · Phage therapy · Poultry · Transduction

Contents

S.A.A. Jassim and R.G. Limoges, *Bacteriophages: Practical Applications for Nature's Biocontrol*, DOI 10.1007/978-3-319-54051-1_2

2.1 Introduction

Industrial agriculture depends heavily on the widespread use of antimicrobials to improve animal health, welfare and productivity. Antibiotics (antimicrobials at large) are also the most successful family of drugs for improving human health. The probability of premature death in humans due to infection would be 40% higher if antimicrobials were non-existent (WHO 2000). It is important to remark that several antibiotics are produced by environmental microorganisms (Waksman and Woodruff 1940), these are used to control and treat diseases as well as in sub-therapeutic doses in animal feed, to promote growth and improve production of animal products (Moore et al. 1946; Jukes et al. 1950). The last 70 years represent a period of drug innovation and implementation in human and animal health as well as in agriculture. Many antibiotic classes were produced and have been used at different periods in the life cycle of poultry, cattle, and swine including: arsenicals, polypeptides, glycolipids, tetracyclines, elfamycins, macrolides, lincosamides, polyethers, beta-lactams, quinoxalines, streptogramins, and sulfonamides (Sarmah et al. 2006). Antibiotics have been a primary defense against bacterial disease in poultry and livestock and have been used in four broad categories:

- Therapeutics: this type of treatment is given to an animal with a diagnosed illness
- Metaphylactics: the presence of clinical illness in one animal triggers drug treatment of the whole herd or flock
- Prophylactics: in the absence of clinical illness, treatment is for prevention of symptoms since both livestock and poultry share water and feed troughs and are in close contact with one another by licking, laying on each other and even rubbing snouts and noses, thus illnesses can spread rapidly. Antibiotics are used to prevent diseases at times when livestock are particularly at risk, like during weaning from the mother
- Growth promotion: antibiotics administered to food animals to enhance their growth rate and production performance.

These treatments were each associated to the emergence of bacteria resistant to antibiotics (Wright 2007; Livermore 2009).

2.2 The Problem

An antibiotic is a substance that kills or prevents bacterial growth. There are hundreds of naturally occurring antibiotics, but only a few are useful in treating conditions in animal agriculture. Bacterial infections with antimicrobial resistance (AMR) are spreading faster than the introduction of new compounds into clinical practice, causing a public health crisis. Synthetic approaches to producing antibiotics have been unable to replace this platform (Ling et al. 2015). AMR is a natural phenomenon that predates the modern selective pressure of clinical antibiotic use (D'Costa et al. 2011). AMR in humans is inter-linked with AMR in other populations, especially farm animals and in the wider environment (Woolhouse et al. 2015). Several studies suggest that antibiotic resistance found in clinical settings is intimately associated with the same mechanisms as those found in the environment (D'Costa et al. 2007). In fact, the animal farms' environment is continually exposed to a wide variety of antibiotics, cleaning agents, detergents, etc., while their metabolites are shared through wastewater treatment plant discharges, agricultural runoff and animal feeding operations. The presence of bacteriophages (or, more simply, phages), in addition to the impact of climate change and increased solar ultraviolet radiation have also contributed to the emergence and spread of antibiotic resistance genes (ARGs) (Jassim and Limoges 2013, 2014; Woolhouse et al. 2015). Moreover, the large-scale mixing of environmental bacteria with exogenous bacteria from anthropogenic sources provides the ideal selective and ecological conditions for the emergence of resistant bacteria (Wellington et al. 2013).

The use of antibiotics for growth promotion is a particularly controversial issue. First described by Moore et al. (1946) when it was observed that chicks improve in growth when fed bacterial shells of *Streptomyces aureofaciens* from which antibiotics had been extracted. In the mid-1950s, when it was discovered that small, 'sub-therapeutic' quantities (1/10 to 1/100 the amount of a therapeutic dose) of antibiotics such as procaine penicillin (a slow-acting antibiotic made from a salt of procaine and a form of penicillin) and tetracycline delivered to animals in feed, could enhance the feed-to-weight ratio for poultry, swine, and beef cattle (Stokestad and Jukes 1950). Non-therapeutic antibiotics have been used extensively on farms since the 1950s and have been reported to increase weight gain by up to 15–20%, a very significant effect (Stahly et al. 1980). The mechanism underlying the growth promotion remains uncertain (Cho et al. 2012); it works for antibacterial but not antifungal or antiviral additives and it works for a variety of animal species including human children (Gough et al. 2014).

Because the dosage of antibiotic that can provide growth enhancement was extremely small, the effect was regarded as largely nutritional by producers and authorities in the food industry (Levy 2002). These pharmaceutical products are known as 'growth promoters' and these substances, as the name indicates, are antibiotics that are used in feed continuously at a low 'sub-therapeutic' level to improve growth and feed conversion. Unfortunately, many antimicrobial agents used for growth promotion are also used for disease prevention, including antibiotic classes ranked by the FDA as critically or highly important to human medicine, such as macrolides, streptogramins, and tetracyclines (United States FDA 2012) whereas this other use has been implicated in emerging AMR bacterial isolates in animals (Da Costa et al. 2013). In order to understand how and why this has occurred, the following example can shed some light on the problem. In 2012 approximately 80% of the overall tonnage of antimicrobial agents sold in the US was for animal use, and approximately 60% of those agents are considered important for human medicine (Paulson and Zaoutis 2015). From the above figures, around 94% of antibiotics were intended to be delivered through animal feed or water, whereas, only 4% of antibiotic drugs sold were intended to be administered by injection for animal therapeutic treatment (US FDA 2012). The administration of an antibiotic via feed or water to an entire flock or herd by farmers has given them less control over the dosage consumed by individual animals (WHO 2012), which has also led AMR to evolve (Emborg et al. 2003; Inglis et al. 2005; Diarra et al. 2007; Aarestrup et al. 2008; Alexander et al. 2008; Silbergeld et al. 2008; Varga et al. 2009; Vieira et al. 2009; Davis et al. 2011; Hammerum 2012; WHO 2012). Furthermore, many researchers have attributed the emergence of AMR on animal farms to low doses of antimicrobial agents given to healthy animals over prolonged periods to promote growth and increase feed efficiency. These antibiotic growth-promoters (AGPs) (Paulson and Zaoutis 2015), are believed to suppress sensitive populations of bacteria in the intestines which compete for food with their host. For example, it has been estimated that as much as 6% of the net energy in the pig diet could be lost due to microbial fermentation in the intestine (Jensen 1998). If the microbial population is better controlled, it is possible that the lost energy could be diverted to growth. Yet the exact modes of action of AGPs are not fully understood as they are probably multi-factorial (Gaskins et al. 2002; Dibner and Richards 2005; Niewold 2007), which justified the link with the increasing incidence of antibiotic resistance among bacterial pathogens (Gyles 2008; Prescott 2008) including bacteria from healthy animals (Yan and Gilbert 2004; Persoons et al. 2010).

In recent years, emerging issues related to methicillin-resistant *Staphylococcus aureus* (MRSA), *Clostridium difficile*, *Escherichia coli* and horizontally transferred genes, indicates that the livestock reservoir has a more significant impact on human health than was estimated 10 years ago, where the focus was mainly on resistance in *Campylobacter* and *Salmonella* (Aarestrup 2015). This has created tremendous worldwide concern over the present state of AMR among animal pathogen zoonotic agents which circulate among food-producing animals including poultry, beef, dairy cattle, goats, sheep as well as aquaculture and their impacts on the health of people on farms and potentially via the food chain (Levy et al. 1976; Holmberg et al. 1984; Hummel et al. 1986; Fey et al. 2000; Gyles 2008; Prescott 2008; Jassim and Limoges 2013, 2014). Furthermore, this has also resulted in the general public's perception that antibiotics used by human beings have been used in food animals and the belief that this has contributed to AMR among foodborne bacteria, which could complicate public health therapies (DuPont 2007).

2.3 AGPs: Challenges and Consequences

Antibiotics have been used as a prophylactic to prevent disease, for treating animal (beef cattle and poultry) and plant infections, as well as in animal farming for promoting growth (McManus et al. 2002; Smith et al. 2002; Singer et al. 2003; Phillips et al. 2004; Shuford and Patel 2005; Cabello 2006; Castanon 2007; Khan et al. 2008). With some exceptions, the antimicrobial classes used in agricultural industries are the same as in human medicine (DuPont 2007; US FDA 2012; Paulson and Zaoutis 2015). At present, sub-therapeutic doses of antibiotics are routinely fed to livestock, poultry and fish on industrial farms to promote faster growth and to compensate for the unsanitary conditions in which they are raised (Emanuele 2010). Among these antibiotics are tetracycline, penicillin, erythromycin and other antimicrobials that are important in human clinical use. They are used extensively, in the absence of disease, for non-therapeutic purposes in today's livestock production (Mellon et al. 2001; Page and Gautier 2012), and can generally be purchased over the counter without veterinary involvement (Khachatourians 1998; Manna et al. 2006; Laxminarayan et al. 2013). Obviously, easy access to antibiotics for animal use has contributed to large amounts being released in natural ecosystems. They are freely used in the agricultural, horticultural and veterinary sectors to keep animals and plants healthy on industrial-scale farms. The worldwide total annual production of antibiotics was estimated between 100,000 and 200,000 tons, where the US used nearly 10 million kilograms (11,023 tons) per year (Mellon et al. 2001; Ternak 2005; Sarmah et al. 2006). The antibiotics fed to hogs in North Carolina each year exceed those that are clinically prescribed for the whole Country (Ternak 2005; Quarles 2006). The human population is also affected, directly or indirectly by this proliferation of antimicrobials which are associated with the increase in human height and the obesity of the population observed since the mass consumption of antibiotics began 40–50 years ago. The association between antibiotic consumption and the increase of human growth and obesity is suspected (Ternak 2005).

Not surprisingly, the current anti-infective drug market for livestock represents one of the largest markets in the world (Page and Gautier 2012). For example, in the US around 8 billion animals, (7.5 billion chickens, 300 million turkeys, and 100 million cattle) are treated by as many as 10 different antibiotics annually or during their shorter lives (Martin 2004; Page and Gautier 2012), for prevention of bacterial diseases via feed or water (Page and Gautier 2012; Laxminarayan et al. 2013). Preventive use can be anything from targeted interventions for controlling the spread of a diagnosed disease in a specific group of animals, to routine treatment of all animals during periods of stress such as weaning, after transportation, when combining new animals with a herd or mixing animals from different sources (Laxminarayan et al. 2013). Overall the use of antibiotics for sub-therapeutic purposes in animals escalated by about 50% between 1985 and 2001 (Gerber et al. 2007), and currently approximately 80% of all antibiotics used in the US are fed to farm animals (US Congress 2011). This was primarily driven by increased use in the poultry industry, where sub-therapeutic antibiotic use increased from 2 million to 10.5 million pounds (907,185 kg–4,762,720 kg) between the 1980s and 2001, which amounted to a dramatic 307% increase on a per-bird basis (Mellon et al. 2001).

In general, it's clear that the excessive use of antibiotics in animal agriculture has contributed a considerable amount of pressure on the natural microbial environment, including beneficial bacteria, human and animal nutrition as well as immunity, resulting in the evolution of dangerous superbugs (Phillips et al. 2004; Yan and Polk 2004; O'Hara and Shanahan 2006; Buffie et al. 2011). As a consequence, over the past 15 years the carbapenemases [are β-lactamases with versatile hydrolytic capacities in which they have the ability to hydrolyze penicillins, cephalosporins, monobactams, and carbapenems (Queenan and Bush 2007)] have increasingly been reported in *Enterobacteriaceae,* which is known as carbapenem-resistant *Enterobacteriaceae* (CRE) (Livermore 2009). The CRE carries a plasmid, or mobile piece of DNA, with an enzyme that breaks down antibiotics. Their ability to transfer that plasmid, and that antibiotic resistance, to normal bacteria present in the environment makes these bacteria even more dangerous. Phages carrying β-lactamase genes were also isolated from sewage, suggesting another vector for transfer of these genes between organisms (Muniesa et al. 2004). The global spread of New Delhi metallo-β-lactamase (NDM) is a significant public health concern. Recent identification of NDM-1 producers, originally in the United Kingdom, India, and Pakistan and now worldwide, is worrisome and indicates that the world is marching at the dawn of a post-antibiotic era (CDC 2013a; Hayden 2015). Detection of infected patients and carriers with carbapenemase producers is necessary for prevention of their spread (Nordmann et al. 2011). The most serious Gram-negative infections are healthcare-associated, and the most common pathogens are *Enterobacteriaceae*, *Pseudomonas aeruginosa, E. coil* and *Acinetobacter* (CDC 2013b; Rashid et al. 2015). *E. coli* was found resistant to antibiotics like ampicillin, tetracycline, and nalidixic acid, and these resistant phenotypes were common in livestock and poultry (Sørum and Sunde 2001; Hasan et al. 2011). Interestingly, multidrug-resistant extended-spectrum β-lactamase (ESBL)-producing *E. coli* isolated from aquatic sources in Bangladesh were found to be resistant to broad-spectrum antimicrobials such as aztreonam, ciprofloxacin, mecillinam, and cefuroxime sodium (Rashid et al. 2015). On the other hand, treating infections of either pan-resistant or nearly pan-resistant Gram-negative microorganisms is an increasingly common challenge in many hospitals, however, this is true as well, but not to the same extent, for some of the Gram-positive infections (e.g., *Staphylococcus* and *Enterococcus*) (CDC 2013b).

Using sub-therapeutic antibiotics in animal agriculture have consequently affected everybody with the increased antibiotic resistant bacteria and in parallel increased frequency of treatment failures (in some cases death), and the increased severity of infections (FAO/OIE/WHO 2003). As stated by the WHO, the increasing emergence of antibiotic resistance in human pathogens is a special concern, not only for treating infectious disease, but also for other pathologies in which antibiotic prophylaxis is needed for avoiding associated infections. In this regard, WHO (2000) reported the following: "the spread of antibiotic resistance within hospitals means that commonplace medical procedures, once taken for granted, could conceivably be consigned to medical limbo; the repercussions are almost unimaginable". Furthermore, in 2011 the Director-General of the WHO, said that "The world is on the brink of losing these miracle cures (antibiotics). In the absence of urgent corrective and protective actions, the world is heading towards a post-antibiotic era, in which many common infections will no longer have a cure" (Liljeqvist et al. 2012). It is also clear that the antibiotics used in livestock are a significant source of antibiotics released into the environment (Dolliver and Gupta 2008).

In an attempt to cure the damage already done, many governments around the world are taking some action to address this issue seriously, with an active role in research (Shute 2013). In this regard, in the European Union (EU), restricted authorisation of antimicrobial types began several decades ago and in 2006 all growth promoting use was abandoned (Laxminarayan et al. 2013; Jassim and Limoges 2014). However, this has not led to any consistent decrease in antibiotic consumption (Woolhouse et al. 2015). Typically, the growth promoter ban has prompted compensatory increases in metaphylactic and prophylactic use. The result is that in Europe, the volume of agricultural usage of antibiotics continues to rival that of medical usage (Woolhouse et al. 2015). In the US, the FDA has released draft guidelines on judicious use of antimicrobials in the rearing of animals for food production. These recommendations aim to reduce the overall use of medically important anti-microbials and include veterinary oversight and consultation. If this guidance is adhered to, a gradual phasing out of growth promoting use is to be expected (Laxminarayan et al. 2013). There have been some localized successes, for example, a more than 50% reduction in the usage of antibiotics (notably macrolides) in pigs was achieved from 1992 to 2008 in Denmark without any loss in productivity (Aarestrup et al. 2010).

2.4 AMR Shared Between Livestock and Humans

Most earlier research on human pathogens led us to ignore for decades the existence of an ecological cycle that does not directly involve humans. The emerging pathogen of growing importance for example *E. coli* O157:H7 was probably created by human activities (Quarles 2006). The author concluded that this is due to saturation of cattle in feedlots with antibiotics putting selective pressure on their microbes. In their frantic scramble to survive, bacteria may increase the frequency of mutation and genetic exchanges. These exchanges include genes for pathogenic activity and antibiotic resistance (Law 2000; Lefebvre et al. 2005) and are encoded on phage, plasmid and chromosomal genes (Law 2000). As in human medicine, the use of antimicrobial drugs in veterinary medicine creates a selective pressure for the emergence of antimicrobial-resistant bacteria, including animal pathogens, human pathogens that have animal reservoirs and commensal bacteria that are present in animals (Singer et al. 2003; Turnidge 2004). This has led to the development of AMR that can be transmitted to humans through the animal meat, milk and egg supply, through direct contact with animals or through environmental contamination (Fig. 2.1). Increasingly, food animals are raised in large numbers under close confinement, transported in large groups to the slaughter and then processed very rapidly. These conditions can cause increased shedding of bacteria and contamination with fecal bacteria from the hide, carcass or meat. The amplified dissemination of pathogens occurs through the food chain via centralized food processing and packaging and worldwide distribution through shipment, wholesale food chains and retailers. The bacterial pathogens are transmitted from contaminated animals, meat and foods to humans through the food supply (Mølbak et al. 1999; White et al. 2001; Manges and Johnson 2012) or through direct exposure from infected or contaminated animals, such as on farms, in slaughterhouse or processing facilities (Lyons et al. 1980; Rowe et al. 1997; Price et al. 2007; Smith et al. 2007; AAP 2012). However, the farmers, farm workers, and farm families as well as casual visitors and travelers to endemic countries are at greater risk to be infected or serve as vehicles to transfer AMR (Levy et al. 1976; CDC 2001; Meltzer and Schwartz 2007; AAP 2012).

Bacteria are found everywhere in the animal farm environment, which means animals are exposed to the potential for disease whether they're raised on open range or indoors. Farm animals are an important component of resistance gene pathogens, normal flora, gut microbiota and vast numbers of soil bacteria in this

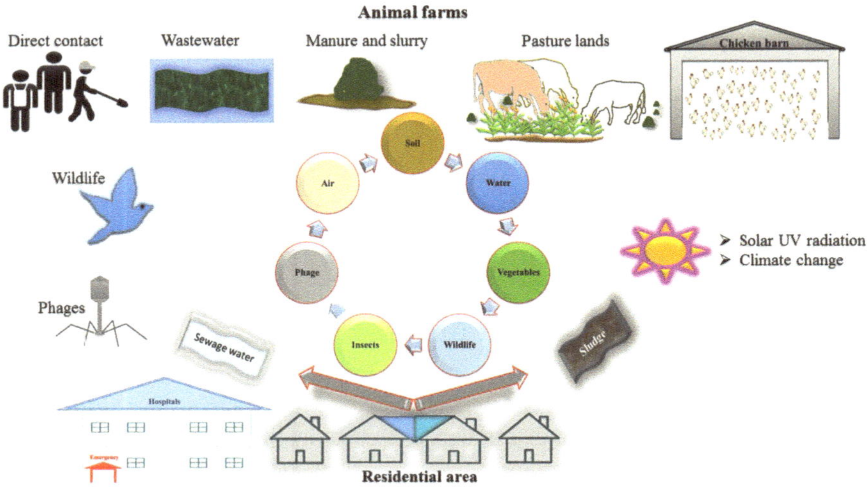

Fig. 2.1 Conceptual model of sharing AMR

complex system. They are exposed to enormous quantities of antibiotics (despite attempts at reduction) and act as another reservoir of resistance genes (Woolhouse et al. 2015). Noteworthy, most antibiotics were produced by screening soil microflora (bacteria, actinomycetes, fungi and algae) which have been producing antibiotics on a global scale for perhaps 2 billion years (D'Costa et al. 2011). They represent the original source of the majority of antibiotics used currently in medicine and veterinary medicine (D'Costa et al. 2007). For example, antibiotic penicillin came from the fungus *Penicillium* found in soil, and vancomycin is made from the *Amycolatopsis orientalis* bacteria found in soil. A newer antibiotic, teixobactin was discovered from a screen of uncultured bacteria in a soil sample. Teixobactin was found to inhibit cell wall synthesis by binding to a highly conserved motif of lipid II (precursor of peptidoglycan) and lipid III (precursor of cell wall teichoic acid) of *Staphylococcus aureus* and *Mycobacterium tuberculosis* without resistance (Ling et al. 2015). It is therefore not surprising that soil is also a major reservoir of AMR. Resistance is likely to be as natural, widespread and ancient as antibiotic production. The relationship between resistance to naturally produced antibiotics in the soil and manufactured antibiotics in the clinic remains unclear (Woolhouse et al. 2015). For instance, a metagenomics study of Fosberg et al. (2012) found multiple examples of resistance genes in the soil that had 100% homology across all major classes of antibiotics to those found in clinical isolates. That study provides clear evidence for horizontal gene transfer between soil bacteria and pathogens; however, it does not show how the gene transfer took place. Woolhouse et al. (2015) explained that, the observation that resistance determinants for synthetic quinolones (qnr genes) can be detected in soil seems to indicate transfer from an unknown source, and the clinic. The soil in farm fields often receives regular application of manure as a fertilizer and can be in direct contact with manure droppings, urine and plants which are also cycled through the animal digestive system, so it is not surprising that soil microflora antibiotics have contributed in part with AMR via vertical and HGT mechanisms for some time. Consider this a 'call to action' for researchers interested in understanding the dynamics of natural antibiotics produced by soil microflora on antibiotic resistance in livestock and poultry, since all livestock are living in an open environment and in direct contact with these soil microflorae most of their life. Researchers should also consider climate change, increased solar ultraviolet radiation (SUR) and the impact of other external forces on phage–host interactions in various ecosystems (Jassim and Limoges 2013).

AMR from livestock and poultry can be spread through fecal material and contaminate foods when manure containing resistant organisms is applied to agricultural soils or through direct and indirect contact. The organisms are then present in farm runoff (Heuer et al. 2011) or wastewater, leading to environmental reservoirs of pathogens and resistance genes (Chapin et al. 2005; Mackie et al. 2006) (Fig. 2.1). Cross-contamination of fruits and vegetables can occur when wastewater or contaminated water is used to irrigate crops and fish raised in contaminated water can also be exposed (Fig. 2.1). Active antimicrobial agents have been detected in surface waters and river sediments (Kümmerer 2004), and the resistance genes identical to those found in swine waste lagoons have been found in groundwater and soil microbes hundreds of meters downstream (Chee-Sanford et al. 2001).

The damage caused to the environment at large and the effect of antibiotics used for treating infections or for farming purposes in AMR, has been studied in more detail (Witte 1998; Ferber 2003; Singer et al. 2003), still little is known on the overall effects of antibiotics on the population dynamics of the micro-biosphere (Sarmah et al. 2006). Antibiotic resistant cross contamination between humans and animals and visa versa is still unclear. There are several studies available that have shown that AMR is widespread in farm animals, for example; apramycin and ampicillin-resistant *E. coli* in newborn calves (Hoyle et al. 2004a, b, 2005; Yates et al. 2004). Nevertheless, there are equally high levels of ampicillin resistance on organic farms (Hoyle et al. 2006), and reports of carbapenem-resistant enterobacteria in livestock (Fischer et al. 2012). However, such studies do not establish the direction of movement, if any, of resistance between livestock populations and humans. Further, carbapenems are not used in livestock so resistance was presumably imported from another agent. The movements of pathogens and AMR are likely from cattle and cattle feces, can travel directly into the food supply from slaughterhouses, or more insidiously contaminate water or food crops (Swerdlow et al. 1992). The pathogens can also be spread by houseflies (*Musca domestica*), cockroaches, and even field slugs the *Deroceras reticulatum* (Rivault et al. 1993; Agui 2001; Kobayashi et al. 2002; Alam and Zurek 2004; Sproston et al. 2006). Antibiotic resistance genes have been found at feedlots, in animals, in the air, in manure, urine, and often held in lagoons and end up in water and in sediments and finally in vegetable crops via contaminated water and fertilizer (Quarles 2006) (Fig. 2.1).

Most antibiotics used for preventing or treating infections in humans or animals, as well as for promoting faster growth of livestock, are only partially metabolized and poorly absorbed in the gastrointestinal tract, resulting in excretion of parent compounds and metabolites (Wegener 2003; Shuford and Patel 2005; Boxall et al. 2006; Dolliver and Gupta 2008; Khan et al. 2008; Chee-Sanford et al. 2009). Since approximately 75% of all antibiotics given to animals are not fully digested, upon excretion these compounds pass through the body and enter the environment beyond feed yard boundaries via application of manure waste onto agricultural fields and runoff (Wegener 2003; Dolliver and Gupta 2008; Chee-Sanford et al. 2009; Roe and Pillai 2003; Chee-Sanford et al. 2009). With huge quantities of manure, 180 million dry tons of livestock and poultry waste are generated every year in the US which are routinely sprayed onto surrounding farm fields. Once in the environment, antibiotics can facilitate de novo mutation of AMR and provide a selective advantage for bacteria that acquire resistance either in treated animals or in the environment (Gilchrist et al. 2007; Silbergeld et al. 2008; Chee-Sanford et al. 2009). They can also encounter new bacteria and create additional resistant strains (Horrigan et al. 2002), also leeching into surface and ground water, contaminating drinking wells and endangering the health of people living nearby (Hrudey and Hrudey 2004; Dolliver and Gupta 2008; Clemans et al. 2011). Therefore, this has resulted in the development of AMR (Lu 2004) which can be spread by animals, birds and insects that come in contact with animal waste (Graham et al. 2009; Page and Gautier 2012) (Fig. 2.1).

The recent advances in molecular biology techniques make it possible to track the movement of bacterial strains and AMR genes through the environment (Choi 2007). Antibiotics released at feeding operations and resistance genes generated there end up in meat products (Sunde and Norstrom 2006). AMR genes have also been found in the air, in manure and in the water at animal feedlots where manure and urine is often held in lagoons (Hrudey and Hrudey 2004; Quarles 2006; Sapkota et al. 2006). Airborne microbes are also thought to have significant impact on human and animal health including threats from pathogenic microorganisms (Behzad et al. 2015) (Fig. 2.1). Pathogens, antibiotics and AMR genes from these lagoons can end up in water and in water sediments (Pei et al. 2006; Schmitt et al. 2006). Where streams or wells are used as drinking water sources, AMR genes can end up in drinking water. They can also be found and transferred by house flies (Macovei and Zurek 2006; Petridis et al. 2006). Conversely, antibiotic resistance genes, acquired by pathogenic bacteria through HGT have been observed in environmental bacteria (Davies 1997), although they can also evolve later on under strong antibiotic selective pressure during the treatment of infections (Martinez and Baquero 2000; Martinez et al. 2007). To understand the development of resistance in full, we need to address antibiotics and their resistance genes, not just in clinics but in natural non-clinical environments (Martinez 2008).

The above findings raise questions and concerns about environmental contamination and the complex interaction that occurs between antimicrobial agents from agricultural and human use, biological vectors such as such as lysogenic phage, flies, birds, etc., and the effect of SUR to provoke direct DNA mutations, along with climate change stimulating rapid bacterial growth. These and other factors exerting selective pressure, could force microbial populations to stimulate horizontal and vertical gene transfer and thus amplify the number and variety of bacteria that are resistant to antimicrobial agents (Jassim and Limoges 2013, 2014) (Fig. 2.2). Consequently, we have the emergence and spread of AMR as a global health threat,

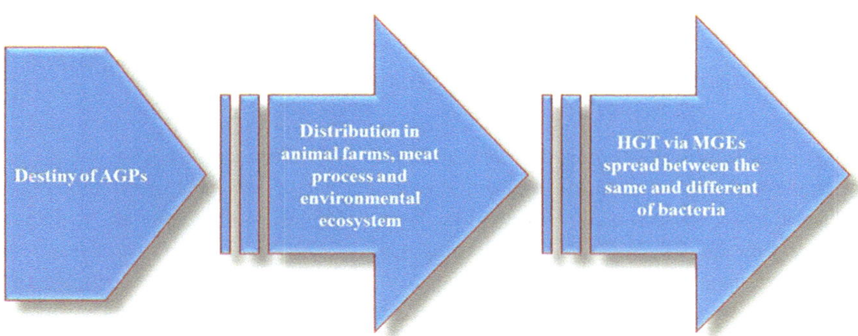

Fig. 2.2 The post-AGP effects

with agricultural use of antibiotics as a contributor to the aggregation of resistance in the environment, coupled with external forces on phage–host interactions in aquatic and other trophic ecosystems (Gilchrist et al. 2007; Levy and Marshall 2004; Jassim and Limoges 2013, 2014) (Fig. 2.2).

2.5 Resistance Genes Not in a Vacuum

AGPs usually contain a small amount of antibiotics with an average of 100 mg of antibiotic substances delivered via animal feed. This sub-therapeutic dose is consumed in animal husbandry in Europe to produce each kg of meat for human consumption (Lu et al. 2004). This regimen would have favored the selection and maintenance of rare bacterial transformants carrying the resistance genes (Lu et al. 2004). These authors have concluded that based on the large numbers of pigs and chickens that were exposed to the antimicrobial agent, the probability of gene pick-up by bacterial commensals in the animal gastrointestinal tract increased. Once incorporated into a gut commensal genome, further dissemination would have followed under antimicrobial selective pressure. Since, farm animals are raised in large numbers under close confinement, they are more likely to shed pathogens in their manures or litter. These conditions would favour such organisms enabling them to pass on their resistance genes to their offspring by replication, or to other related bacteria through "conjugation". Plasmids carrying these genes "jump" from one organism to another or via biological vector through lysogenic phages. This process is a natural, unstoppable phenomenon causing the evolution of many zoonotic pathogens of current concern and is exacerbated by AGPs in animal husbandry. In this regard, the selective effects of antibiotic use can only be understood if considered in the context of environmental routes (e.g., water, air, soil, lagoons, manure and urine, ground water, surface waters, river sediments, soil, wastewater, environmental contact, and biological vectors such as phage vectors, flies, rodents etc.) (Fig. 2.1) and the impact of external forces that enable these bacteria and the genes they carry, to spread between different biomes leading to environmental reservoirs of pathogens and resistance genes (Fig. 2.2).

2.6 Mechanisms of Gene Transfer Between Bacteria

Bacteria have evolved numerous mechanisms to assimilate new genes to help them withstand harsh environmental conditions or settle in a new environmental niche. The same mechanisms now also enable them to develop antimicrobial resistance (Martinez et al. 2009; Wiedenbeck and Cohan 2011). Bacterial resistance to antimicrobial agents develops by a variety of mechanisms: (1) change within the existing genome of a bacterial cell (chromosomal mutations, namely, vertical evolution) and change within the proteome (phase variation) (Crumplin and Odell 1987; Mempel et al. 1994; Martinez and Baquero 2000; Woodford Ellington 2007);

(2) formation of bacterial cell interactions (mixed bacterial biofilms) (Mah 2012); and most importantly; (3) by acquiring new genes from other strains or species through HGT (Boerlin and Reid-Smith 2008; Hegstad et al. 2010a; Palmer et al. 2010).

HGT or sharing of genes between bacteria was thought to be largely, although not exclusively, responsible for the development and intervention of antibiotic-resistant bacteria through various mechanisms called the mobile genetic elements (MGEs) (Hegstad et al. 2010a). MGEs, include phage transduction; conjugation (which involves direct cell-to-cell contact and transfer of plasmids or transposons); or transformation, involving the uptake of free DNA that results from bacterial lysis (Barbosa and Levy 2000; Livermore 2003; Marti et al. 2014a). In some circumstances the presence of low levels of the antibiotic in the environment is the key signal that promotes gene transfer (Jeters et al. 2009), thereby ensuring that the entire microbial community is protected from the antibiotic (Ochiai et al. 1959). HGT may occur relatively infrequently, but once the gene is established in a successful virulent clone, the clone and the carried gene can spread in individual countries and worldwide, such as in the case of multidrug-resistant *Staphylococcus aureus* and pneumococci (*Streptococcus pneumoniae*) (Collignon et al. 2009).

The acquisition of antimicrobial resistance by phage transduction has been demonstrated in chicken meat relevant bacterial species, this mechanism in environmental settings has now been fully explored (Shousha et al. 2015), whereas, the phage can transfer bacterial DNA from one host to the other in a process known as transduction. The co-existence of various resistance genes in the same plasmid or transposon results in the incidental transfer of the whole group, even if the selective pressure is directed towards a specific gene (Summers 2002). This co-selection mechanism impacts the establishment of a linear relationship between the use of a specific antibiotic and the emergence of the corresponding resistance (O'Brien 2002). The fact that the recipient cell receives all the genetic competences mediated by a certain plasmid may result in more complex consequences, such as the transfer of virulence determinants under the selective pressure imposed by the presence of antibiotics or, in opposition, the non-selected transmission of antimicrobial resistance genes driven by the presence of heavy metals or disinfectants (Aiello and Larson 2003; Hegstad et al. 2010b). This dynamic also favours the optimization of these genetic elements, dashing initial hopes of reversing resistance by simply reducing antibiotic use (McEwen and Fedorka-Cray 2002; Enne 2010).

2.7 Phage, AMR and Virulence Factors in Bacteria Sharing

Transduction is a mechanism of genetic exchange, which is mediated by independently replicating phages (Normark and Normark 2002; Frost et al. 2005; Boerlin and Reid-Smith 2008; Michod et al. 2008; Volkova et al. 2014). Phages may typically be grouped into two categories by their life cycle: lytic 'virulent' phages and lysogenic 'temperate' phages. Phages provide one of the most efficient vehicles for moving DNA sequences between bacterial cells by either; (1) multiplying inside the host cell before releasing new phage particles (lytic cycle), or (2) incorporating their viral genome into the host genome, replicating as part of the host (lysogenic 'temperate' cycle) (Fig. 2.3).

Lytic phages infect their host cell and begin the viral replication cycle within a short time frame, around 20 min. At the end of replication and assembly, the host bacterial cell typically lyses and releases the newly formed phage particles (Fig. 2.3), the whole lytic cycle takes around 40 min from start to finish. In contrast to lysogenic phages, lytic phages do not alter the phenotype of the host bacterial cell during a long-term genetic relationship, but they can shape the host population by eliminating susceptible cells in a population. Phage-resistant mutants lose virulence, and thereby these phage-resistant mutants will facilitate genetic exchange by transduction (Michod et al. 2008; McShan and Nguyen 2016). Therefore, transduction is the process by which DNA is transferred from one bacterium to another cell via phage vector 'DNA' (Hartl and Jones 1998), this can be carried-out either largely by the phage lysogenic cycle and to a lesser extent by the phage lytic cycle as described above (Fig. 2.4).

Transduction was first reported by Leonard et al. (1968) when they found both temperate and lytic phages were able to transduce streptomycin resistance. The role of phage in transduction and horizontal transfer is assumed (Christie et al. 2012). This assumption recently has been confirmed by Shousha et al. (2015). They demonstrate that lysogenic phages that infect *E. coli* (coliphages) are able to transduce important antimicrobial resistances in chicken meat. This result suggested that transduction of antimicrobial resistance is not, as has been assumed, a rare event but is rather more common for certain resistance elements. Hence this could

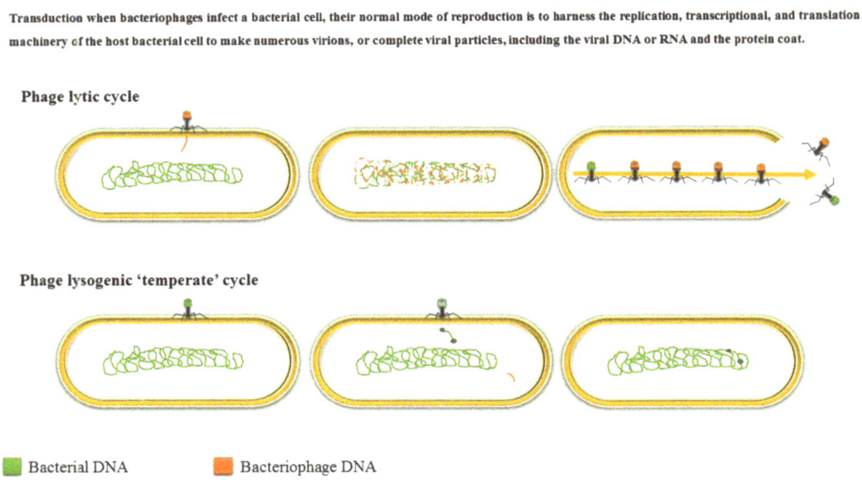

Fig. 2.3 Phage life cycles

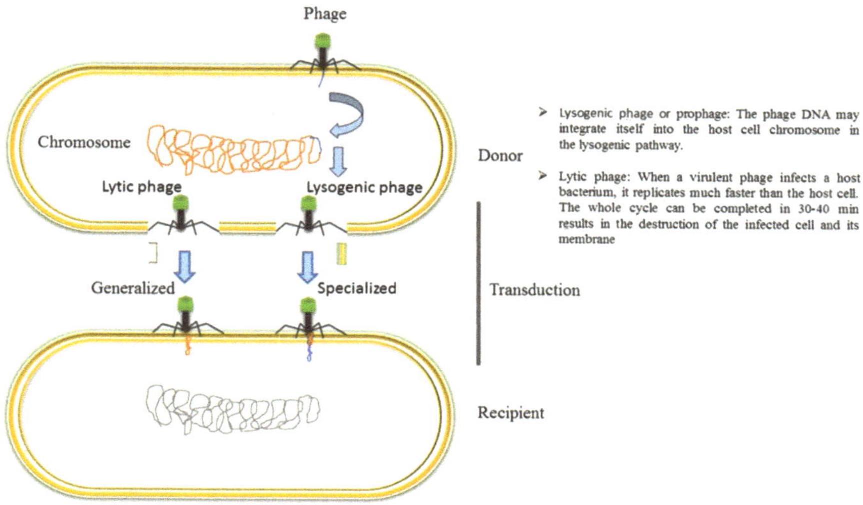

Fig. 2.4 Transfer of DNA between bacteria via phages

suggest that specialized transduction occurs frequently in the environment during lysogeny, when a prophage excises not only its own DNA but also attached bacterial DNA from the host genome. Such a process has been documented in different environments such as in environmental water and on plant surfaces (Fineran et al. 2009; Jassim and Limoges 2013). In this regard, to keep their fitness, the phages can store this DNA in the capsid (encloses the genetic material) and during the next infection cycle the bacterial host might integrate the foreign DNA by homologous recombination (Zhang et al. 2011). In hypothesis, generalized transduction, in which either a lytic or a lysogenic phage packs bacterial DNA instead of phage DNA into its capsid and delivers it to the next bacterial host, may occur without the need for homologous recombination between different species of bacteria (Volkova et al. 2014). Although there has not yet been any definitive proof of the phenomenon, broad-host-range plasmids have been transduced between bacteria of non-related species (Evans et al. 2010).

In general, if the lysogenic cycle is adopted, the phage chromosome is integrated (by covalent bonds) into the bacterial chromosome, where it can remain dormant for thousands of generations (Figs. 2.3 and 2.4). The lysogenic phage or 'prophage' will drive the adaptive evolution of bacteria to achieve more powerful virulence factors inherited from previously infected bacteria via transduction, i.e., the transfer of genetic material to a bacterial cell via phage infection (Campbell 1988; Verheust et al. 2010) (Figs. 2.3, 2.4 and 2.5). Lysogenic phage transduction serves as a driving force in bacterial pathogenesis, acting not only in the evolution of bacterial pathogens through gene transfer, but also contributing directly to bacterial pathogenesis at the time of infection and antimicrobial drug resistance (Blanchard et al. 1986; Blahová et al. 1993; Pereira et al. 1997; Willi et al. 1997; Schmieger and Schicklmaier 1999;

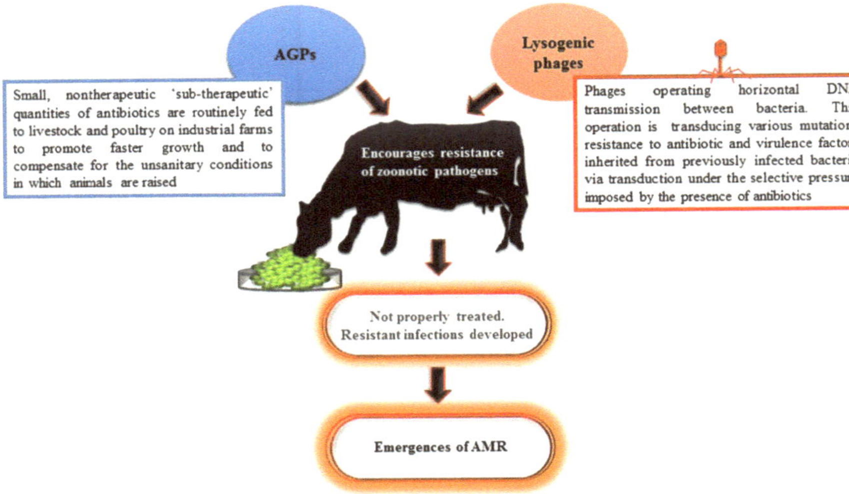

Fig. 2.5 Evolution of AMR in animal agriculture

Wagner and Waldor 2002; Frost et al. 2005; Michod et al. 2008; Verheust et al. 2010). In vitro, phages can transduce resistance to imipenem, aztreonam, and ceftazidime in *Pseudomonas aeruginosa* (Blahová et al. 1993), methicillin in *Staphylococcus epidermidis* (Blanchard et al. 1986), tetracycline in *S. aureus* (Pereira et al. 1997) and Actinobacillus actinomycetemcomitans (Willi et al. 1997). It was found that phages are able to transduce resistance genes from *Salmonella enterica* serovar *typhimurium* DT104 (Schmieger and Schicklmaier 1999). Furthermore, increasing levels of resistance to antimicrobial agents in bacteria, particularly in Gram-negative rods resistant to β-lactam antimicrobial drugs, have become evident (Livermore 1995; Muniesa and Jofre 1998; Muniesa et al. 2004).

The major mechanism of resistance that causes clinically important infection in Gram-negative bacteria is the production of β-lactamases, which includes chromosome and plasmid-encoded enzymes and phages (Livermore 1995; Muniesa and Jofre 1998; Muniesa et al. 2004). The data of Vojtek et al. (2008) has indicated that horizontal transfer of lysogenic phages among group A *Streptococcus* can occur across the M-type barrier due to the extent of horizontal DNA transmission. This data also provided further support for the hypothesis that toxigenic conversion can occur via lysogeny between species (Brabban et al. 2005). Therefore, this mechanism allows more efficient adaptation to changing host challenges, potentially leading to stronger, more specialized and virulent clones (Vojtek et al. 2008) (Figs. 2.3 and 2.4), which may represent a potentially serious hazard to humans, animals and plants (Saunders et al. 2001; Verheust et al. 2010). The influence of phage transduction is evident in localized pathogenicity of E. coli strains endowed with at least six virulence factors: *Shigella* toxin, intimin-mediated adhesion (virulence factor 'adhesion' of Enteropathogenic *E. coli*), haemolysin, serine protease, thermostable enterotoxin and special catalase system (Alexa et al. 2001; Hayashi et al. 2001; Sousa 2006; Vondruskova et al. 2010). This is evidence of the transductive and lysogenic capacities of this class of phages which have contributed to the evolution and shaping of emerging foodborne pathogenic bacteria through the dissemination of virulence and antibiotic resistance genes (Figs. 2.3, 2.4 and 2.5). For example, the genome sequences of *Shigella dysenteriae*, *E. coli* O157:H7 and the Shiga toxin-encoding phages demonstrate the critical role phage-mediated gene transfer events played in the evolution of these high-profile human pathogens (Brabban et al. 2005). These authors also observed the phage's role in the spread of antibiotic resistance between bacteria via interspecies transduction. Such exchanged genes include those encoding for substrate utilization, bacteriocin, exopolysaccharide and biogenic amine production, immunity to phages and antibiotic resistance (Rossi et al. 2014).

Thus through these mechanisms, phages play an important role in the evolution and ecology of bacterial species, as they have the potential to transfer genetic material between bacteria (Figs. 2.3, 2.4 and 2.5). This is evident in the study of Modi et al. (2013) who recently revealed that the viral metagenome (or virome) of antibiotic-treated mice was highly enriched for ARGs compared with that of non-treated control mice. The authors also demonstrated that ex vivo infection of an aerobically cultured naive micro-biota with phages from antibiotic-treated mice resulted in increased bacterial resistance compared to infection with phages from the non-treated control. These findings clearly show that phages have significant implications for the emergence and spread of antibiotic resistance. ARGs may be acquired and transferred among bacteria via MGEs such as conjugative plasmids, insertion sequences, integrons, transposons, and phages (Balcazar 2014). However, the contribution of phages to the spread of ARGs suggests that phages may play a more significant role in the emergence and spread of ARGs than previously expected (Parsley et al. 2010; Colomer-Lluch et al. 2011a, 2014; Balcazar 2014; Marti et al. 2014b; Shousha et al. 2015).

The increasing numbers of resistant bacteria, both commensals and foodborne pathogens, on food is a cause of concern to public health authorities, veterinarians as well as physicians and because it was linked to foods of animal origin, it represents a possible source of antimicrobial resistance in the human environment (Shousha et al. 2015). Unfortunately, the strategies to combat antimicrobial resistance have enjoyed only limited success and there are still many questions relating to how and when resistance transfer occurs (Shousha et al. 2015). Obviously there is an urgent need for new strategies in combating phage transfer of antimicrobial resistance in the food environment. The general setback in phage therapy is laid on the contribution of phages to the spread of ARGs through the horizontal transfer of genetic elements between bacteria, which is critical to the dissemination of resistance, particularly within a mixed bacterial population (e.g., skin, intestine and respiratory mucosa) in the presence of antimicrobial agents (McDermott et al. 2003; Lu and Koeris 2011; Smillie et al. 2011; Sillankorva and Azeredo 2014). This leads to another concern, that since it was found that lytic phages have been used in phage therapy for the treatment of colibacillosis caused by AMR chicken pathogenic *E. coli* strains (Zhang et al. 2013), or for reducing Campylobacter loads in chickens (Kittler et al. 2013), there could be a risk of transduction of antibiotic resistance genes by these phages. Accordingly, this could put phage therapy in jeopardy since phage may be contributing through HGT of AMR genes and virulence factors from other bacteria. For this reason, it was suggested to use purified phage endolysins (lysins) directly rather than the phage itself, thus precluding unintended transfer of genetic material from the phage (Woolhouse et al. 2015). Others have suggested testing the transduction ability of these therapeutic phages as a tool for avoiding the transduction of antibiotic resistance by these phages (Shousha et al. 2015). Thus phage transduction may have a role in the exchange of antimicrobial resistance between microorganisms and thus, between the environments at large (Fineran et al. 2009; Colomer-Lluch et al. 2011a, b; Mod et al. 2013; Balcazar 2014; Shousha et al. 2015). The interactions between phage and bacterial species in the presence of antibiotics are leading to a more highly connected phage–bacterial network for gene exchange (Modi et al. 2013). Therefore, combining antibiotics with phage in animal treatment could lead to more serious AMR outbreaks which can be accelerated by the abundance of phages. With an estimated 10^{31} particles, phages are the most abundant biological entity on earth, and outnumber bacteria 10:1 (Hendrix 2002). It is likely that the phage HGT of AMR genes is on-going and more prevalent in the environmental transfer of genetic material, or transduction and could be responsible for the transfer of pathogenicity determinants and virulence factors, leading to the development of new microbes or even more resistant bacteria (O'Shea and Boyd 2002; Brabban et al. 2005; Maiques et al. 2007). Whole genome sequencing is revealing and beginning to quantify the two-way traffic of AMR bacteria between the farm and the clinic (Woolhouse et al. 2015). This will lead researchers to do further studies in order to predict new challenges to animal food safety (Fig. 2.5).

2.8 Alternative to AGPs

It is a clear fact that today the world has an urgent need for access to new effective treatments for bacterial infections (Laxminarayan et al. 2013) to replace the miracle drugs or 'antibiotics' of the last century. As resistance to antibiotics is becoming increasingly widespread, the sub-therapeutic use of AGPs in animal feeds was discontinued in the EU (European Union 2003). However, the potential consequences of the AGP bans on both animal and human health have been characterized (Casewell et al. 2003; Phillips et al. 2004; Castanon 2007), as follows:

- It is possible to substantially reduce the use of antimicrobial agents in livestock production without compromising animal welfare, health or production. In some cases, this should be done in combination with other measures such as biosecurity and use of vaccines (Aarestrup 2015)

- Resultant increased disease in animals and in particular in pigs, leads to an increased use of therapeutic antibiotics of importance in human medicine (Casewell et al. 2003; OIE 2007; Vigre et al. 2008; ESVAC 2013)
- Demands that farmers make improvements to the farms' hygiene
- Concerns of AMR being transmitted to humans via the food-chain (WHO 2003, 2004)
- Created problems in the international trade of animal meat and in application of the precaution principle allowed by the World Trade Organization (Castanon 2007) because the EU only imports foods obtained from animals that were not fed with antibiotics
- This has created challenges for the animal feed and feed additive industries resulting in the search for "natural" and "safe" alternatives for AGPs to help maintain current animal production levels while reducing feed costs without threatening public health (Ratcliff 2000; Millet and Maertens 2011; Lloyd 2012; Jassim and Limoges 2014)
- Antibiotics are an integral part of industrial agriculture and there are very few alternatives (Woolhouse et al. 2015).

In this regard, a variety of commercial safer non-antimicrobial substances have been studied for use in replacing antibiotics to interact with the intestinal microflora, including immunity modulating agents, phages, phage lysins, peptides, probiotics, prebiotics, synbiotics (combining probiotics and prebiotics), acidification of diets, plant extracts, clay minerals, enzymes, and the inhibitors targeting pathogenicity (bacterial quorum sensing, biofilm, and virulence) (Hooge 1999; Patterson and Burkholder 2003; Ricke 2003; Diebold and Eidelsburger 2006; Hruby and Cowieson 2006; Kocher 2006; Vondruskova et al. 2010; Cheng et al. 2014). Phages might even be able to replace the growth promoting effects of antibiotics by controlling enteric bacteria that compete for food energy thus it was suggested that if phages are to be used to target enteric bacteria, the phage must be able to remain active during the digestive processes (Huff et al. 2006). In this regard two experiments were conducted to determine the effects of dietary supplementation with phage, probiotics and their combination on growth performance, apparent total tract digestibility, fecal bacteria populations and serum immunoglobulins in growing pigs (Kim et al. 2014). The results suggest that phages and probiotics both improve different aspects of a pig's growth performance but that phages are more effective than probiotics and would appear to offer an alternative to antibiotic type growth promoters. The suggested phage level required of 0.5 g kg^{-1} could be used as an antibiotics alternative for growing pigs (Yan et al. 2012). In agreement with previous reports, pigs fed a diet containing anti-*Salmonella typhimurium* phage, at 3×10^9 plaque-forming units (pfu) kg^{-1} of feed, showed improved performance throughout the 28-day study (Gebru et al. 2010). They suggested further work needs to be conducted to better understand their mode of action in this class of pigs.

Nevertheless, phage therapy will not prevent the use of antibiotics as growth promoters, and could only substitute those antibiotics used for animal health purposes, therefore, as such, phage therapy would also be useful in uncoupling medical care and growth promoting (Lorch 1999). Furthermore, without an understanding of the essential problems including phage resistance, phage–host interactions, the microbial ecosystem, and the host animal, this biological pathogen control system will not be used to its fullest potential in improving swine production (Zhang et al. 2015). In spite of all the above alternatives for ARGs Cheng et al. (2014) have summarized that it is hard to conclude that the alternatives might substitute for antibiotics in veterinary medicine in the foreseeable future. The latter authors have also concluded that the prudent use of antibiotics and the establishment of scientific monitoring systems are the best and fastest way to limit the adverse effects of the abuse of antibiotics and to ensure the safety of animal-derived food and environment.

2.9 Phage Therapy

Phage therapy in animal production and in animal experimental models of human infection have been investigated in veterinary therapy since early work, 30 years ago, in cattle (Smith et al. 1987), and the studies have been intensified in the last 10 years due to the increases of AMR strains occurring both among humans and animals (Sulakvelidze 2005; Sulakvelidze and Barrow 2005; Atterbury et al. 2007; Miller et al. 2010; Völkel and Czerny 2011; Abdulamir et al. 2014; Borie et al. 2014; Cheng et al. 2014; Jassim and Limoges 2014; O'Flaherty et al. 2014; Aldoori et al. 2015). Next we review a brief concept for phage therapy; however, this is discussed in greater detail in Chap. 3.

Phages have been extremely effective at treating a number of bacterial infections in controlled animal studies, especially as a biocontrol agent in the prevention of food-borne illnesses, due to its target specificity, rapid bacterial killing and ability to self-replicate (Smith et al. 1987; Biswas et al. 2002; Hawkins et al. 2010). Phages have the potential to treat bacterial infections afflicting animals and in particular to prevent fatal *E. coli* respiratory infections in broiler chickens (Huff et al. 2002a, b, 2003a, b). Aerosol spraying and intramuscular injection have given the best results over using oral delivery of phages via direct administration or addition to drinking water and/or feed (Sillankorva et al. 2012). This may be due to gastric pH levels preventing the proliferation of phages (Spits 2009). Virulent antigen specific phages have been used in an attempt to control *E. coli* O157:H7 in batch culture (Kudva et al. 1999). Loc-Carrillo et al. (2005) and Wagenaar et al. (2005) reported that phage therapy reduces *Campylobacter jejuni* colonization of broiler chickens. Several studies have also addressed the use of phages to decrease *Campylobacter* and *Salmonella* concentrations in poultry (Goode et al. 2003; Atterbury et al. 2007; Allen et al. 2013; Kittler et al. 2013). Phage therapy can be effective, for example against *Salmonella typhimurium* in poultry and pigs, although this requires rapid selection and administration of the phage and high bacterial loads (Allen et al. 2013).

However, despite the reports of the successful use of phages in terms of reducing mortality, the severity of the clinical state and bacterial counts in vitro, in vivo, on tissues, and ex vivo levels, phage therapy remains controversial (Jassim and Limoges 2014). Furthermore, many earlier studies demonstrated that the classical application of phages in bacterial therapy or biocontrol is attainable in theory but in practice, it is not so successful due to the lack of full coverage of target bacteria and the rapid emergence of bacterial mutations leading to complete resistance against phage infection (Barrow and Soothill 1997; Alisky et al. 1998; Carlton 1999; Sulakvelidze et al. 2001; Goodridge and Abedon 2003). Therefore, historically, phage therapy or phage biocontrol were unsuccessful and eventually led to the replacement of phage therapy with antibiotic treatment (Barrow and Soothill 1997). Scientific methodologies could be developed to deal with antibiotic resistance in bacteria using phage, however viral proteins would also integrate into human and animal society with unknown effect. It would be wise to approach such methodologies with caution in order to avoid repeating mistakes that were made with the improper use of antibiotics. Other authors have refuted these assumptions and concluded that the rate of developing resistance against phages can be partially circumvented by using several phages in one preparation or cocktail (much like using two or more antibiotics simultaneously) (Sulakvelidze et al. 2001). More importantly, unlike using trial and error with antibiotics, when resistance against a given phage occurs, the specialists can rapidly select through testing (in a few days or weeks) a new phage that is effective against the phage-resistant bacteria (Sulakvelidze et al. 2001).

Veterinary therapy including both animal and human applications requires the appropriate administration targeting specific bacteria, with a strategy that includes a comprehensive methodology, detailing the phage–host interactions, dose optimization and accounting for all chemical and physical factors (Jassim and Limoges 2013). In general, a deep understanding of intrinsic phage properties is critical to designing therapeutic interventions. The reduction of foodborne pathogens requires a comprehensive phage control program at the farm, where the animals are born, hatched or raised, before shipment to processing plants. Potential pre-harvest sources of foodborne pathogen contamination include breeder herds and flocks, hatcheries, contaminated feed and water, along with environmental sources and vectors, such as litter, animal caretakers, and insects (Bailey 1993; Nayak et al. 2003; this is discussed in greater detail in Chap. 3).

2.10 Phage Therapy for Animals

Phages' specific bactericidal activity gives them special therapeutic advantages over antibiotics (Sulakvelidze et al. 2001; Sulakvelidze and Kutter 2005; Loc-Carrillo and Abedon 2011; Jassim and Limoges 2014), and phages have been reported to be more effective than antibiotics in experimentally infected animals (Smith and Huggins 1982). Like bacteria but unlike antibiotics, phages mutate and therefore can also evolve to counter phage-resistant bacteria (Matsuzaki et al. 2005). Because phages attack bacteria by attaching to receptors on the bacterial cell surface, phage-resistant mutants (which lack these receptors) are often less pathogenic than phage-susceptible bacteria (Inal 2003; Santander and Robeson 2007; Capparelli et al. 2010; Friman et al. 2011; Laanto et al. 2012). The ability of phages to recognize precisely their target hosts, rendered them as favourable antibacterial agents because broad-spectrum antibiotics kill target bacteria along with other beneficial bacteria present on the farm or in the organism, namely, animal intestinal flora (Merril et al. 2003).

2.10.1 Advantages

- Phages are very specific and do not harm the useful bacteria that live in and on the body
- Phages can kill bacteria with no known activity in human or animal cells, making them an attractive alternative to antibiotics
- Because they are harmless, phages can be used for combating harmful bacteria in fattening animals and food
- In comparison to antibiotics, phages have no side effects like diarrhoea or secondary infections
- Phages are abundant in nature; therefore, it is easy to find new phages when bacteria become resistant to them. This means that there should be an 'inexhaustible' supply
- Low cost: the production of phages predominantly involves growth in its host and further purification
- Possible phage transfers between individuals: this is essentially cross infection of phages from treated subjects or environments to untreated individuals, which may be potentially useful in agricultural applications
- Phages are also active against bacteria that have become resistant to antibiotics
- Lysogenic phages can be genetically modified to reverse the bacterial pathogens' drug resistance, thereby restoring their sensitivity to antibiotics
- Purified phage endolysins (lysins) can also be used as antibacterial substances
- Phages are living drugs. They multiply at the site of the infection until there are no more bacteria
- Bacteria that have become resistant to a certain type of phage continue to be destroyed by other types

There is renewed optimism for phages as new 'live drugs' with the hope to overcome the AMR problem. Phages have been approved for use in food and medical industries by several international agencies FDA, GRAS, US-FSIS (Jassim and Limoges 2014), and a phage cocktail that targets *Listeria monocytogenes* contaminants on ready-to-eat foods containing meat and poultry products was granted approval in 2006, which was the first time that the US Food and Drug Administration (FDA) accepted the use of a phage preparation as a food additive (Bren 2007; this is discussed in greater detail in Chap. 3). In the US, the only phages on the market are used in the anti-bacterial treatment of food products

Intralytix (http://www.intralytix.com), and in EU, 2013, launched the first product called 'Phagoburn' to treat burn wounds infected with bacteria *E. coli* and *Pseudomonas aeruginosa* (http://www.phagoburn.eu/). Recently an important extension to phage therapy is the use of phage lytic enzymes that digest the bacterial peptidoglycan, especially of Gram-positive bacteria, as a novel class of alternative antimicrobials (Fischetti 2008; Nelson et al. 2012; this is discussed in greater detail in Chap. 3).

2.10.2 Disadvantages or 'Challenges'

Despite the attractions of phage therapy, scientific and logistical challenges remain. Phages have not yet gained widespread acceptance in therapy treatment as compared to commercially proven pharmaceutical antimicrobial agents. Phage therapy approval could take years and needs substantial investment once if the key obstacles facing phage are solved. The following summary outlines the key issues in phage biocontrol and treatment that scientists have already encountered both in the literature as well as in the laboratories. These can help to frame a platform from which past mistakes with both phages and antibiotics can be avoided.

2.10.3 Obstacles

- Heterogeneity and ecology of both phages and bacteria were not understood
- Need to select highly virulent phages against target bacteria in the patient
- Single phage preparations used to treat mixtures of different bacteria
- Recognition as personalized medicine using a multivalent cocktail carefully selected to treat a patient's specific bacterial infection(s)
- Lack of standardized lytic phages that can target only their host cell without using genetic modification
- Genetically modified phages changing the composition of colonizing bacterial flora in humans, and the risk of subsequent development of new active infections
- Lateral gene transfer of virulence factors and antibiotic resistance
- Restriction, modification, and degradation of phage DNA upon infection
- Resistance mutations in bacterial genes for adsorption, lysogeny and lysogenic conversion
- Strict safety standards for human therapies not met
- Toxigenic conversion via lysogeny between species allowing more efficient adaptation of host, potentially leading to fitter and more virulent clones
- Failure to appropriately characterize or titre phage preparations
- Phages are relatively large in comparison to chemical molecules. For this reason, the sites in the body that can be reached by them must be carefully clarified

- Most of infection agents are hidden in the interior of human and animal cells and this makes it inaccessible to phages
- In the bacterial cell envelope, for example, use of antibiotics in animal production that can cause disruption of microbial cell wall synthesis which will inhibit or defect phage attachment
- Effect of environmental factors which all contribute to the complexity and unpredictability of phage–host interactions in the field, such as SUR and visible direct sunlight, chemical disinfectants, nutrients, pollutants etc.
- Limited knowledge about the kinetics of phage
- Consumer perception problems
- The isolation and the cultivation of phages from natural sources are time consuming and problematic for producing large amounts of active inoculums
- Failure to characterize phage preparation, i.e., to determine the virulence to the target
- Failure to neutralize gastric pH prior to oral administration
- Immunogenicity antibodies developed against phage
- There are a few reports of purified phages used in therapy having side effects
- Presence of endotoxins in phage preparation leading to toxic shock in the patient
- Phages can transfer toxin genes between bacteria via HGT
- Phages are more difficult to administer than antibiotics and physicians will need special training in order to correctly prescribe and use phages
- Phage efficacy in humans or animals is unknown and needs to be tested in the lab prior to use in each treatment
- Phage therapy effectiveness must be tested in an animal model, since each phage can behave differently in vivo
- To indicate the relationship between the phage and the concentration of target bacteria present, the multiplicity of infections must be established before each treatment
- Where multiple infections or the exact species of infecting bacteria is unknown, need cocktail of phages
- Phages are less suitable for acute cases therefore, phage therapy is best suited for infected sites such as wounds, where phages can be easily applied
- Pharmacokinetics of self-replicating agents differs from those of normal drugs
- The shelf life of phages varies and needs to be tested and monitored
- In vivo susceptibility of bacterial pathogens to phages is poorly understood and future research on more phage-host cell interactions is needed to define the requirements for successful phage treatments
- Many phage experiments done in vitro need to be extrapolated to in vivo growth
- Phages can be reproduced from a commercially available phage preparation, a challenge to commercialization
- Each newly isolated phage requires approval; this procedure could become too expensive
- Intellectual property rights are challenging for the use of phage therapy in modern medicine and these can also trigger ethical discussions

- In the healthcare system phage therapy is still seen as a cost and a social program rather than an economic driver
- Phage sectors need more time to develop entrepreneurs and innovation in their sector.

2.11 Discussion and Conclusion

Phage based therapy could potentially lead to bacterial development of phage resistance mutants. It would be wise to approach such methodologies with caution in order to avoid repeating mistakes that were made with the improper use of antibiotics. In this context, this may explain why only a handful of small pharmaceutical companies are turning to phages (Thiel 2004). The grand pharmaceutical companies have shown no interest in phage therapy, in large part because phages cannot be patented and much remains to be done regarding safety, mutations, and efficacy for therapy (Pirnay et al. 2011; Brüssow 2012; Jassim and Limoges 2014). A pharmaceutical drug development strategy for phage therapy is required as described by Jassim and Limoges (2014). The big question is whether phage therapy is ever going to be used to complement or replace traditional antibiotics in western countries.

Obstacles remain before phages can be used as AGP's. Phages have been shown to contribute in HGT of AMR genes and virulence factors from other bacteria as discussed. Insufficiently virulent phages, especially against actual target bacteria, allow bacteria to survive treatment while poorly prepared phage stocks, even if using sufficiently virulent phages, lack the numbers of viable phages required for adequate treatment. Phages converted to temperate phages are problematic due to converts turning phage-sensitive bacteria into insensitive ones, and the encoding of bacterial virulence factors, including bacterial toxins adhesion, haemolysin, serine protease, thermostable enterotoxin and special catalase system (Alexa et al. 2001; Hayashi et al. 2001; Krylov 2001; Skurnik and Strauch 2006; Sousa 2006; Skurnik et al. 2007; Merabishvili et al. 2009; Gill and Hyman 2010; Kutter et al. 2010; Vondruskova et al. 2010). Phages work as biological vehicles to transfer genes between bacteria constantly in the environment. Thus, using and manipulating such phages represents a serious risk to environmentally friendly bacteria and probiotic bacteria in animal intestines. It also potentially provides them with new virulence and aggression/aggressive factors, with resistance to antibiotics which could prompt a new wave of pathogenic bacterial diseases. Therefore, it is necessary to consider fundamentally different strategies to find solutions for the antibiotics dilemma in animal agriculture. This should be an incentive to develop other novel strategies, such as the smart lytic phage which is highly relevant to the practice of phage biocontrol and can pounce on and destroy (lyse) the bacterium quickly. This is discussed in greater detail in Chap. 1. In the next Chapter we describe novel methods and applications that can use smart phages as an optimal complementary tool along with other methods for bacterial growth control including 'safer non-antimicrobial substances'. Smart phages as an alternative to AGPs may restore the balance in microbial environments as well as enhance the farm ecology by reducing animal infections thereby reducing the use of therapeutic antibiotics. In this regard the advantage of using smart phages as a replacement for AGPs will provide the following biological characteristics:

- Improvement in feed efficiency that is economically viable
- Specific antibacterial properties targeting unwanted bacteria without harming beneficial bacteria in animal's intestines
- They become a part of gastrointestinal ecosystems and friendly with probiotic bacteria 'the natural friendly gut bacteria', and
- Prevent the HGT because they are highly lytic and they will replicate in the target host cell to achieve significant phage amplification resulting in greater bacterial killing
- Negligible or no resistance mutations in bacterial genes for adsorption, lysogeny and lysogenic conversion
- Selection of strictly highly virulent lytic phages
- Wide host range: the lytic spectrum of these particles can be much broader than the spectrum of activity of a single phage
- Smart phages with wide host range may not need to be used in combination with other antimicrobial agents, including other antibiotics or chemical disinfectants.

In general, this chapter illustrated the pros and cons of using phage in animal agriculture and is also an introduction for the following chapters, in which we discuss the methods of isolation, identification and analysis of phage to prepare smart phages for phage biocontrol. We go on to discuss various novel applications for phages. Lastly, we discuss practices that can be used to prepare commercial phage biocontrol products. Phage biocontrol describes a complete disinfectant system for every environmental niche, as described in Fig. 2.1. Phage biocontrol can be an alternative to antibiotics or chemical disinfectant agents which have side effects on livestock and are implicated in AMR while increasingly ineffective against pathogenic bacteria. We expect that the use of smart phage biocontrol in animal production may help to decrease the use of antibiotics in animal agriculture, in particular, on those farms adopting better hygiene practices (Castanon 2007).

References

Aarestrup FM (2015) The livestock reservoir for antimicrobial resistance: a personal view on changing patterns of risks, effects of interventions and the way forward. Phil Trans R Soc B 2015370:20140085. doi:10.1098/rstb.2014.0085

Aarestrup FM, Wegener HC, Collignon P (2008) Resistance in bacteria of the food chain: epidemiology and control strategies. Expert Rev Anti Infect Ther 6(5):733–750

Aarestrup FM, Jensen VF, Emborg H-D et al (2010) Changes in the use of antimicrobials and the effects on productivity of swine farms in Denmark. Am J Vet Res 71:726–733. doi:10.2460/ajvr.71.7.726

Abdulamir AS, Jassim SAA, Abu Bakar F (2014) Novel approach of using a cocktail of designed bacteriophages against gut pathogenic *E. coli* for bacterial load biocontrol. Ann Clin Microbiol Antimicrob 13:39. http://www.ann-clinmicrob.com/content/pdf/s12941-014-0039-z.pdf

Agui N (2001) Flies carrying enterohemorrhagic *Escherichia coli* (EHEC) O157 in Japan: A nationwide survey. Med Entomol Zool 52(2):97–103

Aiello A, Larson E (2003) Antibacterial clearing and hygiene products as an emerging risk factor for antibiotic resistance in the community. Lancet Infect Dis 3:501–506

Alam MJ, Zurek L (2004) Association of *Escherichia coli* O157:H7 with houseflies on a cattle farm. Appl Environ Microbiol 70(12):7578–7580. doi:10.1128/AEM.70.12.7578-7580.2004

Aldoori AA, Mahdii EF, Abbas AK, Jassim SAA (2015) Bacteriophage biocontrol rescues mice bacteremic of clinically isolated mastitis from dairy cows associated with methicillin-resistant *Staphyloccocus aureus*. Adv Microbiol 5:383–403

Alexa P, Stouracova K, Hamrik J, Rychlik I (2001) Gene typing of the colonisation factors K88 (F4) in enterotoxigenic *Escherichia coli* strains isolated from diarrhoeic piglets. Vet Med Czech 46:46–49. http://old.vri.cz/docs/vetmed/46-2-46.pdf

Alexander TW, Yanke LJ, Topp E et al (2008) Effect of sub-therapeutic administration of antibiotics on the prevalence of antibiotic-resistant *Escherichia coli* bacteria in feedlot cattle. Appl Environ Microbiol 74(14):4405–4416

Alisky J, Iczkowski K, Rapoport A, Troitsky N (1998) Bacteriophages show promise as antimicrobial agents. J Infect 36:5–15

Allen HK, Levine UY, Looft T et al (2013) Treatment, promotion, commotion: antibiotic alternatives in food-producing animals. Trends Microbiol 21:114–119. doi:10.1016/j.tim.2012.11.001

AAP (American Academy of Pediatrics) (2012) Prevention of infectious disease from contaminated food products. In: Pickering LK, Baker CJ, Kimberlin DW, Long SS (eds) Red Book. Committee on Infectious Diseases. Report of the Committee on Infectious Diseases. Elk Grove Village, IL: American Academy of Pediatrics, Appendix IX, pp 917–920

Atterbury RJ, Van Bergen MAP, Ortiz F et al (2007) Bacteriophage therapy to reduce *Salmonella* colonisation of broiler chickens. Appl Environ Microbio 73(14):4543–4549

Bailey JS (1993) Control of *Salmonella* and *Campylobacter* in poultry production. A summary of work at Russell Research Center. Poult Sci 72:1169–1173

Balcazar JL (2014) Bacteriophages as vehicles for antibiotic resistance genes in the environment. PLoS Pathog 10(7):e1004219. doi:10.1371/journal.ppat.1004219

Barbosa T, Levy S (2000) The impact of antibiotic use on resistance development and persistence. Drug Resist 3:303–311

Barrow PA, Soothill JS (1997) Bacteriophage therapy and prophylaxis: rediscovery and renewed assessment of potential. Trends Genet 5:268–271

Behzad H, Gojobori T, Mineta K (2015) Challenges and opportunities of airborne metagenomics. Genome Biol Evol 7(5):1216–1226. doi:10.1093/gbe/evv064

Biswas B, Adhya S, Washart P, Paul B et al (2002) Bacteriophage therapy rescues mice bacteremic from a clinical isolate of vancomycin-resistant *Enterococcus faecium*. Infect Immun 70: 204–210

Blahová J, Hupková M, Babálová M et al (1993) Transduction of resistance to imipenem, aztreonam and ceftazidime in nosocomial strains of *Pseudomonas aeruginosa* by wild-type phages. Acta Virol 37(6):429–436

Blanchard TJ, Poston SM, Reynolds PJ (1986) Recipient characteristics in the transduction of methicillin resistance in Staphylococcus epidermidis. Antimicrob Agents Chemother 29: 539–541

Boerlin P, Reid-Smith RJ (2008) Antimicrobial resistance: its emergence and transmission. Anim Health Res Rev 9:115–126

Borie C, Robeson J, Galarce N (2014) Lytic bacteriophages in veterinary medicine: a therapeutic option against bacterial pathogens? Arch Med Vet 46:167–179

Boxall ABA, Johnson P, Smith EJ et al (2006) Uptake of veterinary medicines from soils into plants. J Agric Food Chem 54:2288–2297

Brabban AD, Hite E, Callaway TR (2005) Evolution of foodborne pathogens via temperate bacteriophage-mediated gene transfer. Foodborne Pathog Dis 2:287–303. doi:10.1089/fpd.2005.2.287

Bren L (2007) Bacteria-eating virus approved as food additive. FDA Consumer 41:20–22

Brüssow H (2012) What is needed for phage therapy to become a reality in western medicine? Virology 434:138–142

Buffie C, Jarchum I, Equinda M et al (2011) Profound alterations of intestinal microbiata following a single dose of clindamycin results in sustained susceptibility to Clostridium difficile-induced colitis. Infect Immun 80:62–73

CDC (2001) Outbreaks of *Escherichia coli* O157:H7 infections among children associated with farm visits-Pennsylvania and Washington, 2000. Morb Mortal Wkly Rep 50(15):293–297

CDC (2013a) Vital signs: carbapenem-resistant Enterobacteriaceae. Morb Mortal Wkly Rep 62 (9):165–170. http://www.cdc.gov/mmwr/preview/mmwrhtml/mm6209a3.htm

CDC (2013b) Antibiotic resistance threats in the United States. http://www.cdc.gov/drugresistance/pdf/ar-threats-2013-508.pdf

Cabello FC (2006) Heavy use of prophylactic antibiotics in aquaculture: a growing problem for human and animal health and for the environment. Environ Microbiol 8:1137–1144

Campbell A (1988) Phage evolution and speciation. In: Calendar R (ed) The bacteriophages. Plenum Press, New York, pp 1–14

Capparelli R, Nocerino N, Lanzetta R et al (2010) Bacteriophage-resistant *Staphylococcus aureus* mutant confers broad immunity against Staphylococcal infection in mice. PLoS ONE 5(7): e11720. doi:10.1371/journal.pone.0011720

Carlton RM (1999) Phage therapy: past history and future prospects. Arch Immunol Ther Exp 47 (5):267–274

Casewell M, Friis C, Marco E et al (2003) The European ban on growth-promoting antibiotics and emerging consequences for human and animal health. J Antimicrob Chemoth 52:159–161

Castanon JIR (2007) History of the use of antibiotic as growth promoters in European poultry feeds. Poultry Sci 86(11):2466–2471

Cheng G, Hao H, Xie S et al (2014) Antibiotic alternatives: the substitution of antibiotics in animal husbandry? Front Microbiol 5:217. doi:10.3389/fmicb.2014.00217

Chapin A, Rule A, Gibson K et al (2005) Airborne multidrug-resistant bacteria isolated from a concentrated swine feeding operation. Environ Health Perspect 113(2):137–142

Chee-Sanford JC, Aminov RI, Krapac IJ et al (2001) Occurrence and diversity of tetracycline resistance genes in lagoons and groundwater underlying two swine production facilities. Appl Environ Microbiol 67(4):1494–1502

Chee-Sanford JC, Mackie RI, Koike S et al (2009) Fate and transport of antibiotic residues and antibiotic resistance genes following land application of manure waste. J Environ Qual 38 (3):1086–1108

Cho I, Yamanishi S, Cox L et al (2012) Antibiotics in early life alter the murine colonic microbiome and adiposity. Nature 488:621–626. doi:10.1038/nature11400

Choi C (2007) Pollution in solution. Sci American 296(1):22–23

Christie GE, AllisonHE, Kuzio J et al (2012) Prophage-induced changes in cellular cytochemistry and virulence. In: Hyman P, Abedon ST (eds) Bacteriophages in health and disease. CABI, Oxfordshire, UK, ch 4, pp 33–60

Clemans D, Francoeur S, Liggit P, West B (2011) Antibiotic resistance, gene transfer, and water quality patterns observed in waterways near CAFO farms and wastewater treatment facilities. Water Air Soil Pollut 217(1–4):473–489

Collignon P, Powers JH, Chiller TM et al (2009) World Health Organization ranking of antimicrobials according to their importance in human medicine: a critical step for developing risk management strategies for the use of antimicrobials in food production animals. Food Safety 49(1 July):132–141

Colomer-Lluch M, Jofre J, Muniesa M (2011a) Antibiotic resistance genes in the bacteriophage DNA fraction of environmental samples. PLoS ONE 6(3):e17549. doi:10.1371/journal.pone.0017549

Colomer-Lluch M, Imamovic L, Jofre J, Muniesa M (2011b) Bacteriophages carrying antibiotic resistance genes in fecal waste from cattle, pigs, and poultry. Antimicrob Agents Chemother 55:4908–4911

Colomer-Lluch M, Jofre J, Muniesa M (2014) Quinolone resistance genes (qnrA and qnrS) in bacteriophage particles from wastewater samples and the effect of inducing agents on packaged antibiotic resistance genes. J Antimicrob Chemother 69:1265–1274

Crumplin GC, Odell M (1987) Development of resistance to ofloxacin. Drugs 34(Suppl. 1):S1–S8

Davies JE (1997) Origins, acquisition and dissemination of antibiotic resistance determinants. Ciba Found Symp 207:15–27

Davis MF, Price LB, Liu CM-H, Silbergeld EK (2011) An ecological perspective on US industrial poultry production: the role of anthropogenic ecosystems on the emergence of drug-resistant bacteria from agricultural environments. Curr Opin Microbiol 14(3):244–250

Da Costa PM, Loureiro L, Matos AJF (2013) Transfer of multidrug-resistant bacteria between intermingled ecological niches: The interface between humans, animals and the environment. Int J Environ Res Public Health 10(1):278–294. doi:10.3390/ijerph10010278

D'Costa VM, Griffiths E, Wright GD (2007) Expanding the soil antibiotic resistome: exploring environmental diversity. Curr Opin Microbiol 10:481–489

D'Costa VM, King CE, Kalan L et al (2011) Antibiotic resistance is ancient. Nature 477:457–461. doi:10.1038/nature10388

Diarra MS, Silversides FG, Diarrassouba F et al (2007) Impact of feed supplementation with antimicrobial agents on growth performance of broiler chickens, *Clostridium perfringens* and Enterococcus counts, and antibiotic resistance phenotypes and distribution of antimicrobial resistance determinants in *Escherichia coli* isolates. Appl Environ Microbiol 73(20): 6566–6576

Dibner JJ, Richards JD (2005) Antibiotic growth promoters in agriculture: history and mode of action. Poult Sci 84:634–643

Diebold G, Eidelsburger U (2006) Acidification of diets as an alternative to antibiotic growth promoters. In: Barug D, Long J de, A. Kies AK, Verstegen MWA (eds) Antimicrobial growth promoters: where do we go from here. Academic Publisher, Wageningen, The Netherlands, pp 311–327

Dolliver H, Gupta S (2008) Antibiotic losses in leaching and surface runoff from manure-amended agricultural land. J Environ Qual 37:1227–1237

DuPont HL (2007) The growing threat of food-borne bacterial enteropathogens of animal origin. Clin Infect Dis 45:1353–1361

Emanuele P (2010) Antibiotic resistance. AAOHN J 58(9):363–365. doi:10.3928/08910162-20100826-03

Emborg H-D, Andersen JS, Seyfarth AM (2003) Relations between the occurrence of resistance to antimicrobial growth promoters among *Enterococcus faecium* isolated from broilers and broiler meat. Int J Food Microbiol 84(3):273–284

Enne V (2010) Reducing antimicrobial resistance in the community by restricting prescribing: can it be done? J Antimicrob Chemother 65:179–182

ESVAC (2013) Third ESVAC report: sales of veterinary antimicrobial agents in 25 EU/EEA countries in 2011. European Medicines Agency, London, UK. www.ema.europa.eu/docs/en_GB/document_library/Report/2013/10/WC500152311.pdf. Accessed 17 July 2013

European Union (2003) Regulation (EC) No. 1831/2003 of the European Parliament and of the Council on additives for use in animal nutrition. 22 September 2003. Official Journal European Union L 268, 18 October 2003, pp 29–43

Evans TJ, Crow MA, Williamson NR et al (2010) Characterization of a broad-host-range flagellum-dependent phage that mediates high-efficiency generalized transduction in, and between, Serratia and Pantoea. Microbiol 156:240–247

FAO/OIE/WHO (2003) Expert workshop non-human antimicrobial usage and antimicrobial resistance. Geneva, December 1–5, 2003. WHO Publishing Web http://www.who.int/foodsafety/publications/micro/en/amr.pdf. Accessed 1–5 Dec 2003

Ferber D (2003) Antibiotic resistance. WHO advises kicking the livestock antibiotic habit. Science 301:1027

Fey PD, Safranek TJ, Rupp ME et al (2000) Ceftriaxone-resistant salmonella infection acquired by a child from cattle. N Engl J Med 342:1242–1249

Fineran PC, Petty NK, Salmond GPC (2009) Transduction: host DNA transfer by bacteriophages. In: Schaechter M (ed) The desk encyclopedia of microbiology. Elsevier Academic Press, San Diego, CA, pp 666–679

Fischer J, Rodríguez I, Schmoger S et al (2012) *Salmonella enterica* subsp. enterica producing VIM-1 carbapenemase isolated from livestock farms. J Antimicrob Chemother 68:478–480. doi:10.1093/jac/dks393

Fischetti VA (2008) Bacteriophage lysins as effective antibacterials. Curr Opin Microbiol 11:393–400

Fosberg KJ, Reyes A, Wang B et al (2012) The shared antibiotic resistome of soil bacteria and human pathogens. Science 337:1107–1111. doi:10.1126/science.1220761

Friman VP, Hiltuñen T, Jalasvuori M et al (2011) High temperature and bacteriophages can indirectly select for bacterial pathogenicity in environmental reservoirs. PLoS ONE 6(3): e17651. doi:10.1371/journal.pone.0017651

Frost LS, Leplae R, Summers AO, Toussaint A (2005) Mobile genetic elements: the agents of open source evolution. Nat Rev Microbiol 3:722–732

Gaskins HR, Collier CT, Anderson DB (2002) Antibiotics as growth promotants: mode of action. Anim Biotechnol 13:29–42

Gebru E, Lee J, Son J et al (2010) Effect of probiotic-, bacteriophage-, or organic acid-supplemented feeds or fermented soybean meal on the growth performance, acute-phase response, and bacterial shedding of grower pigs challenged with *Salmonella enterica* serotype typhimurium. J Anim Sci 88:3880–3886

Gerber P, Opio C, Steinfeld H (2007) Poultry production and the environment—a review. FAO publishing Web. http://www.fao.org/ag/againfo/home/events/bangkok2007/docs/part2/2_2.pdf

Gilchrist MJ, Greko C, Wallinga DB et al (2007) The potential role of concentrated animal feeding operations in infectious disease epidemics and antibiotic resistance. Environ Health Perspect 115:313–316

Gill JJ, Hyman P (2010) Phage choice, isolation and preparation for phage therapy. Curr Pharm Biotechnol 11:2–14

Goode D, Allen VM, Barrow PA (2003) Reduction of experimental *Salmonella* and *Campylobacter* contamination of chicken skin by application of lytic bacteriophages. Appl Environ Microbiol 69(8):5032–5036

Goodridge L, Abedon ST (2003) Bacteriophage biocontrol and bioprocessing: application of phage therapy to industry. SIM News 53(6):254–262

Gough EK, Moodie EEM, Prendergast AJ et al (2014) The impact of antibiotics on growth in children in low and middle income countries: systematic review and meta-analysis of randomised controlled trials. Br Med J 348:g2267. doi:10.1136/bmj.g2267

Graham JP, Price LB, Evans SL et al (2009) Antibiotic resistant enterococci and staphylococci isolated from flies collected near confined poultry feeding operations. Sci Total Environ 407 (8):2701–2710

Gyles CL (2008) Antimicrobial resistance in selected bacteria from poultry. Anim Health Res Rev 9:149–158

Hammerum AM (2012) Enterococci of animal origin and their significance for public health. Clin Microbiol Infect 18(7):619–625

Hartl DL, Jones EW (1998) Genetics: principles and analysis, 4th edn. Jones and Bartlett Publishers, Boston, Toronto, London (ISBN 0-7637-0489-X)

Hasan B, Faruque R, Drobni M et al (2011) High prevalence of antibiotic resistance in pathogenic *Escherichia coli* from large- and small-scale poultry farms in Bangladesh. Avian Dis 55 (4):689–692

Hayashi T, Makino K, Ohnishi M et al (2001) Complete genome sequence of enterohemorrhagic *Escherichia coli* O157:H7 and genomic comparison with a laboratory strain K-12. DNA Res 8:11–22
Hawkins C, Harper D, Burch D et al (2010) Topical treatment of *Pseudomonas aeruginosa* otitis of dogs with a bacteriophage mixture: a before/after clinical trial. Vet Microbiol 146(3–4): 309–313
Hayden MK (2015) Measuring carbapenem-resistant enterobacteriaceae in the United States a critical step for control. JAMA 314(14):1455–1456
Hegstad K, Mikalsen T, Coque TM et al (2010a) Mobile genetic elements and their contribution to the emergence of antimicrobial resistant *Enterococcus faecalis* and *Enterococcus faecium*. Clin Microbiol Infect 16:541–554
Hegstad K, Langsrud S, Lunestad B et al (2010b) Does the wide use of quaternary ammonium compounds enhance the selection and spread of antimicrobial resistance and thus threaten our health? Microb Drug Resist 16:91–104
Hendrix RW (2002) Bacteriophages: evolution of the majority. Theor Popul Biol 61:471–480
Heuer H, Schmitt H, Smalla K (2011) Antibiotic resistance gene spread due to manure application on agricultural fields. Curr Opin Microbiol 14(3):236–243
Holmberg SD, Wells JG, Cohen ML (1984) Animal-to-man transmission of antimicrobial-resistant Salmonella: investigations of U.S. outbreaks, 1971–1983. Science 225:833–835
Hooge DM (1999) Antibiotics and their alternatives for poultry examined. Feedstuff, 17 May, pp 59
Horrigan L, Lawrence RS, Walker P (2002) How sustainable agriculture can address the environmental and human health harms of industrial agriculture. Environ Health Perspect 110 (5):445–456
Hoyle DV, Knight HI, Shaw DJ et al (2004a) Acquisition and epidemiology of antibiotic-resistant *Escherichia coli* in a cohort of newborn calves. J Antimicrob Chemother 53:867–871. doi:10.1093/jac/dkh177
Hoyle DV, Shaw DJ, Knight HI et al (2004b) Age-related decline in carriage of ampicillin-resistant *Escherichia coli* in young calves. Appl Environ Microbiol 70:6927–6930. doi:10.1128/AEM.70.11.6927-6930.2004
Hoyle DV, Yates CM, Chase-Topping ME et al (2005) Molecular epidemiology of antimicrobial-resistant *Escherichia coli* strains in a cohort of newborn calves. Appl Environ Microbiol 71:6680–6688. doi:10.1128/AEM.71.11.6680-6688.2005
Hoyle DV, Davison HC, Knight HI et al (2006) Molecular characterisation of bovine faecal *Escherichia coli* shows persistence of defined ampicillin resistant strains and the presence of class 1 integrons on an organic beef farm. Vet Microbiol 115:250–257. doi:10.1016/j.vetmic.2006.01.006
Hruby M, Cowieson AJ (2006) The role of enzymes and betaine in antibiotic growth promoter free nutrition. In: Barug D, Long J de, A. Kies AK, Verstegen MWA (eds) Antimicrobial growth promoters: where do we go from here? Academic Publisher, Wageningen, The Netherlands, pp 269–287
Huff WE, Huff GR, Rath NC et al (2002a) Prevention of *Escherichia coli* infection in broiler chickens with a bacteriophage aerosol spray. Poult Sci 81(10):1486–1491
Huff WE, Huff GR, Rath NC, Balog JM et al (2002b) Prevention of *Escherichia coli* respiratory infection in broiler chickens with bacteriophage (SPR02). Poult Sci 81(4):437–441
Huff WE, Huff GR, Rath NC et al (2003a) Evaluation of aerosol spray and intramuscular injection of bacteriophage to treat an *Escherichia coli* respiratory infection. Poult Sci 82(7):1108–1112
Huff WE, Huff GR, Rath NC et al (2003b) Bacteriophage treatment of a severe *Escherichia coli* respiratory infection in broiler chickens. Avian Dis 47:1399–1405
Huff WE, Huff GR, Rath NC (2006) Bacteriophage: a safe and natural alternative to AGPs. In: Barug D, de Jong J, Kies AK, Verstegen MWA (eds), Antimicrobial growth promoters: where do we go from here, vol 978. Wageningen Academic Publishers, The Netherlands, pp 341–360
Hummel R, Tschape H, Witte W (1986) Spread of plasmid-mediated nourseothricin resistance due to antibiotic use in animal husbandry. J Basic Microbiol 26:461–466

Hrudey SE, Hrudey EJ (2004) Safe drinking water: lessons from recent outbreaks in affluent nations. IWA Publishing, London, p 514

Inal JM (2003) Phage therapy: a reappraisal of bacteriophages as antibiotics. Arch Immunol Ther Exp 51:237–244

Inglis GD, McAllister TA, Busz HW et al (2005) Effects of subtherapeutic administration of antimicrobial agents to beef cattle on the prevalence of antimicrobial resistance in *Campylobacter jejuni* and *Campylobacter hyointestinalis*. Appl Environ Microbiol 71 (7):3872–3881

Jassim SAA, Limoges RG (2013) Impact of external forces on cyanophage–host interactions in aquatic ecosystems. World J Microbiol Biotechnol 29(10):1751–1762

Jassim SAA, Limoges RG (2014) Natural solution to antibiotic resistance: bacteriophages 'The Living Drugs'. World J Microbiol Biotechnol 30(8):2153–2170

Jensen BB (1998) The impact of feed additives on the microbial ecology of the gut in young pigs. J Anim Feed Sci 7(1):45–64

Jeters RT, Wang GR, Moon K et al (2009) Tetracycline-associated transcriptional regulation of transfer genes of the *Bacteroides* conjugative transposon CTnDOT. J Bacteriol 191:6374–6382

Jukes TH, Stokstad ELR, Taylor RR et al (1950) Growth promoting effect of aureomycin on pigs. Arch Biochem 26:324–330

Khachatourians GG (1998) Agricultural use of antibiotics and the evolution and transfer of antibiotic-resistant bacteria. Can Med Assoc J 159:1129–1136

Khan SJ, Roser DJ, Davies CM et al (2008) Chemical contaminants in feedlot wastes: concentrations, effects and attenuation. Environ Int 34:839–859

Kim KH, Ingale SL, Kim JS (2014) Bacteriophage and probiotics both enhance the performance of growing pigs but bacteriophage are more effective. Anim Feed Sci Technol 196:88–95

Kittler S, Fischer S, Abdulmawjood A et al (2013) Effect of bacteriophage application on *Campylobacter jejuni* loads in commercial broiler flocks. Appl Environ Microbiol 79 (23):7525–7533. http://dx.doi.org/10.1128/AEM.02703-13

Kobayashi M, Sasaki T, Agui N (2002) Possible food contamination with the excreta of housefly with enterohemorrhagic *Escherichia coli* O157:H7. Med Entomol Zool 53(2):83–87

Kocher A (2006) Interfacing gut health and nutrition: the use of dietary prebiotics and probiotics to maximise growth performance in pigs and poultry. In: Barug D, Long J de, A. Kies AK, Verstegen MWA (eds) Antimicrobial growth promoters: where do we go from here? Academic Publisher, Wageningen, The Netherlands, pp 289–310

Krylov VN (2001) Phagotherapy in terms of bacteriophage genetics: hopes, perspectives, safety, limitations. Genetika 37:869–887

Kudva IT, Jelacic S, Tarr PI et al (1999) Biocontrol of *Escherichia coli* O157 with O157-specific bacteriophages. Appl Environ Microbiol 65:3767–3773

Kümmerer K (2004) Resistance in the environment. J Antimicrob Chemother 54(2):311–320

Kutter E, De Vos D, Gvasalia G et al (2010) Phage therapy in clinical practice: treatment of human infections. Curr Pharm Biotechnol 11:69–86

Laanto E, Bamford JKH, Laakso J, Sundberg L-R (2012) Phage-driven loss of virulence in a fish pathogenic bacterium. PLoS ONE 7(12):e53157. doi:10.1371/journal.pone.0053157

Law D (2000) The history and evolution of *Escherichia coli* O157 and other Shiga toxin-producing *E. coli*. World J Microbiol Biotechnol 16(8–9):701–709

Laxminarayan R, Duse A, Wattal C et al (2013) Antibiotic resistance: the need for global solutions. Lancet Infect Dis 13(12):1057–1098

Lefebvre B, Diarra MS, Giguere K et al (2005) Antibiotic resistance and hypermutability of *Escherichia coli* O157 from feedlot cattle treated with growth-promoting agents. J Food Prot 68(11):2411–2419

Leonard CG, Colón AE, Cole RM (1968) Transduction in group A streptococcus. Biochem Biophys Res Commun 30(2):130–135

Levy SB, FitzGerald GB, Macone AB (1976) Changes in intestinal flora of farm personnel after introduction of a tetracycline-supplemented feed on a farm. N Engl J Med 295(11):583–588

Levy SB (2002) The antibiotic paradox. How the misuse of antibiotics destroys their curative powers, 2nd edn. Perseus Publishing, Cambridge, MA, pp 353 (ISBN 0–7382–0440–4)

Levy SB, Marshall B (2004) Antibacterial resistance worldwide: causes, challenges and responses. Nat Med 10(12 suppl):S122–S129

Ling LL, Schneider T, Peoples AJ et al (2015) A new antibiotic kills pathogens without detectable resistance. Nature 517:455–459. doi:10.1038/nature14098

Livermore DM (1995) β-lactamases in laboratory and clinical resistance. Clin Microbiol Rev 8:557–584

Livermore DM (2009) Has the era of untreatable infections arrived? J Antimicrob Chemother 64: i29–i36. doi:10.1093/jac/dkp255

Liljeqvist T, Andresen D, Zuo Y, Weston C (2012) Antibiotic resistance: moving forward to the past. NSW Public Health Bull 23(1–2):37

Livermore D (2003) Bacterial resistance: origins, epidemiology, and impact. Clin Infect Dis 36: S11–S23

Lloyd DH (2012) Alternatives to conventional antimicrobial drugs: a review of future prospects. Vet Dermatol 23:299–304

Loc-Carrillo C, Abedon ST (2011) Pros and cons of phage therapy. Bacteriophage 1:111–114. doi:10.4161/bact.1.2.14590

Loc-Carrillo C, Atterbury RJ, El-Shibiny A et al (2005) Bacteriophage therapy to reduce *Campylobacter jejuni* colonization of broiler chickens. Appl Environ Microbiol 71(11): 6554–6563

Lorch A (1999) Bacteriophages: an alternative to antibiotics? Biotechnol Dev Monit 39:14–17. http://biotech-monitor.nl/3905.htm

Lu L (2004) Autoinducer 2-based quorum sensing response of *Escherichia coli* to subtherapeutic tetracycline exposure. http://oaktrust.library.tamu.edu/bitstream/handle/1969.1/4198/etd-tamu-2004B-FSTC-Lu.pdf?sequence=1

Lu K, Asano R, Davies J (2004) Antimicrobial resistance gene delivery in animal feeds. CDC 10 (4-April) http://wwwnc.cdc.gov/eid/article/10/4/03–0506_article

Lu TK, Koeris MS (2011) The next generation of bacteriophage therapy. Curr Opin Microbiol 14 (5):524–531. doi:10.1016/j.mib.2011.07.028

Lyons RW, Samples CL, DeSilva HN et al (1980) An epidemic of resistant *Salmonella* in a nursery. Animal-to-human spread. JAMA 243(6):546–547

Mackie RI, Koike S, Krapac I et al (2006) Tetracycline residues and tetracycline resistance genes in groundwater impacted by swine production facilities. Anim Biotechnol 17(2):157–176

Macovei L, Ludek Zurek L (2006) Ecology of antibiotic resistance genes: characterization of enterococci from houseflies collected in food settings. Appl Environ Microbiol 72(6): 4028–4035

Mah TF (2012) Biofilm-specific antibiotic resistance. Future Microbiol 7:1061–1072

Maiques E, Ubeda C, Tormo MA et al (2007) Role of staphylococcal phage and SaPI integrase in intra- and interspecies SaPI transfer. J Bacteriol 189:5608–5616. doi:10.1128/JB.00619-07

Manges AR, Johnson JR (2012) Food-borne origins of *Escherichia coli* causing extraintestinal infections. Clin Infect Dis 55(5):712–719

Manna SK, Brahmane MP, Manna C et al (2006) Occurrence, virulence characteristics and antimicrobial resistance of *Escherichia coli* O157 in slaughtered cattle and diarrhoeic calves in West Bengal, India. Lett Appl Microbiol 43:405–409

Marti E, Variatza E, Balcázar JL (2014a) The role of aquatic ecosystems as reservoirs of antibiotic resistance. Trends Microbiol 22:36–41

Marti E, Variatza E, Balcázar JL (2014b) Bacteriophages as a reservoir of extended-spectrum β-lactamase and fluoroquinolone resistance genes in the environment. Clin Microbiol Infect 20 (7):456–459

Martin R (2004) How ravenous soviet viruses will save the world. Wired Publishing Web. http://www.wired.com/wired/archive/11. 10/phages_pr.html

Martinez JL (2008) Antibiotics and antibiotic resistance genes in natural environments. Science 321:365–367
Martinez JL, Baquero F (2000) Mutation frequencies and antibiotic resistance. Antimicrob Agents Chemother 44:1771–1777
Martinez JL, Baquero F, Andersson DI (2007) Predicting antibiotic resistance. Nat Rev Microbiol 5:958–965
Martinez JL, Fajardo A, Garmendia L et al (2009) A global few of antibiotic resistance. FEMS Microbiol Rev 33:44–65
Matsuzaki S, Rashel M, Uchiyama J et al (2005) Bacteriophage therapy: a revitalized therapy against bacterial infectious diseases. J Infect Chemother 11:211–219
McDermott PF, Walker RD, White DG (2003) Antimicrobials: modes of action and mechanisms of resistance. Int J Toxicol 22(2):135–143
McEwen S, Fedorka-Cray P (2002) Antimicrobial use and resistance in animals. Clin Infect Dis 34 (Suppl. 3):S93–S106
McManus PS, Stockwell VO, Sundin GW, Jones AL (2002) Antibiotic use in plant agriculture. Annu Rev Phytopathol 40:443–465
McShan WM, Nguyen SV (2016) The bacteriophages of *Streptococcus pyogenes*. In: Ferretti JJ, Stevens DL, Fischetti VA (eds) *Streptococcus pyogenes*: basic biology to clinical manifestations. University of Oklahoma Health Sciences Center, Oklahoma City (OK). http://www.ncbi.nlm.nih.gov/books/NBK333409/
Mellon M, Benbrook C, Benbrook KL (2001) Hogging it: estimates of antimicrobial abuse in livestock. Publishing UCS Web. http://www.iatp.org/files/Hogging_It_Estimates_of_Antimicrobial_Abuse_in.pdf. Accessed 4 July 2004
Meltzer E, Schwartz E (2007) Enteric fever: an Israeli perspective. Isr Med Assoc J 9(10):736–741
Mempel M, Feucht H, Ziehbuhr W et al (1994) Lack of mecA transcription in slime-negative phase variants of methicillin-resistant *Staphylococcus epidermidis*. Antimicrob Agents Chemother 38:1251–1255
Merabishvili M, Pirnay JP, Verbeken G et al (2009) Quality-controlled small-scale production of a well-defined bacteriophage cocktail for use in human clinical trials. PLoS ONE 4:e4944. doi:10.1371/journal.pone.0004944
Merril CR, Scholl D, Adhya SL (2003) The prospect for bacteriophage therapy in Western medicine. Nat Rev Drug Discov 2:489–497
Michod RE, Bernstein H, Nedelcu AM (2008) Adaptive value of sex in microbial pathogens. Infect Genet Evol 8(3):267–285
Miller RW, Skinner EJ, Sulakvelidze A et al (2010) Bacteriophage therapy for control of necrotic enteritis of broiler chickens experimentally infected with *Clostridium perfringens*. Avian Dis 54(1):33–40
Millet S, Maertens L (2011) The European ban on antibiotic growth promoters in animal feed: from challenges to opportunities. Vet J 187:143–144
Mølbak K, Baggesen DL, Aarestrup FM et al (1999) An outbreak of multidrug-resistant, quinolone-resistant *Salmonella enterica* serotype *typhimurium* DT104. N Engl J Med 341 (19):1420–1425
Modi SR, Lee HH, Spina CS, Collins JJ (2013) Antibiotic treatment expands the resistance reservoir and ecological network of the phage metagenome. Nature 499(7457):219–222
Moore PR, Evenson A, Luckey TD et al (1946) Use of sulphasuccidine, streptothricin and streptomycin in nutrition studies with the chick. J Biol Chem 165:437–441
Muniesa M, Jofre J (1998) Abundance in sewage of bacteriophages that infect *Escherichia coli* O157:H7 and that carry the Shiga toxin 2 gene. Appl Environ Microbiol 64(7):2443–2448
Muniesa M, García A, Miró E et al (2004) Bacteriophages and diffusion of β-lactamase genes. Emerg Infect Dis 10(6):1134–1137
Nayak R, Kenney PB, Keswani J, Ritz C (2003) Isolation and characterisation of *Salmonella* in a turkey production facility. Br Poult Sci 44:192–202
Nelson DC, Schmelcher M, Rodriguez-Rubio L et al (2012) Endolysins as antimicrobials. Adv Virus Res 83:299–365

Niewold TA (2007) The nonantibiotic anti-inflammatory effect of antimicrobial growth promoters, the real mode of action? A hypothesis. Poult Sci 86:605–609

Nordmann P, Naas T, Laurent Poirel L (2011) Global spread of carbapenemase-producing Enterobacteriaceae. Emerg Infect Dis 17(10):1791–1798

Normark BH, Normark S (2002) Evolution and spread of antimicrobial resistance. J Intern Med 252:91–106

O'Brien TF (2002) Emergence, spread, and environmental effect of antimicrobial resistance: how use of an antimicrobial anywhere can increase resistance to any antimicrobial anywhere else. Clin Infect Dis 34(Suppl. 3):S78–S84

Ochiai K, Yamanaka T, Kimura K, Sawada O (1959) Inheritance of drug resistance (and its transfer) between *Shigella* strains and between *Shigella* and *E. coli* strains. Nippon Iji Shinpo (Jap. M.J.) 1861:34–36

O'Flaherty S, Paul Ross R, Coffey A (2014) Bacteriophage and their lysins for elimination of infectious bacteria. FEMS Microbiol Rev 33(4):801–819

O'Hara A, Shanahan F (2006) The gut flora as a forgotten organ. EMBO Rep 7:688–693

OIE (2007) List of antimicrobials of veterinary importance. http://www.oie.int/fileadmin/Home/eng/Internationa_Standard_Setting/docs/pdf/OIE_list_antimicrobials.pdf. Accessed 17 July 2013

O'Shea YA, Boyd EF (2002) Mobilization of the pathogenicity island between *Vibrio cholera* isolates mediated by CP-T1 generalized transduction. FEMS Microbiol Lett 214:153–157. doi:10.1111/j.1574-6968.2002.tb11339.x

Page SW, Gautier P (2012) Use of antimicrobial agents in livestock. Rev Sci Tech Off Int Epiz 31 (1):145–188

Palmer KL, Kos VN, Gilmore MS (2010) Horizontal gene transfer and the genomics of enterococcal antibiotic resistance. Curr Opin Microbiol 13:632–639

Parsley LC, Consuegra EJ, Kakirde KS et al (2010) Identification of diverse antimicrobial resistance determinants carried on bacterial, plasmid, or viral metagenomes from an activated sludge microbial assemblage. Appl Environ Microbiol 76:3753–3757

Patterson JA, Burkholder KM (2003) Application of prebiotics and probiotics in poultry production. Poult Sci 82:627–631

Paulson JA, Zaoutis TE (2015) Nontherapeutic use of antimicrobial agents in animal agriculture: implications for pediatrics. Pediatrics 136(6):e1670–e1677

Pei R, Kim SC, Carlson KH, Pruden A (2006) Effect of river landscape on the sediment concentrations of antibiotics and corresponding antibiotic resistance genes (ARG). Water Res 40:2427–2435

Pereira MS, Barreto VP, Siqueira-Junior JP (1997) Phage-mediated transfer of tetracycline resistance in *Staphylococcus aureus* isolated from cattle in Brazil. Microbios 92:147–155

Persoons D, Dewulf J, Smet A et al (2010) Prevalence and persistence of antimicrobial resistance in broiler indicator bacteria. Microb Drug Resist 16:67–74

Petridis M, Bagdasarian M, Waldor MK, Walker E (2006) Horizontal transfer of Shiga toxin and antibiotic resistance genes among *Escherichia coli* strains in house fly (Diptera: Muscidae) gut. J Med Entomol 43(2):288–295

Phillips I, Phillips I, Casewell M, Cox et al (2004) Does the use of antibiotics in food animals pose a risk to human health? A critical review of published data. J Antimicrob Chemoth 53:28–52

Ricke SC (2003) Perspectives on the use of organic acids and short chain fatty acids as antimicrobials. Poult Sci 82:632–639

Pirnay JP, De Vos D, Verbeken G et al (2011) The phage therapy paradigm: Prêt-à-Porter or Sur-mesure? Pharm Res 28:934–937

Prescott JF (2008) Antimicrobial use in food and companion animals. Anim Health Res Rev 9:127–133

Price LB, Graham JP, Lackey LG et al (2007) Elevated risk of carrying gentamicin-resistant *Escherichia coli* among US poultry workers. Environ Health Perspect 115(12):1738–1742

Quarles W (2006) Feedlots, pathogens, and antibiotic pollution. In: Common sense pest control quarterly: special double issue. Feedlot antibiotics produce pathogens? XXII (2–3):5–7. http://www.birc.org/SprSummer2006.pdf

Queenan AM, Bush K (2007) Carbapenemases: the versatile β-Lactamases. Clin Microbiol Rev 20 (3):440–458

Rashid M, Rakib MM, Hasan B (2015) Antimicrobial-resistant and ESBL-producing *Escherichia coli* in different ecological niches in Bangladesh. Infect Ecol Epidemiol 5. doi:10.3402/iee.v5.26712

Ratcliff J (2000) Antibiotic bans: a European perspective. In: Proceedings of the AFMA symposium on: improving animal performance through nutrition 2000, AFMA 2000, Pretoria, South Africa, p 128

Rivault G, Cloarec A, Le Guyader A (1993) Bacterial contamination of food by cockroaches. J Environ Health 55(8):21–22

Roe MT, Pillai SD (2003) Monitoring and identifying antibiotic resistance mechanisms in bacteria. Poult Sci 82(4):622–626

Rossi F, Rizzotti L, Felis GE, Torriani S (2014) Horizontal gene transfer among microorganisms in food: current knowledge and future perspectives. Food Microbiol 42:232–243

Rowe B, Ward LR, Threlfall EJ (1997) Multidrug-resistant *Salmonella typhi*: a worldwide epidemic. Clin Infect Dis 24(Suppl 1):S106–S109

Sapkota AR, Ojo KK, Roberts MC, Schwab KJ (2006) Antibiotic resistance genes in multidrug-resistant Enterococcus spp. and Streptococcus spp. recovered from the indoor air of a large-scale swine-feeding operation. Lett Appl Microbiol 43(5):534540

Sarmah AK, Meyer MT, Boxall AB (2006) A global perspective on the use, sales, exposure pathways, occurrence, fate and effects of veterinary antibiotics (VAs) in the environment. Chemosphere 65(5):725–759

Santander J, Robeson J (2007) Phage-resistance of *Salmonella enterica* serovar enteritidis and pathogenesis in *Caenorhabditis elegans* is mediated by the lipopolysaccharide. Electron J Biotechnol 10(4):627–632. doi:10.1098/rspb.2001.1945

Saunders JR, Allison H, James CE, McCarthy AJ, Sharp R (2001) Phage-mediated transfer of virulence genes. J Chem Technol Biotechnol 76(7):662–666

Schmieger H, Schicklmaier P (1999) Transduction of multiple drug resistance of *Salmonella enterica* serovar Typhimurium DT104. FEMS Microbiol Lett 170:251–256

Schmitt H, Martinali B, Stoob K et al (2006) Antibiotics as environmental pollutants: effects on soil microorganisms. Umweltwiss Schadst Forsch 18(2):110–118

Shousha A, Awaiwanont N, Sofka D et al (2015) Bacteriophages isolated from chicken meat and the horizontal transfer of antimicrobial resistance genes. Appl Environ Microbiol 81(14): 4600–4606

Shuford JA, Patel R (2005) Antimicrobial growth promoter use in livestock-implications for human health. Rev Med Microbiol 16:17–24

Shute J (2013) Antibiotic resistance: death of a wonder drug. The Telegraph. http://s.telegraph.co.uk/graphics/projects/antibiotic-resistance/

Silbergeld EK, Graham J, Price LB (2008) Industrial food animal production, antimicrobial resistance, and human health. Annu Rev Public Health 29:151–169

Sillankorva SM, Oliveira H, Azeredo J (2012) Bacteriophages and their role in food safety. Int J Microbiol 2012: Article ID 863945. doi:10.1155/2012/86394

Sillankorva S, Azeredo J (2014) Bacteriophage attack as an anti-biofilm strategy. Methods Mol Biol 1147:277–285

Singer RS, Finch R, Wegener HC et al (2003) Antibiotic resistance-the interplay between antibiotic use in animals and human beings. Lancet Infect Dis 3:47–51

Skurnik M, Strauch E (2006) Phage therapy: facts and fiction. Int J Med Microbiol 296(1):5–14. doi:10.1016/j.ijmm.2005.09.002

Skurnik M, Pajunen M, Kiljunen S (2007) Biotechnological challenges of phage therapy. Biotechnol Lett 29:995–1003. doi:10.1007/s10529-007-9346-1
Smillie CS, Smith MB, Friedman J et al (2011) Ecology drives a global network of gene exchange connecting the human microbiome. Nature 480(7376):241–244
Smith DL, Harris AD, Johnson JA et al (2002) Animal antibiotic use has an early but important impact on the emergence of antibiotic resistance in human commensal bacteria. Proc Natl Acad Sci USA 99:6434–6439
Smith HW, Huggins MB (1982) Successful treatment of experimental *Escherichia coli* infections in mice using phages: its general superiority over antibiotics. J Gen Microbiol 128:307–318
Smith HW, Huggins MB, Shaw KM (1987) The control of experimental *Escherichia coli* diarrhoea in calves by means of bacteriophages. J Gen Microbiol 133(5):1111–1126
Smith JL, Fratamico PM, Gunther NW (2007) Extraintestinal pathogenic *Escherichia coli*. Foodborne Pathog Dis 4(2):134–163
Sousa CP (2006) The versatile strategies of *Escherichia coli* path types: a mini review. J Venom Anim Toxins Incl Trop Dis 12:363–373
Spits HB (2009) Bacteriophages for improvement of intestinal health in pigs and poultry. Master thesis, Faculty of Medicine, University of Utrecht. http://dspace.library.uu.nl/handle/1874/36365
Sproston EL, Macrae M, Ogden ID et al (2006) Slugs: potential novel vectors of *Escherichia coli* O157. Appl Environ Microbiol 72(1):144–149
Stahly TS, Cromwell GL, Monegue HJ (1980) Effects of dietary inclusion of copper and(or) antibiotics on the performance of weanling pigs. J Anim Sci 51:1347–1351
Sørum H, Sunde M (2001) Resistance to antibiotics in the normal flora of animals. Vet Res 32(3–4):227–241
Stokestad ELR, Jukes TH (1950) Further observations on the "animal protein factor". Proc Soc Exp Biol Med 73:523–528
Sulakvelidze A (2005) Phage therapy: an attractive option for dealing with antibiotic-resistant bacterial infections. Drug Discovery Today 10:807–809
Sulakvelidze A, Kutter E (2005) Bacteriophage therapy in humans. In: Kutter E, Sulakvelidze A (eds) Bacteriophages: biology and application. CRC Press, Boca Raton, pp 381–436
Sulakvelidze A, Alavidze Z, Morris JG Jr (2001) Bacteriophage therapy. Antimicrob Agents Chemother 45(3):649–659
Sulakvelidze A, Barrow P (2005) Phage therapy in animals and agribusiness. In: Kutter E, Sulakvelidze A (eds) Bacteriophages: biology and application. CRC Press, Boca Raton, pp 335–380
Summers AO (2002) Generally overlooked fundamentals of bacterial genetics and ecology. Clin Infect Dis 34(Suppl 3):S85–S92
Sunde M, Nordstrom M (2006) The prevalence of, associations between and conjugal transfer of antibiotic resistance genes in *Escherichia coli* isolated from Norwegian meat and meat products. J Antimicrob Chemother 58(4):741–747
Swerdlow DL, Woodruff BA, Brady RC et al (1992) A waterborne outbreak in Missouri of *Escherichia coli* O157:H7 associated with bloody diarrhea and death. Ann Intern Med 117(10):812–819
Ternak G (2005) Antibiotics may act as growth/obesity promoters in humans as an inadvertent result of antibiotic pollution? Med Hypotheses 64(1):14–16
Thiel K (2004) Old dogma, new tricks-21st century phage therapy. Nat Biotechnol 22:31–36
Turnidge J (2004) Antibiotic use in animals-prejudices, perceptions and realities. J Antimicrob Chemoth 53:26–27
US Congress (2011) Preservation of antibiotics for medical treatment act of 2011. H.R.965 112th: Congress, 2011–2013. Publishing Govtrack Web. https://www.govtrack.us/congress/bills/112/hr965/text. Accessed 9 Mar 2011
US FDA (2012) FDA annual summary on antimicrobials sold or distributed for use in food-producing animals. Food and Drug Administration Department of Health and Human

Services. September, 2014. http://www.fda.gov/downloads/ForIndustry/UserFees/AnimalDrugUserFeeActADUFA/UCM416983.pdf
Varga C, Rajic A, McFall ME et al (2009) Associations between reported on-farm antimicrobial use practices and observed antimicrobial resistance in generic fecal *Escherichia coli* isolated from Alberta finishing swine farms. Prev Vet Med 88(3):185–192
Verheust C, Pauwels K, Mahillon J et al (2010) Contained use of bacteriophages: risk assessment and biosafety recommendations. Appl Biosaf 15(1):32–44
Vieira AR, Houe H, Wegener HC et al (2009) Association between tetracycline consumption and tetracycline resistance in *Escherichia coli* from healthy Danish slaughter pigs. Foodborne Pathog Dis 6(1):99–109. doi:10.1089/fpd.2008.0152
Vigre H, Larsen PB, Andreasen M et al (2008) The effect of discontinued use of antimicrobial growth promoters on the risk of therapeutic antibiotic treatment in Danish farrow-to-finish pig farms. Epidemiol Infect 136(1):92–107. doi:10.1017/S095026880700814X
Vojtek I, Pirzada ZA, Henriques-Normark B et al (2008) Lysogenic transfer of group A Streptococcus superantigen gene among streptococci. J Infect Dis 197(2):225–234
Volkova VV, Lu Z, Besser T, Gröhn YT (2014) Modeling infection dynamics of bacteriophages in enteric *Escherichia coli:* estimating the contribution of transduction to antimicrobial gene spread. Appl Environ Microbiol 80:4350–4362
Völkel I, Czerny CP (2011) Phage therapy in the fields of veterinary medicine: an overview. Berl Munch Tierarztl Wochenschr 124(7–8):303–312
Vondruskova H, Slamova R, Trckova M et al (2010) Alternatives to antibiotic growth promoters in prevention of diarrhoea in weaned piglets: a review. Vet Med Czech 55(5):199–224
Waksman SA, Woodruff HB (1940) The soil as a source of microorganisms antagonistic to disease-producing bacteria. J Bacteriol 40:581–600
Wagenaar JA, Van Bergen MA, Mueller MA et al (2005) Phage therapy reduces *Campylobacter jejuni* colonization in broilers. Vet Microbiol 109(3–4):275–283
Wagner PW, Waldor MK (2002) Bacteriophage control of bacterial virulence. Infect Immun 70(8):3985–3993
Wegener HC (2003) Antibiotics in animal feed and their role in resistance development. Curr Opin Micobiol 6:439–445
Wellington EMH, Boxall ABA, Cross P et al (2013) The role of the natural environment in the emergence of antibiotic resistance in Gram-negative bacteria. Lancet Infect Dis 13:155–165
WHO (2000) Overcoming antimicrobial resistance. Report on infectious diseases. World Health Organization, Geneva, Switzerland, No. WHO/CDS/2000.2
WHO (2003) First joint FAO/OIE/WHO expert workshop on non-human antimicrobial usage and antimicrobial resistance: scientific assessment. Geneva, Switzerland. http://whqlibdoc.who.int/hq/2004/WHO_CDS_CPE_ZFK_2004.7.pdf. Accessed June 2007
WHO (2004) Second joint FAO/OIE/WHO expert workshop on non-human antimicrobial usage and antimicrobial resistance: management options. Oslo, Norway. http://whqlibdoc.who.int/hq/2004/WHO_CDS_CPE_ZFK_2004.8.pdf. Accessed Jun. 2007
WHO (2012) Antibiotic resistance in foodborne pathogens: evidence of the need for a risk management strategy. NW. https://cspinet.org/new/pdf/abrupdate.pdf
White DG, Zhao S, Sudler R et al (2001) The isolation of antibiotic-resistant salmonella from retail ground meats. N Engl J Med 345(16):1147–1154
Wiedenbeck J, Cohan FM (2011) Origins of bacterial diversity through horizontal genetic transfer and adaptation to new ecological niches. FEMS Microbiol Rev 35:957–976
Willi K, Sandmeier H, Kulik EM, Meyer J (1997) Transduction of antibiotic resistance markers among Actinobacillus actinomycetemcomitans strains by temperate bacteriophages Aa phi 23. Cell Mol Life Sci 53:904–910
Witte W (1998) Medical consequences of antibiotic use in agriculture. Science 279:996–997
Woodford N, Ellington MJ (2007) The emergence of antibiotic resistance by mutation. Clin Microbiol Infect 13:5–18

Woolhouse M, Ward M, van Bunnik B, Farrar J (2015) Antimicrobial resistance in humans, livestock and the wider environment. Phil Trans R Soc B 370:20140083. doi:10.1098/rstb.2014.0083

Wright GD (2007) The antibiotic resistome: the nexus of chemical and genetic diversity. Nat Rev Microbiol 5:175–186. doi:10.1038/nrmicro1614

Yan SS, Gilbert JM (2004) Antimicrobial drug delivery in food animals and microbial food safety concerns: an overview of in vitro and in vivo factors potentially affecting the animal gut microflora. Adv Drug Deliv Rev 56:1497–1521

Yan F, Polk B (2004) Commensal bacteria in the gut: learning who our friends are. Curr Opin Gastroenterol 20:565–571

Yan L, Hong SM, Kim IH (2012) Effect of bacteriophage supplementation on the growth performance, nutrient digestibility, blood characteristics, and fecal microbial shedding in growing pigs. Asian-Aust J Anim Sci 25:1451–1456

Yates CM, Pearce MC, Woolhouse MEJ, Amyes SGB (2004) High frequency transfer and horizontal spread of apramycin resistance in calf faecal *Escherichia coli*. J Antimicrob Chemother 54:534–537. doi:10.1093/jac/dkh353

Zhang Z, Kottadiel VI, Vafabakhsh R et al (2011) A promiscuous DNA packaging machine from bacteriophage T4. PLoS Biol 9(2):e1000592. doi:10.1371/journal.pbio.1000592

Zhang C, Wenli L, Wenhua L et al (2013) T4-like phage Bp7, a potential antimicrobial agent for controlling drug-resistant *Escherichia coli* in chickens. Appl Environ Microbiol 79:5559–5565. doi:10.1128/AEM.01505-13

Zhang J, Li Z, Cao Z et al (2015) Bacteriophages as antimicrobial agents against major pathogens in swine: a review. J Anim Sci Biotechnol 6:39. doi:10.1186/s40104-015-0039-7

Chapter 3
Bacteriophage Biocontrol in Poultry

Abstract This chapter focuses on two foodborne pathogens of contemporary importance in poultry industries: *Campylobacter* and *Salmonella*. These two foodborne diseases associated with the consumption of tainted poultry meat and processed products are of public health significance worldwide. We also discuss the development of innovative bacteriophage applications to control foodborne pathogens in the poultry food chain and using novel phage technologies to replace or enhance the effectiveness of antibiotics in food production applications and in poultry feed industries.

Keywords Animal farm · Bacteriophage · Bacterial detection · Biocontrol · *Campylobacter* · Environment · Foodborne pathogens · Livestock · Multidrug-resistant bacteria · Poultry · *Salmonella*

Contents

S.A.A. Jassim and R.G. Limoges, *Bacteriophages: Practical Applications for Nature's Biocontrol*, DOI 10.1007/978-3-319-54051-1_3

3.1 Introduction

3.1.1 Poultry as a Source of Food

Global poultry meat and egg production has rapidly grown in recent years. The current worldwide forecasts of world poultry meat production suggest that it could rise by 29% over the coming decade, averaging 2.4% growth per year to reach almost 118 million tons. The corresponding output of eggs shows that between 2000 and 2010, global egg output expanded by more than two per cent a year from 51 to 63.8 million tons (FAO 2010). Experts from the International Egg Commission expect that within a few years the production volume of eggs will be greater than that of beef and veal (assuming the growth rates remain fairly constant) (FAO 2010). Chicken accounts for about 86% of all poultry worldwide and hen eggs represent 92% of the global primary production of eggs (FAO 2010). Poultry consumption today is at an all-time high and this trend is expected to continue, driven by the combination of poultry's low production costs, high nutritional value and projected economic growth throughout the world.

3.1.2 Poultry Safety

In terms of food safety, poultry ranks first or second in foods associated with disease in Australia, Canada, England, and Wales while in the US poultry is the food vehicle in 8% (ranked third) of the reported food-borne disease outbreaks (Conner et al. 2001). Epidemiological reports indicate that more than 95% of all food-borne illnesses are the result of activities occurring after the product has left the plant; that is, illness is generally the result of temperature abuse and improper handling or preparation (Conner et al. 2001). However, when contamination and illness occur, investigators tend to look at raw product (how it was produced, processed, and handled) and press for elimination of pathogens before the product reaches the consumer. This creates a challenge to the poultry industry to improve the microbiological safety and quality of its products (Conner et al. 2001).

3.1.3 Foodborne Bacterial Pathogens

Salmonella and *Campylobacter* are two of the most prevalent foodborne bacterial pathogens and are believed to cause approximately 81% of all bacterial based foodborne illness (FAO-WHO 2009). These two pathogens are having an increasing and significant impact on human health, the food industry and government regulations regarding food safety. During the past several years, both pathogens demonstrated increased antibiotic resistance and globally the number of infections from these pathogens is increasing. Salmonellosis, the disease caused by *Salmonella*, is a global public health problem. Infections that spread to the bloodstream can result in a severe illness leading to disability or death. Antibiotics are essential for the treatment of patients with these invasive infections. However, the choice of such antimicrobial agents has become severely limited because of the emergence of antibiotic-resistant *Salmonella,* possibly or at least partly resulting from the overuse of antibiotics in food animals (Jassim and Limoges 2014).

The possible sources *Salmonella* infection within broiler flocks include water, feed, litter, farm staff and the environment, both inside and outside the broiler house (FAO-WHO 2009). Furthermore, hatcheries are possible sources of infection, as is vertical transmission (FAO-WHO 2009). Therefore, many of the current risk assessment models today start by estimating the prevalence of contaminated *Salmonella*-positive birds as the birds enter the slaughterhouses, and this means that on-farm control strategies are very poorly investigated at the present time (FAO-WHO 2009). For this reason, *Salmonella* levels have been chosen by the USDA as one of the quantitative standards used to indicate the level of bacterial contamination of meat and poultry products. Pursuant to the current USDA Hazardous Analysis Critical Control Points (HACCP)/Pathogen Reduction (PR) Act (HACCP/PR), performance standards have been set that allow inspectors to monitor whether processing plants are using adequate measures to reduce pathogen contamination of poultry (Dreyfuss et al. 2007). A maximum acceptable limit was set based on baseline surveys for each class of animal and food product.

The contamination or infection/colonisation of food producing poultry with organisms pathogenic to man is well recognised as resulting in exposure of these pathogens to the human population. The use of probabilistic quantitative risk assessment (QRA) methodologies in the modelling of food-borne bacterial pathogens in primary food production is so far fairly limited and often exists as the first module of a 'Farm-to Fork' human exposure QRA focussing on the risk of infection from food producing animals. Such QRA's include the modelling of *Campylobacter* spp. in broiler chickens and *Salmonella* spp. in egg production (European Commission 2002). Modern intensive animal husbandry practices introduced to maximize production seem to have led to the emergence and increased prevalence of *Salmonella* serovars and/or *Campylobacter* in herds or flocks of all the most important production animals (poultry, cattle, pig). For example, in the USA in 1969, 470,832 layer-hen farms with an average of 632 hens per farm produced 67 billion eggs per year; by 1992, the number of farms dropped by 85% to 70,623, the number of hens per farm increased by 470% to 2985 and annual production rose to 70 billion eggs (Sobel et al. 2002). In addition, the conditions and stress associated with transporting animals to slaughter and dietary changes prior to slaughter can increase carriage rates and shedding (WHO 2001). Poultry and poultry meat, often carrying millions of bacteria internally and externally, is an excellent vehicle for foodborne pathogens, particularly *Campylobacter* and *Salmonella* (Izat et al. 1988; Corry and Atabay 2001; Capita et al. 2004). In some circumstances *Listeria*, *Clostridium perfringens*, *Bacillus anthracis*, and the Proteobacteria (classes Gammaproteobacteria and Epsilonproteobacteria) are a major group (phylum) of Gram-negative bacteria. They include a wide variety of pathogens, such as, *Salmonella*, *Campylobacter*, *Escherichia*, *Vibrio*, *Helicobacter*, *Yersinia*, *Acinetobacter*, *Pseudomonas*, *Serratia* species, *Aeromonas* and many other notable genera also have the potential to be important pathogens in poultry products. However, *Salmonella* and *Campylobacter* are considered to be the major food-borne pathogens in the poultry industry and can be caused by consumption of eggs or poultry that are not adequately cooked.

Since *Campylobacter* and *Salmonella* are the most challenging bacteria to control and because their infection is the most frequent bacterial cause of enteric disease worldwide, we will focus on these two bacteria as a model for others in treating food-borne gastroenteritis in humans via phage biocontrol.

3.2 Sources of *Campylobacter* Infection for Poultry

The *Campylobacter* genus is currently composed of 25 species (Man 2011). They are spiral shaped Gram-negative cells and they are typically micro-aerobic, but have been found to also grow in aerobic as well as anaerobic environments (Newell and Fearnley 2003). They are thermophilic and have an optimum growth temperature of 42 °C but they can grow at 30 °C. Because avian species have a higher body temperature than mammals, *Campylobacter* is particularly suited to colonise them. However, *Campylobacter* resides in a commensal relationship with chickens. Infected chicken feces may contain up to 10^9 bacteria per 25 g, and due to the installations, the bacteria are rapidly spread to other chickens and can easily spread from bird to bird through a common water source or through contact with infected feces (Nyati and Prasad 2014). When an infected bird is slaughtered, *Campylobacter* can be transferred from the intestine to the meat. This colonization vastly exceeds the infectious dose of 10^3–10^4 bacteria for humans (Humphrey et al. 2007).

These bacteria have been implicated as a major cause of bacterial food-borne infection in the industrial world. For example, approximately 80% of raw chicken meat sold in the United Kingdom has been found to be contaminated with *Campylobacter* (Jorgensen et al. 2002). Up to 30% of human cases are due to handling, preparation and consumption of raw or undercooked poultry, although the routes of transmission from chickens to humans are not well understood (Humphrey et al. 2001; Kapperued et al. 2003; EFSA 2011). Most human Campylobacteriosis are due to *C. jejuni* (90%) with some due to *C. coli* (10%) (Friedman et al. 2000; Wilson et al. 2008). Other species such as *C. upsaliensis*, *C. lari*, and *C. fetus*, have also been associated with human diarrhea, particularly in the developing world (Coker et al. 2002; Man 2011). Their reservoirs are warm-blooded animals such as cattle, sheep, pigs, poultry, wildlife and domestic pets. Animal carriers are generally asymptomatic, but the species *C. fetus* can cause diarrhea and aseptic abortions in large animals (Moore et al. 2005). Contamination occurs through animal faeces and sewage discharge (Newell 2002). The bacterium distributed in farms by horizontal transmission from the environment of the broiler house is the most important route for flock colonization, which often occurs via human traffic (Newell and Fearnley 2003; Callicott et al. 2006). Avian hosts constitute a natural reservoir for thermophilic *Campylobacter* species, primarily *C. jejuni* and *C. coli*, and poultry flocks are frequently colonized in the intestinal tract with high numbers of the organisms as indicated. The prevalence rates in poultry, especially in slaughter-age broiler flocks, could reach as high as 100% on some farms (Sahin et al. 2015). *Campylobacter* has been demonstrated to be able to form biofilms on a variety of surfaces (Joshua et al. 2006) which increase its chances of survival in the poultry production environment (Trachoo et al. 2002).

The epidemiology and distribution of this bacterium occur following the exposure as follows: *Campylobacter* colonization of these birds develops rapidly, so that high levels of the organism may be present in faeces and cecal droppings by the time the flock is cleared, typically 5–8 days following the thinning process (Newell and Fearnley 2003; Allen et al. 2008). This has been supported with the comparison between *Campylobacter* present on vehicles, transport crates and modules arriving on a broiler farm at thinning and those subsequently recovered from the remaining birds (Allen et al. 2008). Moreover, there was preliminary evidence that particular strains may have spread between farms owned by the same company (Ridley et al. 2011a). Since the thinning crews and associated vehicles often travel from farm-to-farm with their own equipment, boots and working clothes and are likely to have had recent exposure to heavily contaminated environments, a requirement for heightened hygiene control in the thinning process is indicated (Ridley et al. 2011a, b; Sahin et al. 2015). The risk of transmitting *Campylobacter* to the flock can be reduced by good hygienic practices, including well maintained poultry houses with restricted access and effective use of physical hygiene barriers that involve dedicated footwear and clothing and/or disinfectant footbaths (Berndtson et al. 1996; van de Giessen et al. 1998; Hald et al. 2000). To prevent flock colonization by *Campylobacter* on poultry farms and in slaughterhouses, sufficient biosecurity measures are needed (Newell et al. 2011; Ridley et al. 2011a). The upgrading of biosecurity has successfully produced *Campylobacter*-free chickens, but these measures are expensive and difficult to maintain. This is largely due to the susceptibility of chickens to infection by *Campylobacter* and its ubiquity in the environment (Newell and Fearnley 2003). Moreover, unless increased biosecurity is applied globally, the benefits of reducing *Campylobacter* on some farms will be negated by cross-contamination from positive flocks at the abattoir (Herman et al. 2003).

3.2.1 Campylobacteriosis

Campylobacteriosis is one of the most important bacterial food-borne illnesses in humans worldwide (WHO 2013). Most people who become ill with Campylobacteriosis get diarrhea, cramping, abdominal pain and fever within two to five days after exposure to the organism. The diarrhea may be bloody or dysentery syndrome and can be accompanied by nausea and vomiting. The illness typically lasts about 6–7 days. Some infected persons do not have any symptoms. In persons with compromised immune systems, *Campylobacter* occasionally spreads to the bloodstream and causes a serious life-threatening infection.

The global epidemiology of *Campylobacter* spp. appears to depend on the geographical locations and animal habitats as the frequency and number of Campylobacteriosis cases are different from one continent to another. Campylobacteriosis is the most frequently reported zoonotic disease in the EU, with poultry meat being the major source for human cases (EFSA 2011). In the US *Campylobacter* species are found to cause the most illnesses attributed to animal contact (Hale et al. 2012) and is a common commensal of the gastrointestinal tract of various mammals and birds (Lee and Newell 2006). It is frequently found in chicken farms where the bacteria may spread rapidly (Humphrey et al. 2000). Contaminated birds are shedding up to 10^8 *Campylobacter* per gram of cecal contents. Although *Campylobacter* is insignificant for poultry health, it is a leading cause of food-borne gastroenteritis in humans worldwide and contaminated poultry meat is recognized as the main source for human exposure.

Campylobacter is consistently the most commonly reported gastrointestinal bacterial pathogen in humans in the European Union (EU) since 2005. The estimate of the true incidence rate for *Campylobacter*-associated infections is 9.2 million, with a total annual cost of €2.4 billion (EFSA 2011). The incidence of human *Campylobacter* infection in the EU is reflected by 200,507 confirmed cases of Campylobacteriosis in 2007 (EFSA 2009a), whereas, in only England and Wales, with case estimates numbering over 359,000 for the year 2000 (Adak et al. 2000) and for 2008–2009 the incidence of Campylobacteriosis was estimated to be 9.3 per 1000 person-years in the United Kingdom, and 5.8 per 1000 person-years in the Netherlands (for 2009). Only 1 out of every 9.3 cases in the United Kingdom and 1 out of 12 in the Netherlands are reported to national surveillance bodies, which explains the variance. In the USA, it is estimated that 1 out of 30.3 cases is reported by FoodNet sites, and that national incidence was 1.3 million cases in 2006 or 4.4 per 1000 (WHO 2013), with an estimated annual cost of illness to be $1.7 billion (Hoffmann et al. 2012). These studies also indicate that 1 out of 7 patients with Campylobacteriosis in the United Kingdom, and 1 out of 4 in the Netherlands, consulted their doctor; reflecting the severe nature of Campylobacteriosis (WHO 2013). The consumption of undercooked poultry meat and handling of raw poultry account for up to 41% of reported human infections with *Campylobacter* spp. (Havelaar et al. 2007; Stafford et al. 2007; Tam et al. 2009). Therefore, considerable research efforts have been devoted to the development of interventions to diminish *Campylobacter* contamination in poultry, with the intention to reduce the burden of food-borne illnesses (Sahin et al. 2015). For this reason, billions of dollars have been spent researching food safety, animal healthcare, environmental sanitation and human healthcare to find a cure, prevent infection or to control these bacteria without success (Crutchfield et al. 1997). In general, the issue of foodborne Campylobacteriosis is well recognized by WHO, FAO and OIE. Since *Campylobacter* is a zoonotic hazard, with both food and waterborne routes of transmission and in light of the challenges it presents in terms of control, it is clear that, both locally and globally, the issue needs to be addressed in a multidisciplinary manner (WHO 2013).

3.3 Sources of *Salmonella* Infection for Poultry

Salmonella are short bacilli, 0.7 × 2.5 μm, Gram-negative, aerobic or facultative anaerobic and are widely distributed in nature. The natural habitat of *Salmonella* may be divided into three categories based on the specificity of the host and clinical pattern of the disease:

(1) highly adapted to men: *Sal. typhi* and *Sal. paratyphi* A, B and C,
(2) agents of typhoid fever; highly adapted to animals: *Sal. dublin* (bovines), *Sal. choleraesuis* and *Sal. typhisuis* (swine), *Sal. pullorum* and *Sal. gallinarum* (poultry), responsible for animal paratyphoid and
(3) includes most of the serovars that affect men and animals, called zoonotic *Salmonella*, responsible for worldwide foodborne diseases, and detected in most species of animals used for human consumption, both wild and domestic animals (Gantois et al. 2009). A number of these are host specific (e.g., *Salmonella* serotype Pullorum in chickens, *Salmonella* serotype Cholera-Suis in pigs) (Behling et al. 2010). They attach to and colonize the intestines of chickens. It is a hearty organism that may remain active on dry fluff for weeks.

In many countries, poultry remains the dominant reservoir, although pork, beef, and mutton have served as vehicles of infection (Behling et al. 2010). Contamination of poultry carcasses and parts with *Salmonella* organisms is well documented and data are available for many parts of the world (Simmons et al. 2003). Transmission of *Salmonella* to men generally occurs by means of contaminated food and water, although person-to-person transmission may take place, mainly in hospitals. Contamination of poultry products (meat and eggs) destined for human consumption may occur at the slaughterhouse, during food preparation, or by cross-contamination with material from poultry with intestinal and systemic infections. *Salmonella* in poultry meat is transmitted through the handling of raw poultry carcasses and products, together with the consumption of undercooked poultry meat (Panisello et al. 2000). As for poultry meat, even a small number of infected birds at the slaughterhouse may contaminate the whole slaughter line if the carcasses are not correctly processed, multiplying the chances of occurrence of foodborne disease and a threat to public health.

Transmission of *Salmonella* in birds may occur vertically, via eggs, (this is unlike *Campylobacter* mentioned above) with the birth of infected chicks; horizontally, by means of ingestion of water, feed, fecal material, contaminated bedding material or dust; or by oral, nasal, conjunctival, cloacal and umbilical routes (Cox et al. 1996). Many *Salmonella* serotypes may survive for weeks or months in manure or bedding material, in equipment, in empty sheds, in the dirt around sheds that have been cleaned and disinfected, in feces of wild poultry, in dust particles, and in bird feeders. According to the above authors, *Salmonella* may survive in contaminated feed for at least 26 months, in feces of infected birds for more than 11 days when inside of sheds, or for 9 days in open spaces. Domestic and wild animals may also be carriers of *Salmonella*, spreading the microorganism in the environment where they live. These bacteria may cause acute and/or chronic disease in susceptible animals.

3.3.1 Salmonellosis

In the United States and EU, salmonellosis is the second most frequently reported zoonotic disease in humans, and the great majority of these infections are associated with the consumption of products such as meat, poultry, eggs, milk, seafood, and fresh produce contaminated with *Salmonella* (Foley et al. 2008; Bardina et al. 2012). The majority of human-infecting *Salmonella* strains originate from poultry and poultry-derived food products (Guard-Petter 2001). *Salmonella enterica* serovars, *enteritidis* and *Sal. typhimurium*, are responsible for the majority of *Salmonella* outbreaks and most events are reported to be due to consumption of contaminated eggs and poultry (Foley et al. 2008; EFSA 2011). *Salmonella* is also a known spoilage bacterium in processed foods. Evidence from many studies indicates that from 1 to 10 cells may constitute an infectious dose in some circumstances (Behling et al. 2010). Once ingested, it can cause fever, diarrhea, abdominal cramps, and even life-threatening infections (Pickering 2006). Incubation period is generally 6–72 h (Behravesh 2008) and can be longer than 10 days. In the US, there are more than 800,000 notified cases of infections caused by *Salmonella* spp., with an average of 500 deaths a year. Worldwide occurrence of salmonellosis is calculated in 1.3 billion cases and 3 million deaths (Thong et al. 1995). In 1988, there were 4 million cases of foodborne disease in the US and Canada, representing an estimated cost of US$4.8 billion, including losses in commercialization, productivity and labour (Todd 1989). In a 5-year period (1985–1989), there were 189 outbreaks in the US caused by *Sal. enteritidis*, with 6604 people involved, and 43 deaths. In Canada, the discovery of *Salmonella* from abattoir chickens increased from 16% in 2002 to 28% in 2008 (Agunos et al. 2012). Similarly, in the US, the prevalence of *Salmonella* also increased between 2007 and 2009 in retail chicken (Agunos et al. 2012), but it is unclear if the increases seen in both countries are coincidental or if the exchange of poultry products played a role (Nesbitt et al. 2011). Vertical transmission of *Sal. enteritidis* in commercial poultry was responsible for the increased number of cases of human infection in Europe, North America and other parts of the world (Humphrey et al. 1988; ICMSF 1998; Behling et al. 2010).

Salmonella spp. has a long history of food contamination and have caused illness from ingestion of contaminated eggs (Tessari et al. 2012). The vast majority of the Salmonellosis outbreaks that have resulted in death were also attributed to eggs (Tessari et al. 2012). The egg-borne pandemic of *Salmonella* serotype *enteritidis* phage type 4 and *Sal. typhimurium* in Europe and *Salmonella* serotype *enteritidis* phage type 8 in North America illustrates the importance of poultry products as vehicles of human salmonellosis (Behling et al. 2010; Tessari et al. 2012). Together, they account for approximately 80% of all human isolates with *Sal. enteritidis* being the most important. Other serovars do not individually exceed 1%. Bacteria can be on the outside of an eggshell because the egg exits the hen's body through the same passageway as faeces; thus, the reason for washing and sanitizing eggs at the processing plant. Bacteria can also be inside an un-cracked, whole egg. Contamination of eggs may be due to bacteria such as *Sal. enteritidis* in the hen's ovary or oviduct before the shell forms around the yolk and white.

Sal. enteritidis does not make the hen sick. It is also possible for eggs to become infected by *Sal. enteritidis* faecal contamination via the pores of the shells after they have been laid. Using epidemiological surveys, the CDC identified the consumption of eggs or egg-based foods as responsible for most of the outbreaks involving specific phagotypes (PT) of *Sal. enteritidis* PT-4 in the EU. Furthermore, *Sal. enteritidis*, *Sal. typhimurium*, *Sal. infantis*, *Sal. hadar and Sal. virchow* are considered the most important serovars in terms of public health in the EU (EFSA-ECDC 2007). On the other hand, the PT-8 and PT-13a were found in the US (Perales and Audicana 1989). The serovar types most frequently isolated from human sources in the US were *Sal. typhimurium*, *Sal. enteritidis*, *Sal. newport*, *Sal. heidelberg and Sal. javiana* (CDC 2007). The predominant serotypes involved in foodborne diseases changed, during the past decades, from *Sal. agona*, *Sal. hadar* and *Sal. typhimurium* to *Sal. enteritidis*, which is the predominant cause of salmonellosis in several countries (Suresh et al. 2006). Changes in the predominance of serotypes reflect changes in animal raising practices and dissemination of new serotypes due to increased international trade. At the present time, the main concern is the emergence of *Salmonella* serotypes that are multidrug-resistant bacteria (MDRB) (Huang 1999). Since 1980 human outbreaks caused by *Sal. enteritidis* have shown common sources in the US and EU (CDC 2005, 2013a, b). It is noteworthy that salmonellosis in chickens is rare and if birds are infected with certain serovars and phage types, disease that is clinically similar to colibacillosis (disease caused by the bacteria *Escherichia coli*) may occur (Gast 2008).

3.4 Preventing *Campylobacter* and *Salmonella* in Poultry

Over the last 25 years the world became more interrelated and interdependent on international trade, along with the current facilities for travelling which have increased dissemination of pathogenic agents from contaminants in foodstuffs. This has been associated with many new waves of diseases, thus local foodborne disease outbreaks have become a potential threat for the whole world. The increase in international trade has three main consequences:

(i) the rapid transfer of microorganisms from one country to another,
(ii) the time between processing and consumption of food is increasing, leading to increased opportunity for contamination and time/temperature abuse of the products, and
(iii) the population is more likely to be exposed to a higher number of different strains/types of foodborne pathogens (Rocourt et al. 2003).

The current world globalization, commercialization and distribution make it possible to transfer a contaminated product from one continent to another one via air freight, to affect the health of people in several countries at the same time. Thus a single contaminated foodstuff may lead to discarding literally tons of food, with considerable economic losses and resultant restrictions to trade and tourism (Tauxe et al. 2010). The modernization of poultry farms and globalization of the poultry breeding trade have also played a role in infection (Rocourt et al. 2003; Velge et al. 2005; Behling et al. 2010), with several serotypes being isolated from retail poultry products from many years back in various parts of the world (Rindhe et al. 2008; Behling et al. 2010).

3.4.1 Campylobacter

In some European countries, keeping *Campylobacter* colonized and non-colonized flocks separate during processing is seen as a promising control strategy (Wagenaar et al. 2006), but this may not be case in the US where prevalence of *Campylobacter* in broiler flocks is very high. Freezing of contaminated poultry can also be utilized to significantly reduce *Campylobacter* populations as was recently demonstrated in Iceland (Georgsson et al. 2006). However, this was not seen as a cost-effective strategy in the US market, since unlike other foodborne pathogens, *Campylobacter* does not grow effectively in the environment or under normal food storage conditions (Parka et al. 2002). Nevertheless, all types of poultry can become colonized with *Campylobacter* at any stage on the farms (Newell and Wagenaar 2000). Yet no evidence was found for vertical transmission of *Campylobacter* to progeny parent breeders that were hatched from eggs coming from *Campylobacter*-positive grandparent flocks (Callicott et al. 2006).

There have been many published reports of interventions with the potential to control *Campylobacter* in chickens on the farm (pre-harvest) and in the slaughter/processing facility (post-harvest) (Fig. 3.1). However, in commercial poultry settings, few of these interventions have been demonstrated to be as effective in lowering either the number of positive carcasses or the number of *Campylobacter* (load) on each carcass as in the research laboratory (Hue et al. 2011; WHO 2013). This may be because *Campylobacter* is easily spread within the processing plant environment from previous positive poultry entering the facility or the actual load on each bird is so high that the interventions in the processing plant environment are ineffective in reducing the contamination load (Hansson et al. 2010). In addition, the benefit of an intervention on the farm may be lost if there are not concurrent interventions in the transport from the farm to the processing facility to reduce cross-contamination (Berrang et al. 2003; Hansson et al. 2005).

3.4.2 *Salmonella*

Prevalence of *Salmonella* in poultry meat has been reported worldwide from retail outlets, retail markets and processing plants. It has been reported to be as low as 1.56% from a Morocco poultry processing plant (Cohen et al. 2007) and as high as

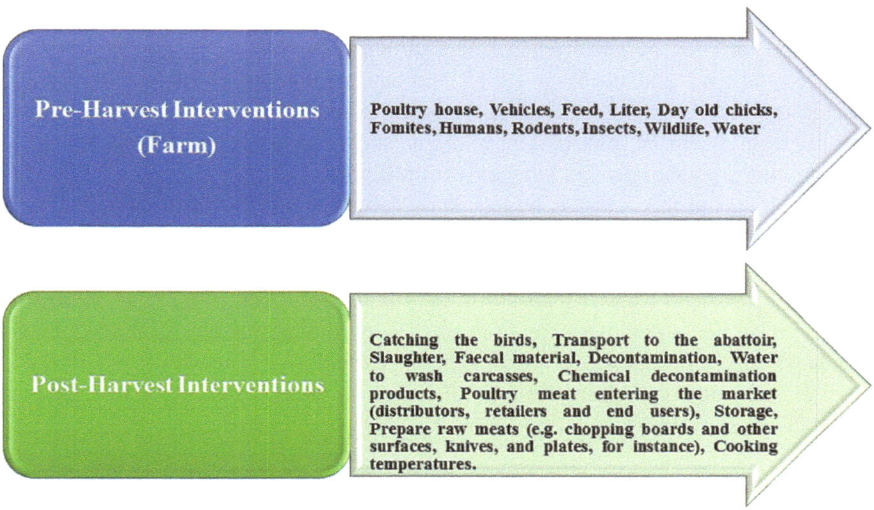

Fig. 3.1 Chemical and physical treatment in the farm and plant environment

15% from a poultry processing plant in USA (Russell 2009). In egg production, there is a linear relationship between the flock prevalence and the number of eggs contaminated with *Sal. enteritidis* (Gast et al. 2002; FAO-WHO 2009). As stated before, the epidemiological complexity of the disease, which involves vertical transmission, fecal excretion, horizontal transmission, environmental contamination and presence of carriers in different species, make salmonellosis control difficult. Multiplication outside the body of the host is facilitated by high temperatures and the presence of protein (for example, in residual waters). Therefore, the most important points of transmission of *Salmonella* are tropical and subtropical regions, as well as places where there is a large concentration of animals and people. *Salmonella* may also be found in products refrigerated at 2 °C; the microorganism is able to remain viable in frozen products for long periods. After entering the digestive system together with contaminated food and water, *Salmonella* reach the intestines, where they attach to intestinal cells and multiply.

Salmonella are distributed all over the world, countries that produce poultry on a large scale have evolved different methods of production, processing, and testing especially with regard to controlling and testing for *Salmonella*. For example, in the EU, some countries test all breeder flocks for *Salmonella* and destroy any breeder flock that is found to be positive (Russell 2009). In this way, they have significantly reduced *Salmonella* to 3–6% on birds coming into the processing facility. This approach is impossible in the US since the US produce twice as much poultry in Athens, Georgia than is produced in all of Sweden, where these practices are common (Russell 2009). Furthermore, some countries in the EU use competitive exclusion to prevent *Salmonella* colonization of baby chicks. This approach is illegal in the US because the bacterial cultures used to inoculate chicks, which occupy the attachment sites in the intestine and produce by-products that kill *Salmonella*, are undefined—the particular species used are not specifically identified (Russell 2009). Clearly it appears that salmonellosis epidemiology and control are highly complex, where hygienic and sanitary standards vary with the region, based on feeding and cooking habits and animal raising practices. In general, considering that the main route of transmission is in the food chain, the presence of this microorganism in production animals shows that *Salmonella* is the most incident and relevant etiological agent of intestinal infections. It causes millions of dollars in losses to the industry, mainly in cattle, swine and poultry production, both in local and international trade. In some countries, rigid food inspection is a constant need in order to produce foodstuffs of high quality. Besides the importance of preventive measures against the risk of *Salmonella* infection in humans, control of salmonellosis has a positive economic impact in countries where outbreaks occur. Estimated costs of medical expenses, sick leaves and loss of productivity related to the high incidence of salmonellosis in the US range from US$1.3 to US$4.0 billion a year (Taitt et al. 2004).

3.5 Control Measures for *Campylobacter* and *Salmonella*

In poultry and meat industries there are number of techniques to control *Campylobacter* and *Salmonella* as reported in comprehensive detail by Behling et al. (2010). These include washer and product temperature controls, chemical interventions, water replacements, counter-flow technology in the scalder and chiller, equipment maintenance, HACCP plans, training and awareness programs for employees, environmental monitoring, air sampling programs, plant sanitation, good manufacturing practices (GMP)/procedural analysis, and state-of-the-art testing. In spite of these interventions, *Campylobacter* and *Salmonella* have been known to persist on chicken skin during poultry processing because of their ability to attach to skin and become entrapped in deeper skin layers, crevices, or feather follicles (Notermans and Kampelmacher 1975; McMeekin et al. 1984; Lillard 1986). These sites may provide a suitable microenvironment for bacteria to lodge (McMeekin et al. 1979; Chantarapanont et al. 2003) and physical protection from antimicrobial agents such as chlorine, acidified sodium chlorite, and peracetic acid (Yang et al. 2001; Chantarapanont et al. 2004). Thus more vigilant intervention strategies to reduce the microbial contamination on poultry meat are necessary to lower the potential of foodborne illness. It has been reported that a high number of bacteria were recovered after 40 consecutive rinses of a single carcass due to a firm attachment to poultry skin, some of which were extremely difficult to eliminate using a conventional washing method Lillard (1989). The report of Zhang et al. (2011) shows that the 12% incidence of *Salmonella* and 92% prevalence of *Campylobacter* on eviscerated broiler carcasses decreased to nil and 4%, respectively, after in-line spraying of disinfectant. However, the rate increased to 13% for *Salmonella* and 52% for *Campylobacter* after carcass chilling, indicating that the chemical spray before chilling did not eliminate pathogens sufficiently from the broiler surface.

Hot water treatments have also been used to improve the safety of food products. Several researchers have had some success in decontaminating poultry meat using either hot water sprays/immersion or steam exposure (Cox et al. 1974a; James et al. 2000, 2007; Purnell et al. 2004). It appears that these treatments are more effective at higher temperatures and longer exposures but such treatments caused the skin to shrink and change colour (James et al. 2007). However, immersing a pre-chilled carcass in 60 °C water for 10 min, the number of *Salmonella* was reduced 100-fold without adversely affecting carcass appearance (Morrison and Fleet 1985). The early study of Cox et al. (1974b) reported that by immersion in water at 71 °C for 1 min, the total aerobic bacterial count on the carcass was lowered by almost 2 log unit, however, a partially cooked appearance was produced. Using a hot water spray at 65.6 or 71.1 °C reduced microbial counts on skin approximately 1 log cycle, and shelf life was subsequently extended for 1–2 days (Thomson et al. 1974). In general Li et al. (2002) found that the inside-outside spraying of carcasses at 55 and 60 °C significantly reduced the inoculated *C. jejuni* by more than 0.78 log cycle/carcass compared with a water spray at 20 °C. It was concluded that a 90% reduction in the risk to public health could be obtained by reducing the number of *Campylobacter* in broiler intestines by 3 logs, and even a 1 log reduction on the carcasses would reduce the risk between 50 and 90% (EFSA 2011).

Carcass rinsing is another approach (Capita et al. 2004; Cox et al. 2010). The rinsing technique has been used by several researchers who found that removal of bacteria depended on the force used. It was found that gentle rinsing may not be sufficient to remove microorganisms on carcasses or from selected samples (Notermans and Kampelmacher 1974; McMeekin et al. 1984; Lillard 1986; Hinton and Cason 2008). Using a stomach bag technique has removed more bacteria that are attached tightly to deeper layers of skin (Avens and Miller 1970; Notermans et al. 1975), although several bacteria still remain in the skin (Hannah et al. 2011). Even after 5 consecutive stomaching procedures were used, the total bacteria on skin were similar with one term washing (Lillard 1988). Therefore, it appears that these types of bacterial cells that are attached tightly or internally may recover when the whole sample is minced (Avens and Miller 1970).

Poultry are thus considered one of the most important *Salmonella* and *Campylobacter* reservoirs, a public health concern and a major foodborne pathogen affecting humans. Control of these bacterial diseases is a challenge to public health because of their emergence/re-emergence and high mutation rate, with antibiotic resistance in developing and developed countries. Shipping goods between continents and countries is implicated as an important epidemiological factor because of the lack of symptoms in animals infected and the technical difficulty in detecting them before or during the inspection of foods of animal origin. This has helped the spread of *Salmonella* and *Campylobacter* infection to various parts of the globe which can now occur simultaneously in various countries. Therefore, the control measures for these bacteria needs to be re-enforced and reviewed at each step of processing. Thus, with current control measures on farms and production facilities, there will always be a risk that some of those bacteria will contaminate equipment, employees, and carcasses. The strategies to prevent transmission of *Salmonella* and *Campylobacter* to humans must focus on the whole production chain of broiler meat and on the subsequent storage and handling of meat from 'Farm-to-Fork'. Only a high level of biosecurity (cleaning, disinfection and biocontrol) will help to reduce the incidence of diseases on poultry farms and in food processing plants. Poor levels of biosecurity can cause disease outbreaks requiring the slaughter of the entire flock along with other drastic measures.

3.6 Critical Control Points for Poultry Farms

There are many variables that need to be controlled for poultry farms to be free of *Sal. enteritidis, Sal. Typhimurium* and other *Salmonella* serotypes along with *Campylobacter*. The most important measure to be taken in poultry farms is to reduce the risk of *Salmonella* spp. and *Campylobacter* spp. contaminations (FAO-WHO 2009). It is imperative to disinfect or control *Salmonella* spp. and

Campylobacter spp. on poultry farms and in the feed process as well as on personnel (moving in or out the farm) in addition to environmental aspects (FAO 2010; Newell et al. 2011).

Risk factors include:

(i) Waste and wastewater: Water contamination has become a major issue confronting industrial poultry operators.
(ii) Chicken manure: Almost 18 billion birds are raised each year worldwide and produce more than 22 million tons of manure. Poultry manure contaminates groundwater and surface waterways.
(iii) Chicken processing: Chicken guts, heads, feathers, blood and wastewater resulting from the processing are rendered down to their essence before being hauled away as sludge to fertilize farmlands in the area.
(iv) Water use: Modern broiler houses (e.g. typically 500 ft long) require almost 38 l of water per minute. Treated wastewater is released into nearby streams or sprayed on farmlands in the area.
(v) Feeding: Feed is usually offered to birds via the chain system. This system transports feed into the metal feeder at precise times during the day.
(vi) Egg collection: Hens lay eggs, often already infected, onto an angled wire floor and the eggs roll towards the front of the cage and onto a belt. The belt transports eggs out of the house either to the egg packaging facility or storage cooler.
(vii) Egg shell cleaners and sanitizers: This step is very important since the egg exits the hen's body through the same passageway as faeces, thus *Sal. enteritidis* can be on the outside of an eggshell. Thus once the eggs enter the egg packaging centre, they are washed by a detergent solution within minutes or no later than 12–14 h post-lay (FAO 2010).

The detergent solutions and compounds used to wash and de-stain egg shells are potential food additives, therefore, they are regulated by the FDA. It is noteworthy that the FDA does not have any published regulations dealing with egg shell cleaning and de-staining compounds (USDA 2013). Nevertheless, FDA regulates antimicrobial agents that are used on raw agricultural commodities, such as shell eggs, wherever food is prepared, packaged, or held for commercial purposes. The EPA regulates antimicrobial agents that are applied in the field, at treatment facilities and during transportation between the field and such facilities. Sanitizers formulated in compliance with Title 21 of the US Code of Federal Regulations (CFR), Section 178.1010, are acceptable for use in sanitizing shell eggs provided they are registered with EPA for this use. The FDA has provided US-Food Safety and Inspection Service (US-FSIS) with some general guidelines for chemicals acceptable for use in formulating products for cleaning and de-staining shell eggs. Substances that are GRAS for use in food: (Title 21 of the CFR), Parts 182 and 184 (USDA 2013).

3.7 Ecology of *Campylobacter* and *Salmonella* Bacteriophages

Bacteriophages (or, more simply, phages) are viruses that infect and kill bacterial cells. Generally, phages are found in the same environments that their bacterial host (s) inhabit (Son et al. 2010), and only recognize specific bacterial hosts (Sulakvelidze et al. 2001). Phages were found readily abundant in foods and have been isolated from a wide variety of raw products (e.g., beef, chicken, roast turkey, pies, biscuit dough, cheese, yoghurt, mussels and oysters) (Kennedy et al. 1986; Croci et al. 2000; Hsu et al. 2002; Suárez et al. 2002; Atterbury et al. 2003a). Therefore, they are consumed daily by humans (Sillankorva et al. 2012). They are common commensals of the human gut and play an important role in regulating the diversity and population structure of various bacteria in the human gastrointestinal tract. For example, phages are capable of infecting various *Salmonella* serotypes, *E. coli*, and *Bacteroides fragilis* which have been isolated from human fecal specimens in concentrations as high as 10^5 plaque-forming units (pfu) 10^{-2} g of feces (Calci et al. 1998; Furuse et al. 1983; Havelaar et al. 1986). It has been suggested that phages are the second most abundant category, after bacteria in fecal matter (Breitbart et al. 2003). *Salmonella* phages were also found in very high numbers (10^2–10^5 pfu ml^{-1}) in all types of wastewater (Havelaar 1987).

Since pathogenic bacteria with increasing resistance to currently used antimicrobial agents is growing, phages have now resurfaced as potential biocontrol and therapeutic agents (Chhibber et al. 2008; Parisien et al. 2008, O'Flaherty et al. 2009; Coffey et al. 2010; Mahony et al. 2011; Nelson et al. 2012; Jassim and Limoges 2014). Phage therapy is described as the use of phages as biocontrol (prophylaxis) agents of bacteria. Phage applications include different approaches in various fields as follows: phage therapy 'treatment of infections in people' (Abedon 2010; Gill and Hyman 2010; Kutter et al. 2010; Jassim and Limoges 2014), veterinary phage biocontrol for 'treatment or prevention of infections in animals' (Atterbury 2009; Sillankorva et al. 2012; Adulamir et al. 2014; Jassim and Limoges 2014; Aldoori et al. 2015), phage treatment of plants (Balogh et al. 2010; Jones et al. 2012), phage treatment of foods (Hagens and Loessner 2010; Sillankorva et al. 2012), phage treatment of aquaculture and fish (Park et al. 2000; Park and Nakai 2003), phage bioprocessing (Goodridge and Abedon 2003; Jassim et al. 2012), phage pharmacology (Abedon and Thomas-Abedon 2010), wastewater treatment (Withey et al. 2005), and phage modification (Goodridge 2010; Jassim et al. 1995, 2010). *Sal. typhimurium* phages were also proposed to be used as indicator organisms for human pathogenic viruses in the evaluation of disinfection processes for water treatment plants (Havelaar 1987).

3.8 Phages for Biocontrol of *Campylobacter* and *Salmonella*

Phages have several characteristics that make them attractive as therapeutic agents or agents of biocontrol (Jassim and Limoges 2014); these include their effectiveness in killing their target bacteria, their specificity, natural residence in the environment and the fact that they are self-replicating and self-limiting (Sulakvelidze et al. 2001; Jassim and Limoges 2014). Well-controlled animal models have demonstrated that using smart phages can prevent or treat animals infected with certain pathogenic bacteria and may be good alternatives for the treatment of infections with MDRB (Abdulamir et al. 2014; Jassim and Limoges 2014; Aldoori et al. 2015). In general veterinary therapy and biocontrol applications require appropriate administration, targeting specific bacteria, with a strategy that includes a comprehensive methodology, detailing the phage-host interactions, dose optimization and accounting for all chemical and physical factors (Jassim and Limoges 2013). Poultry products have arguably been the most widely-used meats to study the efficacy of phage-mediated biocontrol in foods. Members of the *Campylobacter* and *Salmonella* genera have been the most frequently targeted pathogens on chicken meat because of their significance as contributors to food-borne disease.

3.8.1 *Campylobacter* Phages

Extensively reared birds are more likely to be colonized by *Campylobacter* and thus are also a source of companion bacteriophages. Not surprisingly, several *Campylobacter*-specific bacteriophages have been isolated from broiler chickens (Grawjewski et al. 1985; Connerton et al. 2004), retail poultry (Atterbury et al. 2003a, b), and the rearing cycle of free-range and organic chickens from hatching until slaughter (El-Shibiny et al. 2005). Furthermore, *Campylobacter*-specific phages were also found in pig manure, abattoir effluents and sewage (Salama et al. 1989; Khakhria and Lior 1992). These *Campylobacter* phages survive the conditions and processing in poultry production (chilling and freezing) and on retail chicken under commercial storage conditions at 4 °C (Atterbury et al. 2003a).

Campylobacter phages were used to treat infected poultry, however *Campylobacter* which were resistant to phage infection were recovered from phage-treated chickens at a frequency of <4% (Loc-Carrillo et al. 2005). The incidence of phage-resistant phenotypes following phage treatment has always been perceived as a potential drawback to phage biocontrol in animals (Smith and Huggins 1982; Smith et al. 1987; Sklar and Joerger 2001; Atterbury et al. 2003a, b; Wagenaar et al. 2005; Loc-Carrillo et al. 2005; El-Shibiny et al. 2009; Carvalho et al. 2010; Kittler et al. 2013) and are correlated with reduced virulence (Smith et al. 1987; Park et al. 2000). To overcome this setback of using phage for therapy/biocontrol, it has been suggested to administer a cocktail of phages

(Carvalho et al. 2010; Kittler et al. 2013). These phage-resistant types were compromised in their ability to colonize experimental chickens and rapidly reverted to a phage-sensitive phenotype in vivo (Loc-Carrillo et al. 2005). This may suggest that resistant bacteria lose fitness, notwithstanding being sensitive to the phage. However, the results of Carvalho et al. (2010), disputes results described by Loc-Carrillo et al. (2005) whereas, the resistant phenotype did not lose the ability to colonise the chicken gut and did not completely revert to sensitive type. In fact, the phage cocktail was able to reduce *C. coli* and *C. jejuni* in infected poultry by only 2 $\log_{10}$ colony-forming unit (cfu) g^{-1}. In parallel, *Campylobacter* strains resistant to phage infection were recovered from phage-treated chickens at a frequency of 13% (Carvalho et al. 2010). Similar results were reported with *Campylobacter* strains resistant to phage infected broiler chickens, where only 2 log cfu declined (El-Shibiny et al. 2009); a 3 log cfu decline in *C. jejuni* counts was initially observed in the therapeutic group, however, after 5 days, bacterial counts stabilized at a level 1 log cfu lower than that of the broilers not receiving phage therapy (Wagenaar et al. 2005). Under commercial conditions, a phage cocktail for *Campylobacter* reduction was tested on three commercial broiler farms, each with a control and an experimental group. *Campylobacter* count reductions of >$\log 10^{3.2}$ cfu g^{-1} cecal content compared to the control was still detected (Kittler et al. 2013). The essential parameters of the infection model were set to mimic natural *Campylobacter* colonisation, with the colonisation level to be between 1×10^6 and 1×10^9 cfu g^{-1} of faeces, the number found in commercial broiler flocks (Rosenquist et al. 2006). Therefore, since the above studies achieved only 2–3 log cfu g^{-1} reduction from the initial bacterial number inoculums in each bird from $\log 10^{6-9}$ cfu/bird, the results are negligible as *Campylobacter* infection was reduced by <50%. This can also be pointed out as a major drawback of phage therapy/ biocontrol. In order to overcome this problem, it was suggested that the precise timing of treatment may be important to achieve optimal reduction of *Campylobacter* colonisation pre slaughter (Carvalho et al. 2010), and it was found that the best results were obtained between 1 and 4 days prior to slaughter (Kittler et al. 2013). Furthermore, the same authors have also recommended that when selecting the phages that will compose the cocktail, an additional criterion should be the ability to infect other phage resistant *Campylobacter* phenotypes. In conclusion, several small-scale studies have been performed to evaluate the potential of phages for a reduction of *Campylobacter* in the broiler gut and on chicken skin (Atterbury et al. 2003a, b; Goode et al. 2003; Loc-Carrillo et al. 2005; Wagenaar et al. 2005; El Shibiny et al. 2009; Carvalho et al. 2010; Hertwig et al. 2013; Hammerl et al. 2014). In 2013 the first field trial with *Campylobacter* phages (campylo-phages) was carried out in commercial broiler flocks (Kittler et al. 2013). The majority of these studies used single phages or cocktails composed of several phages for therapy, resulting in reduced *Campylobacter* counts in cecal content between 2 and 3 log10 cycles. In general, the key elements for the success of phage biocontrol against *Campylobacter* in broiler chickens are a proper selection of phage, the dose of phage applied and the time elapsed after administration (Loc-Carrillo et al. 2005).

On the basis of *Campylobacter* phage morphology, genome size and endonuclease restriction profile, the currently known lytic *Campylobacter* phages are divided into three groups (Javed et al. 2014) of which, members of group II and group III may be useful in such applications. It was found that Group II frequently infect both *C. jejuni* and *C. coli* strains making them better suited for phage biocontrol and therapy. Some phages isolated from poultry in group III phages also showed a strong lytic activity on certain *C. jejuni* strains (Loc-Carrillo et al. 2005). As phage resistance has been perceived as a potential drawback to phage applications, it is of great importance that group II and group III phages apparently use multiple host cell receptors for binding (Coward et al. 2006; Sorensen et al. 2011). The selection of the appropriate phage and their dose optimization are key elements for the success of phage therapy to reduce *Campylobacter* in broiler chickens. However, notwithstanding their potential, there is little information available regarding the prevalence and influence of phages on *Campylobacter* colonizing extensively reared poultry flocks. In general, experimental studies from different research groups have shown the effectiveness on *Campylobacter*. However, due to their short-term effect, the application would be limited to a small time window just before slaughter. The preliminary conclusion was only proof of principle, and on-farm studies had not yet been performed in a large scale (Kittler et al. 2013). Practical application might be hampered by lack of broad-host phages, resistance and environmental contamination issues. They concluded that novel intervention strategies employing phage may be useful in the future; however, they are not yet commercially available (Joerger 2003) and development and application of a commercial product would need extensive research (Atterbury et al. 2007; Kittler et al. 2013). It is clearly necessary to use phages that have strong virulent characteristics for phage biocontrol to become a sustainable biological control measure and to be harnessed for the reduction of *Campylobacter* emanating from poultry sources. Hence, to overcome the problem of phage resistance in *Campylobacter*, it is advisable to apply phage programming technology to different groups of type II and III *Campylobacter* phages in order to limit the entry of these pathogens into the human food chain.

3.8.2 *Salmonella Phages*

Salmonella phages have been used to control *Salmonella* contamination in a variety of foods. Goode et al. (2003) used phages to control *Sal. enteritidis* on chicken skin that had been inoculated with commercially relevant numbers of bacteria. The *Salmonella* phage (Felix-O1), which has a broad host range within the genus *Salmonella*, demonstrated an approximately 2 log units' reduction in *Sal. typhimurium* DT104 inoculated on chicken frankfurters (Whichard et al. 2003; Fig. 3.2). Phage therapy can be effective, for example against *Sal. typhimurium* in poultry and pigs, although this requires rapid selection and administration of the phage and high

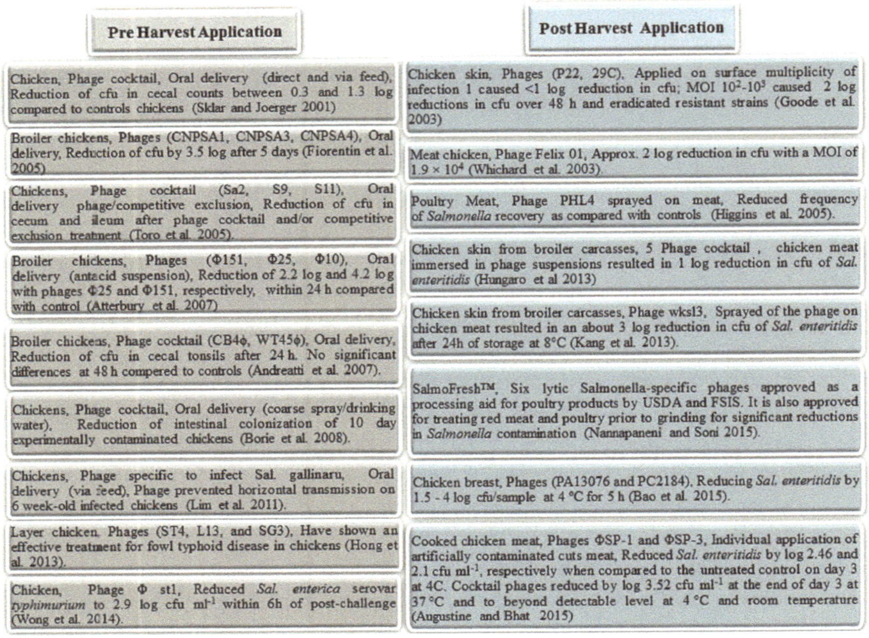

Fig. 3.2 *Salmonella* phages for pre-harvest and post-harvest biocontrol applications

bacterial loads (Allen et al. 2013). According to Bardina et al. (2012) findings, the frequent treatment of chickens with the phage cocktail, including prior to *Salmonella* colonization of the intestinal tract, is necessary to achieve a significant reduction of bacterial cell numbers. In general, the reduction of foodborne pathogens requires a comprehensive phage control program at the farm, where the animals are born, hatched or raised, before shipment to processing plants. The potential pre-harvest sources of foodborne pathogen contamination include breeder herds and flocks, hatcheries, contaminated feed and water, along with environmental sources and vectors, such as litter, animal caretakers, and insects (Bailey 1993; Nayak et al. 2003). Interventions are also necessary to reduce cross-contamination in post-harvest processes in the farm and plant environments. In this regard, numerous studies were conducted to prevent such poultry infections by using phage biocontrol in the pre-harvest and post-harvest applications and have demonstrated their effectiveness to prevent or reduce colonization and diseases in poultry. Recent studies are summarized in Fig. 3.2, concluding that experimental results of phage biocontrol for *Salmonella* looked very promising, but there had been no published field trials. There were commercial companies investigating this, but it was too early to recommend it as an effective intervention in a commercial setting (Atterbury et al. 2007).

3.8.3 Summary

In spite of modern technologies, good manufacturing practices, quality control and hygiene, changes in animal husbandry, agronomic processes, and advances in food or agricultural technology, food safety is continuously challenged by changes in lifestyle, consumer demands (e.g., ready-to-eat products) as well as by the increase of international trade (Rocourt et al. 2003; Velge et al. 2005). The most efficient means for limiting the growth of microbes are good production hygiene, a rational running of the process line and a well-designed use of biocides and disinfectants (Maukonen et al. 2003). However, even when acceptable cleaning procedures are applied, bacteria are found in foods and food contact surfaces (Holah et al. 2002). Food products may become contaminated at different stages along the food chain, from growth or production until final consumption. Furthermore, the inherent ability of pathogens to attach to living and inert surfaces, where they start living in microbial communities known as biofilms and become highly tolerant to varied antimicrobial agents (Lewis 2008; Nannapaneni and Soni 2015) also contributes to the pathogen prevalence in foods and food contact surfaces. Therefore, new food preservation techniques must be continually developed in order to control the emerging pathogens and their impact on a global scale and to meet the primary goal of any food safety program, 'consumer protection'.

The proof of principle of phage biocontrol in poultry was already established for several pathogens (detailed description above). In general, it is estimated that with phage, a reduction of 2 log on the *Campylobacter* loads in poultry intestines was surprisingly considered sufficient (Rosenquist et al. 2003). As yet, these are the only phage proof of principle, and phages on-farm studies had not yet been performed, since practical application might be hampered by lack of broad-host phages, resistance and environmental contamination issues. Furthermore, application of a commercial product would need extensive research (Joerger 2003; Atterbury et al. 2007; Janež and Loc-Carrillo 2013). Therefore, effective intervention on the farm will likely require a multifaceted approach and in light of bacterial resistance to phages, it is unlikely that there will be any one "silver bullet" approach to eliminate *Campylobacter* and *Salmonella* in poultry processing.

3.9 Approval for Direct Use on Food Products

Phage is already used in animal meat industries, USDA-FSIS (2014) have approved spraying with phages for treating ready-to-eat poultry and meat products for direct human use (Bren 2007). Spraying or washing with phages is currently being used more widely at pre-harvest as more companies develop a marketing strategy for pre-harvest applications of their products. In the context of food safety, the phage-based product Agriphage™, produced by OmniLytics Inc. to treat bacterial spot disease on crops, was the first phage-based product formally approved for use in agriculture by the US regulatory agencies (Endersen et al. 2014). ListShield (a phage preparation targeted against *Listeria monocytogenes*), created by Intralytix to control *Listeria* on foods and processing surfaces was the first approval granted by the FDA and UDSA for a phage-based food additive. Furthermore, the successful use of coliphages has led to the development of a phage-product (EcoShield, Intralytix Inc.) which also received regulatory approval from the FDA in 2011 (Jassim and Limoges 2014). Since then, ListShield has been registered for use as an organic food additive in Europe and has also received approval by Food Standards Australia and New Zealand (Hodgson 2013). This product can be used on red meat parts and trims, intended to be ground (Food Contact Notification no. 1018) and has proved to eliminate from 95 to 100% of *E. coli* O157:H7. Finalyse® (Elanco Food Solutions) is another product on the market using naturally occurring phages specific for *E. coli* O157:H7. Finalyse is sprayed on cattle to reduce the load of *E. coli*, prior to entering the beef packing facility (pre-harvest strategy) (Sillankorva et al. 2012; USDA-FSIS 2014). Phages are FDA approved for use in or on live cattle as a treatment or for control of *E. coli* O157:H7 shedding in cattle (USDA-FSIS 2014). In 2012, the FDA approved LISTEX (developed and produced by Micreos) using phages on cheese to kill *L. monocytogenes* bacteria, giving them generally recognized as safe (GRAS) status (U.S. FDA/CFSAN 2006). In July 2007, the same phage was approved for use on all food products (U.S. FDA/CFSAN 2007; USDA-FSIS 2014). In February 2013, Salmo Fresh™ (Intralytix Inc.) received regulatory approval for use in eliminating *Salmonella* in poultry products and other foods (http://www.intralytix.com). In 2011 USDA confirmed that LISTEX is a clean label processing-aid and is included in USDA (2011). Research in the field of food safety is continuing to see if lytic phages are a viable option to control other food-borne pathogens in various food products. Recently, several phage-products are available:

(1) OmniLytics Inc. which received, in 2007, USDA's Food Safety and Inspection Services (USDA-FSIS) approval to use anti-*E. coli* and anti-*Salmonella* phage-based preparations applied as a electric mist fan, spray, or wash on live animals prior to slaughter (Garcia et al. 2010; USDA-FSIS 2014).
(2) BIOTECTOR S1, a phage product from CheilJedang Corporation that is to be applied on animal feed to control *Salmonella* in poultry (Sillankorva et al. 2012).
(3) Preparations of phages from South Korea are commercially available in many parts of the world to reduce mortality in poultry due to *Sal. gallinarum* and *Sal. pullorum* (Seal et al. 2013).
(4) The EU launched in 2013 the first project in this area, called "Phagoburn" to test the phages against resistant bacteria *E. coli* and *Pseudomonas aeruginosa* addressing major burn wounds (http://www.phagoburn.eu/).
(5) The Institute for Environmental Science and Research (ESR) has concluded field trials of a phage cocktail-STECleanz™ targeting the control and prevention of Shiga-Toxin producing *E. coli* (STEC) (ESR 2016).

3.10 Phage-Derived Enzymes

With the increasing worldwide prevalence of antibiotic resistant bacteria, phage endolysins (peptidoglycan hydrolases) 'lysins' represent a very promising novel alternative class of antibacterial for controlling bacterial infection or other unwanted contaminations in various fields, particularly in light of the increasing frequency of drug-resistant pathogens worldwide (Fischetti 2008; Fenton et al. 2010; Schmelcher and Loessner 2016). Lysins are phage-encoded peptidoglycan hydrolases which, when applied exogenously (as purified recombinant proteins) to Gram-positive bacteria, bring about rapid lysis and death of the bacterial cell (Young 1992; Fischetti 2008). They are particularly effective when applied exogenously to Gram-positive bacteria, resulting in immediate bacteriolysis. Their potential as biocontrol agents in the food industry is increasingly becoming a focus of research interest (O'Flaherty et al. 2009; Garcia et al. 2010; Shakeeba et al. 2010, Callewaert et al. 2011; Coffey et al. 2011). For example, in the area of the poultry industries, the endolysin Ply3626 was found active against all 48 strains of *Clostridium perfringens* tested and this enzyme has been proposed as an effective application for the control of *C. perfringens* in poultry (Zimmer et al. 2002). The potential application of endolysins for use in animal and carcass sanitation prior to slaughter and processing has also been suggested by Fenton et al. (2013), and it may be possible to use purified phage lysins directly rather than the phage itself (Woolhouse et al. 2015).

However, Gram negative phages are unable to synthesise endolysins (lysins) during the phage lysis cycle. Therefore, instead of this enzyme Waseh et al. (2010) conducted studies with a purified truncated phage tailspike endoglycosidase (P22sTsp) from the phage P22 targeting *Sal. Typhimurium*. Oral administration of P22sTsp to *Sal. Typhimurium* infected chickens resulted in a significant reduction of the pathogen in the gut and prevented further penetration into internal organs. The authors suggested that the ability of tailspike proteins to resist the acid-induced denaturation and digestion by proteases in the gastrointestinal tract make them attractive for use as oral therapeutics against bacterial Gram-negative pathogens in food-producing animals. Further studies are needed for the use of endoglycosidase (P22sTsp) as a rapid biocontrol for poultry and food industries and it could also be developed for use in rapid bacterial detection.

3.11 Challenges: Phage Biocontrol in Food Industries

Notwithstanding all the advantages stated above, one of the major obstacles of phage biocontrol which attracts most of the concern is the potential emergence of phage-resistant bacteria mutants, as in the case of antibiotics, which has been observed in foodborne pathogen and spoilage bacteria (Sklar and Joerger 2001;

Greer and Dilts 2002; Jassim and Limoges 2014). Typically, the resistance developed towards a particular phage occurs when the bacteria surface protein is lost which prevents the phage from infecting its host due to the absence of the receptor for attachment. The ecology of both phages and bacteria were also not understood, resistance, failure to neutralize gastric pH prior to oral administration, inactivation of phages by host immune responses and environmental contamination issues are other obstacles (Dabrowska et al. 2005; Sulakvelidze 2011; Jassim and Limoges 2014). It was also suggested that changes of the bacterial hosts used for maintenance of phages must be avoided as these can drastically modify the parameters of the phage preparations, including host range and lytic activity (Levin and Bull 2004; Sulakvelidze 2011). The generally poor efficacy of commercial phage preparations led to widespread criticism and disagreement about the effectiveness of phages in treating disease (Atterbury 2009).

The lysogenic phage provides axes force in bacterial pathogenesis and contributing in the bacterial pathogens evolution through horizontal gene transfer. The horizontal acquisition of a restriction enzyme system which allows the bacteria to degrade foreign DNA with endonucleases is another route of acquiring resistance towards phage in bacteria (Verheust et al. 2010; Jassim and Limoges 2014; Ly-Chatain 2014). This will eventually lead to the evolution of pathogenic bacteria (Verheust et al. 2010). For this reason, cocktails of phages have been used to help in resolving the issues of bacterial resistance (Tanji et al. 2004). Furthermore, reduced efficacy of phages has been observed in several studies on the phage treatment for pathogens that colonized farm animals and also on the surface of food products. It has been suggested that the decreased efficacy of the phage in livestock may be due to exposure to the constantly changing physical and chemical factors associated with the micro-environment's food matrix and with the conditions of application, including environmental factors, the distinct properties of the phage itself and the interaction with the immune system of a living animal (Greer 2005; Jassim and Limoges 2014). The various components in a food matrix such as carbohydrates, proteins and fats also have slight effects on the ability of the phage to interact with its specific pathogens (Greer 2005).

In general, the phage could become a bio-decontamination or useful biocontrol tool through understanding how to control the phage-resistant bacteria (Mizoguchi et al. 2003; Fischer et al. 2004). Subsequent studies revealed that not all phages replicate similarly and that there are important differences in the replication cycles of lytic and lysogenic phages (Sulakvelidze et al. 2001; Jassim et al. 2010). In this regard, several suggestions were shared in order to point a direction for overcoming common obstacles in applied phage technology (Jassim and Limoges 2014) and are discussed below. The key to successful use of phages in modern farm, food processing and clinical applications is to understand the common obstacles as well as best practices and to develop answers that work in harmony with nature.

3.12 Phage Biocontrol from Farm-to-Fork

3.12.1 Background

Any change in the conditions or practices associated with the production of animal agricultural commodities can affect the safety of the food supply. For example, introducing antibiotics in livestock and poultry for more than half a century to control diseases and in sub-therapeutic doses in animal feed, to promote growth and improve production of animal products, has resulted in the development of antibiotic-resistant bacteria. The consequences affect everybody in the world and access to effective treatment for bacterial infections is urgently needed (Jassim and Limoges 2014). Feedlot animals and large-scale poultry rearing with large processing facilities have made it difficult to control bacterial pathogens and have been implicated in the increasing incidence of human pathogens, such as *Salmonella*, in domestic animals over the past 50 years. The efforts to develop new methods for controlling microbial contamination in food and the food processing environment are extremely important (Endersen et al. 2014). By the same token, the current control strategies for reducing *Campylobacter* in poultry production on farms propose the use of biosecurity measures to exclude the organisms from the flock and/or complementary, non-biosecurity-based approaches, such as antibacterial treatments, probiotics or vaccination, to prevent or reduce levels of colonization. As yet no comprehensive measures have been found effective and feasible to use on-farm as interventions against *Campylobacter* (Newell et al. 2011).

There is an urgent need to rapidly identify any microbial risk factors on farms or in meat industries, then to apply a rapid and simple protocol to control the pathogens thereby preventing widespread contamination, which occurs rapidly in the farm environment i.e., within 3–5 days of the time when a flock becomes positive (Ridley et al. 2011b). Without longitudinal sampling of both the flock and the external environment, there is no way of knowing the direction of transmission (chicken to environment or vice versa) (Ridley et al. 2011b). Recovery from *Campylobacter* infection from the environment is generally poor because the stresses on the organism are high and their viability is compromised (Ridley et al. 2008). Ideally, improved farm biosecurity as a potential intervention against flock colonization, is essential to obtain optimal control of flock infections and preventing vertical transmission. Vertical transmission is usually taken to include the internal contamination of the egg within the reproductive tract before completion of shell deposition. Horizontal transmission after chicks are placed on a farm appears to be the normal route of infection for intensively reared flocks. The sources of such contamination are multiple and various. As a consequence of their fastidious growth requirements, *Campylobacter* can naturally multiply only within a warm-blooded host, usually within the gastrointestinal tract. Since all mammals and birds (wild and domestic) are considered potential hosts, the environment (soil, water, pasture, etc.) must be contaminated frequently with these organisms (Newell et al. 2011).

The best use of phage applications in a farm-to-fork context are discussed below in terms of current uses and future potential as biocontrol encompassing areas such as primary production, post-harvest processing, bio sanitation, and bio-detection.

3.12.2 Phage Conceptual Applications from Farm-to-Fork

3.12.2.1 Phage Isolation

Phage Isolation is based on using naturally occurring phages. Wild phages are isolated from various poultry environmental samples such as poultry litter or broiler litter, a mixture of poultry excreta, spilled feed, feathers, material used as bedding in poultry operations, water, sewage, soil, etc., (this is discussed in greater detail in Chap. 1).

3.12.2.2 Phage Programming

(i) *Background*

Phages have several potential replication cycles. The two most dominant are the lytic (virulent) and the lysogenic (temperate) cycles (Jassim and Limoges 2014). The lytic cycle is typically considered the main method of viral (phage) nucleic acid replication and synthesis of virus encoded proteins resulting in the destruction of the infected cell and the release of newly-formed phages from the lysed (destroyed) host cell. The lysogenic cycle is the phenomenon in which a bacterium is infected by a temperate or dormant phage, where the viral DNA is integrated into the bacterial chromosome, becoming a part of the host cell and replicated along with the host chromosome for many generations, coexisting as opposed to lysing the host cell.

(ii) *Programming*

The isolated phages will be optimized to produce smart phages via phage programming technology (this is discussed in greater detail in Chap. 1). In brief, phage programming describes non-genetic manipulation biotechnology that can enhance phage-host interaction and phage reproduction while avoiding host resistance (Jassim et al. 1995, 2010). The technology is able to control phage mutation and also convert lysogenic (temperate) phages into virulent phages. These new smart phages have overlapping host ranges and are necessary to overcome obstacles faced by wild phages, if the phage is to be used as a biocontrol. We overcome the mechanisms of resistance and reactivate lysogenic phages by developing an inventive phage programming technology, briefly illustrated in Fig. 3.3, which is capable of enhancing the phage lytic infectivity resulting in death of their host bacteria. The programmed phage is capable of converting naturally occurring wild

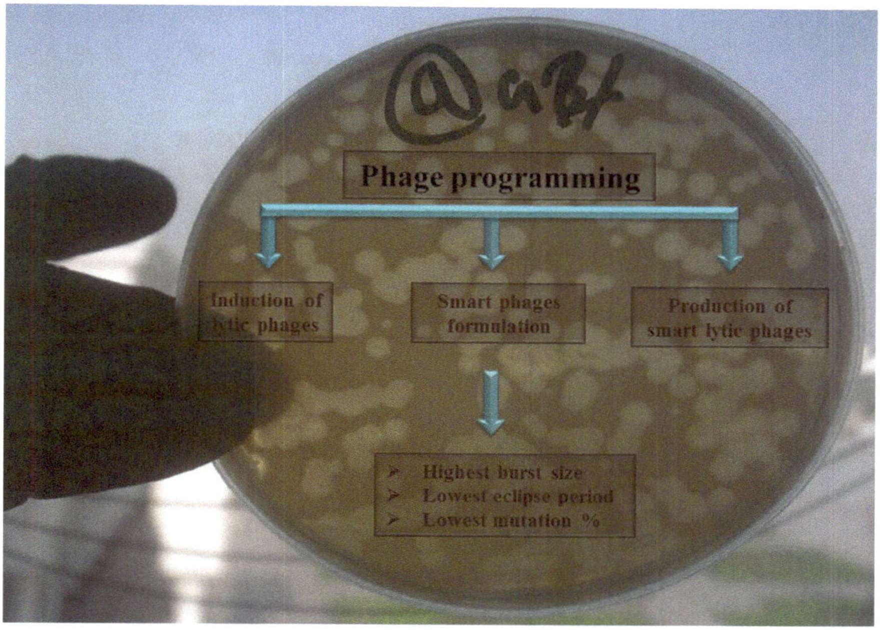

Fig. 3.3 Phage programming to produce smart lytic phage

phages (or any laboratory phages) to smart phages with a broader range of host specificity that can also overcome a bacterium's resistive defense mechanisms and completely destroy the target bacterial cell. These smart phages will be utilized to develop poultry bio-control, feed, and bio-processing to prevent foodborne pathogens.

The smart phages will be prepared for use either in a single phage application or cocktails of multiple phage strains to target specific bacterial cells. If a bacterial strain developed resistance to one specific phage in the cocktail, the other phages will compensate and subdue the resistance at its very initial step. This is the same principle used for treating patients using multi-drug chemotherapy against serious infectious diseases, such as bacterial septicaemia. The phages naturally multiply in the presence of their target pathogens as they replicate through the process of lysing or killing off the target bacteria.

3.12.3 Phage Biocontrol Applications

Phages as therapeutic tools to control major foodborne pathogens such as *Campylobacter* and *Salmonella* was also reported (Carvalho et al. 2012). Phages can be administered orally, incorporated in drinking water or food to control

Salmonella and *Campylobacter* in poultry, or by spray to target avian pathogenic *E. coli* in poultry, and orally/rectally to control *E. coli* in ruminants (Sillankorva et al. 2012; USDA-FSIS 2014). A comprehensive description of several proprietary technology platforms follows. The smart phages can be applied singularly or in a cocktail for poultry processing. We also describe various devices for topical, oral or systemic poultry treatment applications which will reduce and/or eliminate major bacterial pathogens resident in the animals or their environment. Applications include control of antibiotic resistant bacteria in various facilities such as egg and poultry barns, egg collection and sanitizing, during breeding, hatching, and grow-out as well as in meat processing.

3.12.3.1 Phage Egg Sanitizing

Smart phages can be used to decontaminate *Campylobacter* and *Salmonella* on the outside of an eggshell during the cleaning process using a phage spray sanitizing chamber or rinse cycle connected to or part of egg washing machines as illustrated in Fig. 3.4. The phage can be supplied inexpensively to the egg cleaning facility with the sprayer (Fig. 3.4). A concentrated appropriate phage stock suspension in Lambda (λ)-buffer [6 mmol l^{-1} Tris pH 7.2, 10 mmol l^{-1} $Mg(SO_4)_2 \cdot 7H_2O$, 50 μg ml^{-1} gelatin (Oxoid, UK)] titre 10^{10} pfu ml^{-1} diluted in warm (30–37 °C) clean fresh water to give a final phage titre of 10^7 pfu ml^{-1}. Most data available to date comes from experimentally inoculated foods in laboratories and in many of the experiments, optimum control of pathogens was achieved at high multiplicity of infection values (high ratio of phage to target bacteria) (Higgins et al. 2005; EFSA 2009b; Abdulamir et al. 2014). The belt transports eggs to the cleaning section. First the egg is washed by water to remove dirt, soil, stains, etc., then on to the phage sanitizing section (Fig. 3.4). The phage treatment by sprays will cover the whole eggshells with small droplets of water containing millions of phage particles. If the target bacterial pathogen is present on the eggshell, the phage will infect the target bacteria within 3–5 min (this is of course after egg leaves the washing facility) the phage will complete the kill of their target cells within 45 min. As a result of this process, the phage populations will increase on the infected eggshells due to replication and transmission from neighbouring organisms thereby increasing the phage yield, adding more protection against any further

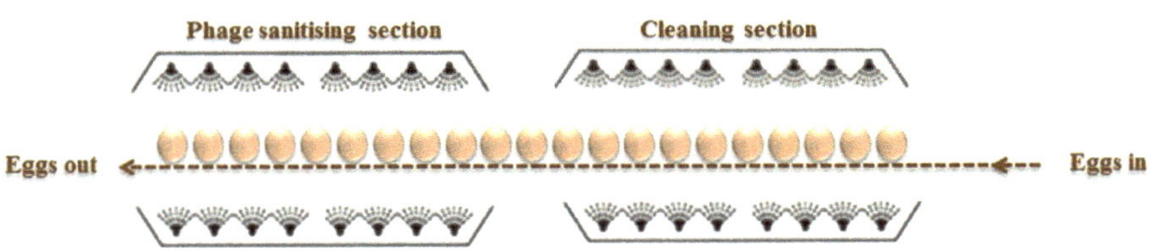

Fig. 3.4 Egg phage spray sanitizing

contamination by the target bacteria after the eggs leave the egg cleaning and phage disinfection facility. The phage will stay active on eggshells for only a few days during egg packaging, shipping and storage. It is important to mention that although phages were sometimes not inactivated, they were apparently immobilized relatively soon after addition to non-liquid forms and therefore became inactive by limited diffusion (EFSA 2009b). Furthermore, the phage survival and persistence may be affected by a combination of physical factors such as pH, temperature, water content etc., (this is discussed in greater detail in Chap. 1).

3.12.3.2 Bird Barns

Cleaning of manure from soiled surfaces is quite difficult with chicken feathers on the walls and dust covering everything in the barn. These conditions are quite challenging but since broiler birds and egg production are enhanced in a clean environment, and for economic reasons, they are cleaned after each flock, before the next flock is introduced into the installation. The cleaning methods used by farmers include scraping and brushing manure from the surfaces or spraying and soaking with water and a cleaning solution. Scraping and brushing manure from the surfaces by workers is a dangerous practice since many pathogenic bacterial diseases can be inhaled with the dust, which can cause a serious health problem and many bacterial foodborne pathogens can be transferred from one farm to another as well as to their families and communities at large. Levels of hygiene vary significantly between farms, companies and countries. Industry recommendations are relatively standard and include the provision of appropriate hygiene barriers, hand washing facilities, boot dips/changes, etc. The use of house-specific boots and clothes, the use of overshoes, and the effective use of boot dips are all associated with a reduced risk of flock infection (Newell et al. 2011). However, the implementation of such procedures is variable. The relative efficacy of house specific boots compared with boot dips remains unclear from available intervention studies, and a combination of both methods might even be beneficial (Newell et al. 2011). Furthermore, the same authors have reported the purpose of a hygiene barrier should be to physically separate the "dirty" outside environment from the "clean and protected" inside environment. In the most basic situations, a hygiene barrier for a poultry house is a physical point (possibly as simple as a line drawn on the floor) at which there is a boot dip and/or a change of footwear. Hygiene barriers may also enable/ensure a change of outer clothes (overalls) and provide hand washing facilities. These hygiene procedures should be undertaken both going into a poultry house and upon leaving a poultry house, nevertheless, under practical conditions, it has proved extremely difficult, though not impossible, to provide the stringent biosecurity measures required to effectively and consistently exclude infection from such poultry houses (Newell et al. 2011). In spite of the rigorous general biosecurity measures on poultry farms to reduce infection, sustaining such measures on the farm appears to be extremely difficult. It is becoming increasingly vital to reduce the costs of follow-up cleaning, decontamination, cross infections and quality control. For this reason, phage applications can be introduced in poultry farms and poultry meat industries as a possible means to enhance the farm hygiene.

Smart phage biocontrol at the farm can be used to decontaminate the bird barn and egg production facilities. Phage sanitisation is proposed for decontaminating poultry manure, bird barns and egg production. The phage solution is comprised of smart phages (cocktail) at titre of 10^7 pfu ml^{-1} in clean warm water (30–37 °C). The water is supplemented with Lambda-buffer [6 mmol l^{-1} Tris pH 7.2, 10 mmol l^{-1} $Mg(SO_4)_2 \cdot 7H_2O$, 50 μg ml^{-1} gelatin (Oxoid, UK)] (see Sect. 3.12.3.1). By using a hose nozzle with multi spray settings, similar to that for a car wash, or by using a sprayer or an electric mist fan with ultrasonic vibration, we atomize the water to 1–5 μm ultra-particles and spread the phage mist (Fig. 3.5). The phage can disinfect directly on floors as well as walls and equipment. After 2–3 h, decontaminated manure and dust are removed. Phage barn cleaning can be implemented after each flock and as needed to ensure that each flock starts its life in a clean environment.

3.12.3.3 Feed and Water

Barns have automatic feeding systems so the birds can eat whenever they want. Broiler birds also have constant access to fresh, clean water. Chicken producers are required to clean and disinfect their water lines before each new flock and test their water annually for the presence of bacteria. The water is delivered to the birds through a nipple drinking system. Many poultry producers are interested in phage as an alternative feed additive to dietary antibiotics for controlling *Salmonella* infection. A wide variety of phage applications such as air sac inoculation, drinking supply, aerosol spray and their combinations have been reported (Kim et al. 2013). A phage feed additive will reduce the carriage of pathogens (e.g. *Salmonella*, *Campylobacter* and *E. coli* O157:H7) in poultry, thereby lowering the dependence

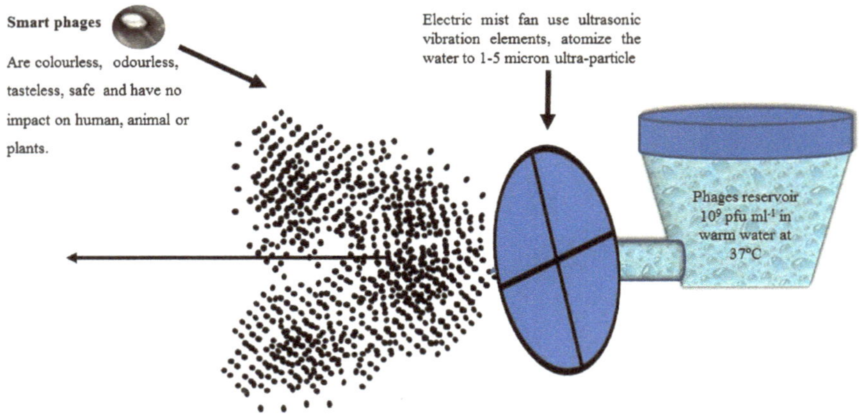

Fig. 3.5 Concept for phage electric mist fan

on antibiotics in poultry farming. Programmed smart phage cocktails for feed products can be integrated into normal chicken feeding practices (Fig. 3.6) to control pathogens, thereby increasing their efficacy and reducing contamination on poultry products which are ready for sale. There are two ways the phages can be administered orally, (1) incorporated in drinking water at phage titre of 10^7 pfu m^{-1} or (2) supplemented in animal feed at 10^7 pfu g^{-1} (Fig. 3.6). Phage concentration of 10^7 pfu m^{-1} is considered optimal for phage feeding via drinking water (Abdulamir et al. 2014).

Despite these advances, inherent challenges remain in the practical use of oral phage biocontrol. It is well-known that phages typically lack stability in the extremely acidic environment of the chicken stomach (Jończyk et al. 2011), and their residence time in the intestinal tract is very short (Bardina et al. 2012). These problems translate into a low efficiency of phage infection that can only be compensated for by very frequent treatment, which is costly and time-consuming (Bardina et al. 2012). In this regard, the optimal use of phages for controlling *Salmonella* infections in broilers will require phages formulations thereof that are more acid stable and have longer residence times. There are 3 protocols which have been developed for stabilizing phages in acidic media: (i) using alkali to phage suspensions (Abdulamir et al. 2014), (ii) the addition of antacids (e.g., $CaCO_3$) to phage suspensions (Atterbury et al. 2007), and (iii) the encapsulation of phage approach has been attempted to improve the efficacy of phage treatment using

Fig. 3.6 Phage feed for poultry as a growth promoter (phage probiotics supplement) and to reduce *Salmonella* and *Campylobacter* infections

polymeric microcapsules, including those comprising alginate-pectin (Dini et al. 2012) (size, 1 mm), alginate-chitosan (size, 780 m) (Ma et al. 2008), methacrylate (Stanford et al. 2010), alginate-$CaCO_3$ (size, 900 m) (Ma et al. 2012) and liposome (Coloma et al. 2015). Phage capsules were also developed by SAA Jassim (unpublished observation) (Fig. 3.7). Albino rats were fed with a designed phage cocktail (Jassim et al. 2010) using absorbable vegetable capsules administered orally to deliver the phage cocktail to the tested animal. The capsules were manually filled with 0.5 ml of the 14 smart phage cocktail 10^5 pfu ml^{-1} in lambda buffer. Rats were given 3 capsules per a day at 2 h intervals for 6 days. It was found that the lowering in *E. coli* output started in day 2 and maximum log reduction yielded 4.63 log units in 6 days. Achieving the greatest reduction in *E. coli* O157: H7 counts (log_{10} 4.63) at the experimental end point of 6 days without intercept by animal immunity demonstrated that phage capsules provide an excellent route for animal biocontrol.

The bacterial level cfu ml^{-1} of *E. coli* O157:H7 (VTEC O157) fecal output (red line) in phage-treated rats compared to that of non-phage-treated rats (dashed black line) (A), and log_{10} reduction (B) of *E. coli* fecal output in phage-treated rats to that of non-phage-treated rats (Jassim SAA and Abdulamir AS, unpublished observation).

Due to the possibility of infection of animals with *Salmonella*, this organism should not be present in animal feed (EFSA 2008). Considering that the prevalence of *Salmonella* contamination in certain feed ingredients is found to be high and identified as a risk for contamination of the subsequent feed chain, therefore the

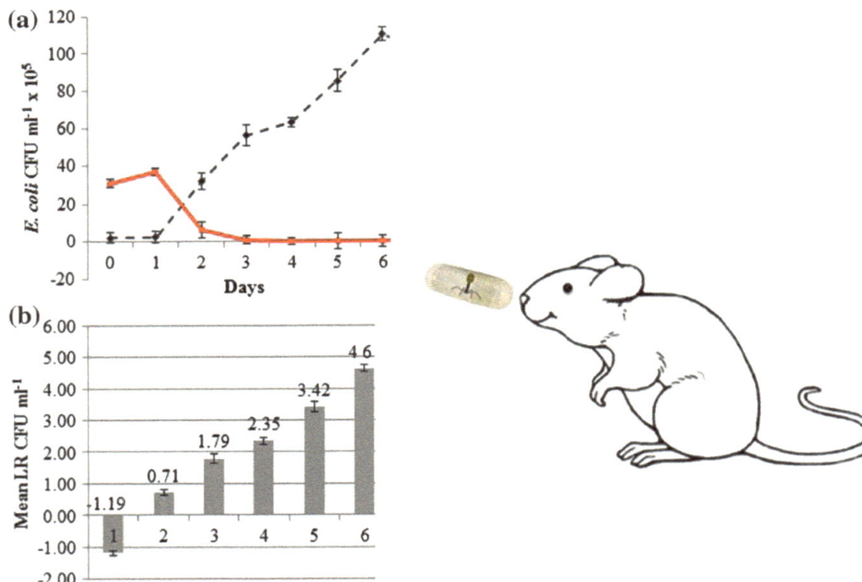

Fig. 3.7 Phage feeding *via* oral vegetable capsules

objective should be to prevent the *Salmonella* contamination as early as possible in the feed chain (EFSA 2008). Using phage in animal feed and drink as a biocontrol measure with the aim to prevent, reduce or limit the multiplication, colonisation and penetration of *Salmonella* in the gut, will have the greatest effect on probiotic bacteria (the indigenous intestinal flora) which are likely to be antagonistic to pathogens such as *Salmonella* and VTEC O157. In summary, long term studies are needed to demonstrate the best industrial scale methods for the production of *Salmonella* free feed material in spite of an often relatively high initial *Salmonella* contamination.

Numerous animal studies returned results on the safety of phages as biocontrol agents. Carlton et al. (2005) demonstrated that no abnormal physical or behavioral changes were observed in albino rats administered orally with repeated high doses of phage P100 for 5 consecutive days. Kang et al. (2013) also observed no effects on the physical appearance and behavior of the 8 male BALB/c mice in the oral toxicity test after a high titer (1.1×10^{11} pfu kg^{-1} body weight) of wksl3 phage was administered orally. Convincing evidence was shown by a study which involved oral administration of *E. coli* phage T4 into human volunteers (Skurnik et al. 2007). They found that neither the low (10^3 pfu ml^{-1}) nor the high (10^5 pfu ml^{-1}) phage dosage resulted in any adverse effects (Skurnik et al. 2007). These studies suggest, as expected, that the intake of phage with food is safe and can be used as an additive for bio-preservation of foods.

3.12.3.4 Direct Use on Poultry and Meat Processing

Poultry provides an excellent medium for the growth of certain pathogenic (disease-causing) bacteria, such as *Salmonella* and *Campylobacter*. Potential causes of contamination of poultry during the slaughtering and processing procedures include contact of the carcass with body parts that contain a high microbial load (e.g., feathers, feet, intestinal contents), use of contaminated equipment and physical manipulation of the meat (e.g., de-boning, grinding). In broiler production, both hatchery and feed contamination are implicated in *Salmonella* strains seen at slaughter (EFSA 2008). *Salmonella* may be transferred to carcasses during processing from three major sources:

(i) Batch of birds being slaughtered,
(ii) Cross contamination between a positive batch of birds and subsequent carcasses from negative batches,
(iii) Inefficient initial disinfection will cause the establishment of resident populations of *Salmonella* in biofilms associated with the processing equipment, environment and recycled water supplies. Once inaccessible biofilms have formed they can be extremely difficult to eliminate.

Therefore, the prevention of microbial contamination involves careful regulation and monitoring of the slaughtering and processing plants, proper handling and storage and adequate cooking of raw and processed poultry products. Experimental studies from different research groups suggested that reduction of pathogens colonization in animals during primary production (phage pre-harvest application) is a strategy that can be followed in avian farms just before slaughter or during animal growth to reduce the probability of cross-contamination with the poultry feces during food processing (Fig. 3.2). Such treatments should render effective protection for foodstuffs against the small number of pathogenic bacteria that have escaped the food processing regimen, or have become contaminated post-sale from improper food handling. They also prevent the formation of biofilms which normally resist treatment (Jassim et al. 2012; Hibma et al. 1997; Lewis 2008; Nannapaneni and Soni 2015).

Following is an example of using phage adaptation to new or modified hosts in food samples. Jassim et al. (2012) have used phage biotechnologies which do not use genetic manipulation. Techniques called phage design (2010) and phage breeding technology (Jassim et al. 1995; Hibma et al. 1997) were used to explore fast new approaches to phage-based bio-processing for pathogenic *E. coli* bacteria in food products. These approaches remain the most advanced technologies published in phage biocontrol. One hundred and forty highly lytic, designed coliphages were used on naturally contaminated *E. coli* and experimentally inoculated enterohemorrhagic *E. coli* (EHEC) samples of lettuce, cabbage, meat and egg. In addition, experimentally produced biofilms of *E. coli* were tested. A phage concentration of 10^3 pfu ml^{-1} was used for food product immersion by applying a 20-sec optimal dipping time in a phage cocktail and for spraying of food products, 10^5 pfu ml^{-1} of a phage cocktail was used. Food samples were cut into pieces and were either sprayed with or held in a bag immersed in lambda buffer containing a cocktail of 140 phages. Phage bio-processing was successful in completely eliminating *E. coli* in all processed samples after 48 h stored at 4 °C. Partial elimination of *E. coli* was observed in earlier storage periods (7 and 18 h) at 24 and 37 °C. Moreover, *E. coli* biofilms were reduced >3 log cycles upon using this phage bio-processing. The use of a phage cocktail of 140 highly lytic designed phages proved highly effective in suppressing *E. coli* contaminating food products. Proper decontamination and prevention methods of pathogenic *E. coli* achieved in this study can replace the less effective current chemical decontamination methods (Jassim et al. 2012). Phage application has also been studied as a method to control the presence of biofilms in the food-processing environment. Such a treatment is useful in the control of *L. monocytogenes*, which is very effective at forming biofilms in drainage systems, making decontamination difficult (Hibma et al. 1997).

3.12.3.5 Biosensors

(i) *Climate control*

Most poultry meat processors have computer systems that are used to control the temperature in the factory, keep the air fresh and control the lighting systems. The air can be a serious contamination source; it is therefore essential for food a processing facility to regularly evaluate the quality of the environment. Similarly, a smart biohazard detector 'Phage Alarm and Detector' (PAD; Fig. 3.8) can be designed to alert the meat producer if the pathogenic bacteria are present in a meat processing room so that precautions can be implemented to prevent food/meat contamination. Furthermore, it can be connected to the dry fog phage system (DFPS; discussed in greater detail in Chap. 4) (Fig. 3.9). The system is suggested to improve monitoring and bio-control of *Campylobacter* and *Salmonella* bacteria. We can also add any other important bacterial pathogens for broiler production and poultry meat industries to be detected in the same PAD. The need for such systems is vital and important to control foodborne and airborne pathogens since the air in laying hen houses contains high concentrations of airborne *Salmonella* bacteria (Ahmed et al. 2013). Numbers of *Campylobacter* increase during feather removal, as mechanical feather picking causes a mist as well as particles of dust, feathers, feces and other matter to become airborne around the machine (Berrang et al. 2004). Therefore, there is a need for a sensitive detection method for *Campylobacter* in the air at broiler farms and industries (Olsen et al. 2009).

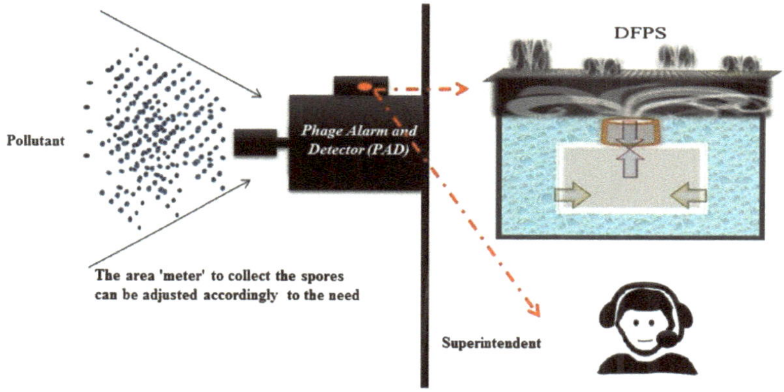

Fig. 3.8 Illustration shows smart PAD can be connected to DFPS and superintendent telephone

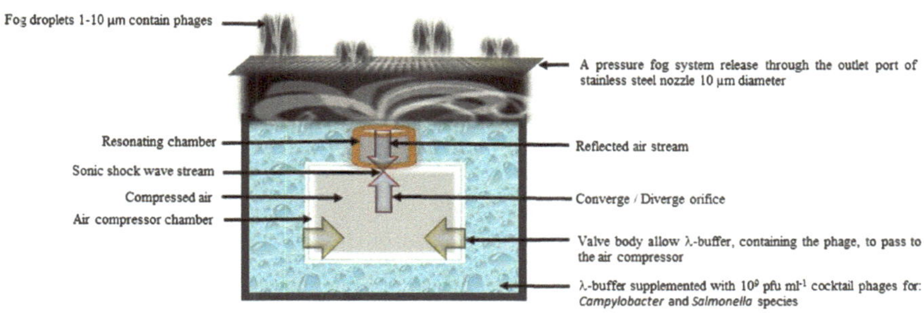

Fig. 3.9 Drawing of dry fog phage system (DFPS) illustrate operation

(ii) *Biohazard detector 'PAD'*

A brief description of a PAD 'air monitoring system': Transmission of pathogenic bacteria from one host to another is primarily through the air. The bacteria are shed from an infected bird in the nasal and or ocular secretions, fecal material and feather dust. The organism remains stable outside the host body and dries as a dusty substance. This dust or aerosol contaminates the air that is then inhaled by another possible host.

Verification of airborne contamination can be set to open automatic once or twice a day or as needed by pulling air into the PAD from a poultry house, hatchery, slaughterhouse, meat processing room, etc. Air is pulled into the air collector through a stainless steel grid with a flow capacity of approximately 120 ml of air per min. The air is transferred to a 1 cm^2 thin chip collection pad containing smart phage. Approximately 120 ml of air (around 1 min) on 1 cm^2 thin chip is required.

(iii) *Reporter phage signal*

Adenosine triphosphate (ATP) bioluminescence has been shown by Jassim and co-workers to be extremely sensitive to levels of ATP isolated from bacterial lysis (Jassim et al. 1990, 1993; Denyer et al. 1991; Walker et al. 1992; Stewart et al. 1993; Hibma et al. 1996). Using phages as probes to bind and lyse 10^4 cfu ml^{-1} *E. coli* or *Salmonella* in the sample where the released ATP can be detected after 60 min (Blasco et al. 1998). The sensitivity of the assay was as low as 10^3 cfu ml^{-1}, following a 120 min incubation post infection (Loessner et al. 1996). The designed phages have substantially improved test sensitivity for the ATP assay, having a detection limit of $<10^2$ cfu ml^{-1} at <70 min (Jassim et al. 2010).

Since the thin chip contains 10^9 pfu phages on 1 cm^2, if the target bacteria are present in the air, the phage will rapidly infect bacterial cells and lyse them within 60 min. Then after 61 min (time for 120 ml air collection 1 min + the phage lysis time 60 min) the chip can be subjected to an auto detector to measure the release of intracellular materials ATP with ATP-monitoring reagent (firefly luciferase and D-luciferin) as described by Jassim and Griffiths (2007) and Jassim et al. (2011). In this regard, the smart chip can be supplemented with luciferin-luciferase enzymes. The luminescent signal can be auto measured kinetically for a period of 10 s (from the start of bacterial infection by phage to the bacterial lysis) using a luminescence reader to be placed and connected with the PAD. Bioluminescent reactions are catalysed by firefly luciferase (Fig. 3.10). Firefly luciferase, using ATP, catalyses the two-step oxidation of luciferin to oxyluciferin, which yields light at 560 nm. The light output is given as the integral relative light units (RLUs). ATP measurements are carried out at room temperature.

Since the PAD collects 120 ml of air per min, the number of the bacteria collected from 120 ml on thin phage chip will be 10^4 cfu placed in 1 cm^2 and by using smart phage the detection time can be reduced by half (approximately 35–40 min) since ATP of $<10^2$ cfu ml^{-1} has been detected <70 min. Therefore, it is

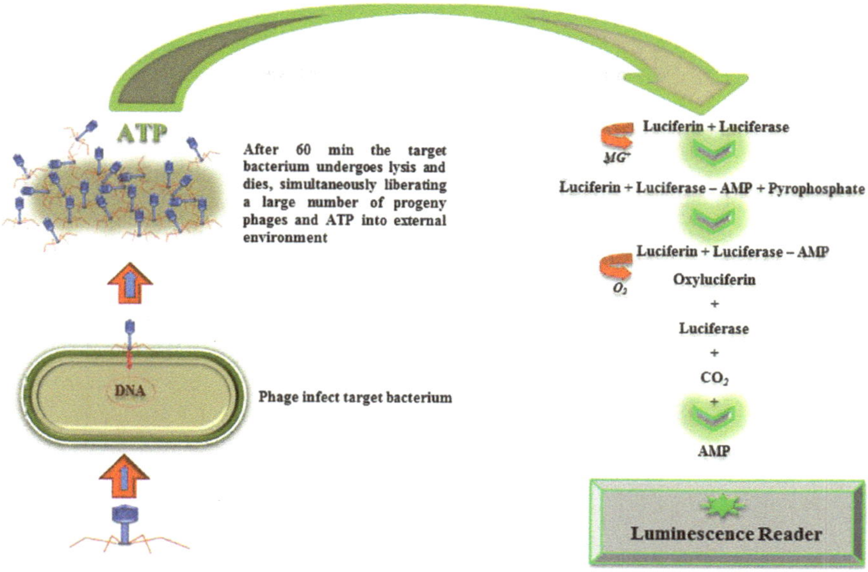

Fig. 3.10 ATP bioluminescence assay coupled with ADP to quantify bacterial contamination in air

possible for the PAD-ATP bioluminescence (Figs. 3.8 and 3.10) to detect a single (1) bacterial cell in one (1) ml of air within 70 min qualitative and quantitative. However, the detection time can be reduced from 70 to 15 min by introduction of *lux AB* genes into phages as rapid reporter bacteria. The principle of using *lux AB* genes in phages was developed by Ulitzur and Kuhn (1987, 1989) by introducing a detectable marker enzyme, fire-fly luciferase, into phages which can then be used for bacterial detection. If a target bacterial host is present, phage nucleic acid will infect that host and will be expressed in the bacteria. When the modified phage carries the marker luciferase, the presence of the bacteria can be determined by the emission of light by luciferin added exogenously, which can be easily measured or visualised under the photon imaging camera (Denyer et al. 1998), or using the luminometer measurements. The so-called 'Bronx Box' has been developed as a low cost means of detecting a light signal from mycobacteria infected with LRP using photographic film (Riska et al. 1999). Using a similar principle, the fluorescent protein reporter phage (Oda et al. 2004) can also be adopted with the thin phage chip in the PAD.

In general, the novel PAD-ATP bioluminescence can be linked via computer network to the superintendent telephone and/or to the meat producer on his home or cell phone if there is a foodborne contamination problem in the factory. He can thus stop the production and save millions of dollars in waste along with the reputation of his facility while he addresses the problem before resuming production. If the power should go out, a generator can keep the systems running and ensure the processing is continued. The PAD-ATP bioluminescence can also be connected to DFPS where as soon as the PAD-ATP bioluminescence system detects the target pathogen in the air, PAD-ATP bioluminescence will instantly send a signal to DFPS in the meat processing room or the plant environment to open and release the phage fog for DFPS. The phage fog can rapidly disinfect and kill the target bacterial pathogens in air and on all surfaces such as; meat, walls, floor, equipment, tools, etc. This improvement in the ATP test detection limits would subsequently enable institution of proactive measures for quality assurance, i.e., implementing HACCP programs. The system allows for variable detection limits as well as specific identification. Many other measurements of released intracellular materials after phage lysis of bacteria can also be developed with the smart phage chip, such as the detection of endotoxin Gram-negative cells after lysis (Jassim et al. 2011), matrix-assisted laser desorption/ ionization, time-of-flight mass spectrometry resulting in a visible signal following phage infection (Rees and Voorhees 2005), amperometric detection of enzymatic activity following phage infection (Yemini et al. 2007) and fluorescent probes, Live/Dead fluorescent stain (Jassim and Griffiths 2007).

3.12.3.6 Phage Biosecurity

Broiler producers follow strict biosecurity protocols that are designed to reduce the ability of *Salmonella* and *Campylobacter* to enter the barns and affect the broiler flocks they care for. These two bacterial pathogens and other important foodborne bacterial pathogens can be carried into the barns on equipment, clothing, footwear and hands. Diseases can be easily transferred from one poultry farm to the next just by the traffic that is on the farm everyday as reported above (Cox et al. 1996; Rocourt et al. 2003; Velge et al. 2005; Behling et al. 2010). Persistent environmental contamination is a major factor in the infection of poultry flocks. All vehicles used for the transport of incoming raw materials and compound feeding stuffs must be subject to regular cleaning and sanitizing programs to ensure that they are clean. Several studies have pointed out that condensation in ingredient storage containers produces an environment capable of supporting the growth of *Salmonella* spp. (EFSA 2008). Housing and feeding systems as well as flat storage for straight ingredients are often open to intrusion by wild birds, rodents, cats, dogs, foxes and badgers which may defecate in the feed and for that reason effective control measures are important to prevent contamination (EFSA 2008). It is important that silos are cleaned regularly as well as flat storage areas (EFSA 2008). Surveys of the feed chain have shown high *Salmonella* levels detected in dust samples from the pre and post-heating treatment area of the feed mill, as well as inside the pellet cooling systems (EFSA 2008). Recontamination can be prevented through cleaning and disinfecting unsanitary equipment surfaces and dust that are in contact with feed. Farmers could adopt phage farm food safety programs in place to reduce the risk of *Campylobacter* and *Salmonella* disease transfer. The phage poultry biosecurity program can be developed that builds on the above National requirements. The program also requires growers to have phage wash stations at the gate so vehicles can be cleaned upon entering and exiting the farm. There are also phage hand-sanitizing stations the phage-based hand rubs (PBHR; discussed in greater detail in Chaps. 4 and 7) and boot dips at the entrance to each barn to stop bacterial pathogens from entering the barn on people's hands and feet (Fig. 3.11).

To control *Salmonella* and *Campylobacter* and most bacterial foodborne diseases, we need comprehensive control measures using lytic and virulent smart phages which can be delivered using a phage sprayer (Fig. 3.4), phage electric mist fan (Fig. 3.5) and DFPS (Fig. 3.9). They are user-friendly and deal with multiple threats in feed/food processing, farm environments, contaminated animals, silos, tools, equipment, feeding systems and with the community at large (Fig. 3.12).

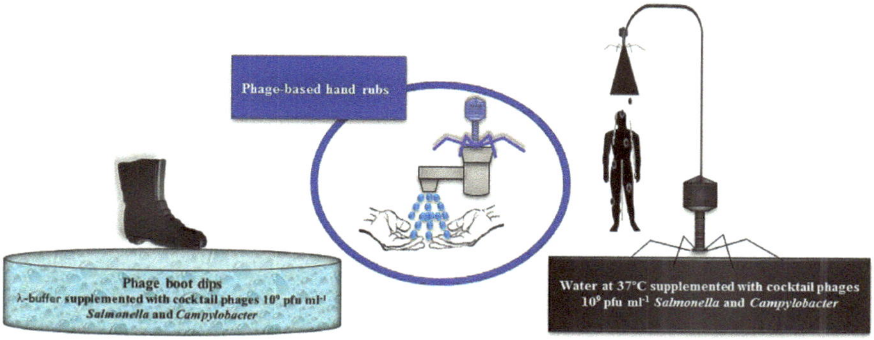

Fig. 3.11 Smart phages for personal hygiene: phage sanitizing stations

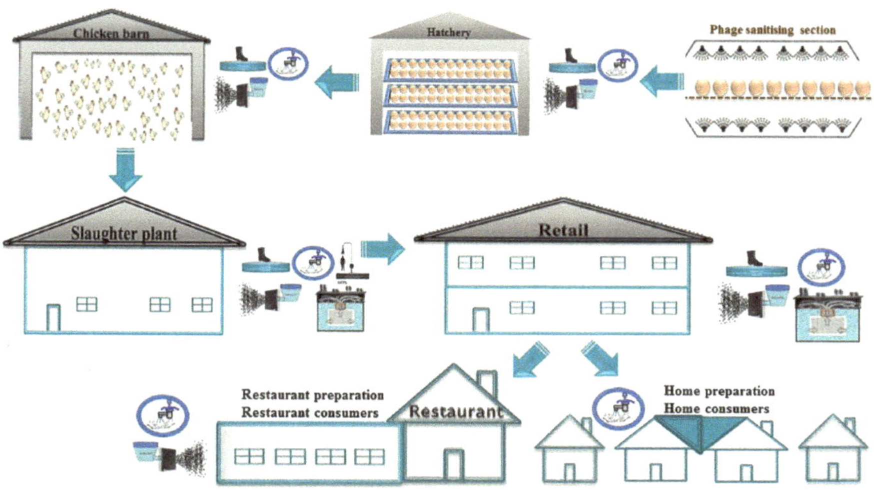

Fig. 3.12 Schematic diagram illustrating comprehensive phage biocontrol measures from farm-to-fork

The conventional decontamination techniques employed in food-processing facilities such as chemicals, physical disrupting agents and irradiation have significant drawbacks. They usually cause corrosion of the instruments used in food processing, damage foods and cause the toxic effects of chemical residues (Sulakvelidze 2013). These deleterious effects are not seen in bacteriophages which are non-toxic and do not interfere with mammalian cells (Bruttin and Brüssow 2005; Skurnik et al. 2007), while also environmentally friendly (Loc-Carrillo and Abedon 2011). The phage based biocontrol/disinfectant treatment of broiler and laying hen flocks is safe to use in all poultry industrial levels including programs associated with the higher prevalence and/or risk of bacterial infection. They are an effective measure to decrease the shedding and prevalence of *Salmonella*, *Campylobacter* or any other important bacterial foodborne pathogens in poultry populations, and have a beneficial effect on the balance of commensal intestinal organisms. Such treatment gradually reduces the percentage of positive flocks in order to reach an acceptable level and in parallel reducing the need for antibiotics. A preventive phage based biocontrol/disinfectant agent will then maintain the safety level from farm-to-fork.

3.13 Conclusion and Future Perspective

Phages are the most ubiquitous microbes on Earth. They are found in dental plaque, gastrointestinal tracts, skin and other organs, as well as in food and drinking water, which are loaded with these microorganisms, while humans and animals live harmoniously with them (Sulakvelidze 2011). Phages have long been recognized for their potential as therapeutic agents. The recent approval for the use of phages as safe 'GRAS' by US FDA for food safety purposes has increased the impetus of phage research to uncover phage-mediated applications with activity against other food pathogens.

The lytic cycle or 'virulent phages' fit in the class of 'natural antimicrobial controlling agents' while their high specificity and rapid killing kinetics against their target species and their safety profile make them particularly attractive natural antibacterial products (Sulakvelidze et al. 2001; Skurnik et al. 2007; Courchesne et al. 2009). Antibiotic resistance of bacterial pathogens is presently a source of major concern to modern medicine and its incidence has grown sharply in recent years (Sillankorva et al. 2008; Jassim and Limoges 2014). As a result, improving phage biocontrol may represent an important step in the fight against these resistant bacteria. There are many advantages in the use of phage for biological control. The success of phage biocontrol to greatly reduce harmful bacteria entering the food chain at farm level requires the production of virulent phages that can survive in extreme environments and having a broad host range for the target genus, while also lacking bacterial virulence genes (Jassim and Limoges 2014).

Veterinary biocontrol applications require the appropriate administration targeting specific bacteria, with a strategy that includes a comprehensive methodology, detailing the phage-host interactions, dose optimization and accounting for all chemical and physical factors (Jassim and Limoges 2013, 2014). In general, a deep understanding of intrinsic phage properties is critical to designing biocontrol interventions. The use of phage as a pre-harvest strategy is to prevent animal illness and also to reduce the pathogens carried by livestock via direct administration of phages. Meanwhile, post-harvest strategies involve the use of phages on animal carcasses to remove unwanted contaminants on the products. In the pre-harvest interventions, the phages are usually administrated directly to the live 'food animals' before being processed into meat. The purpose of such an approach is that phages may eliminate or reduce the colonization of the pathogenic bacteria on the livestock prior to slaughter and carcass processing to ensure that the processed meat is free from those pathogens. Meanwhile, the post-harvest intervention is to improve the food safety by applying phages on the surface of foods, hence eliminating or reducing the contamination of foods with foodborne bacterial pathogens, making the foods safe to consume. A combination of pre-and post-harvest intervention strategies should ultimately provide consumers with safer meat and poultry products, which would significantly reduce the medical and productivity costs associated with salmonellosis and Campylobacteriosis.

Using phages and in particular properly programmed phages in biocontrol applications offers advantages for several reasons:

(i) a single bacterial species, or even strain allows safe targeting to the pathogenic bacteria without causing any harm to the normal flora when used for biocontrol purposes,
(ii) self-replication and self-limiting, meaning that low or single dosages will multiply as long as there is still a host threshold present, multiplying their overall antimicrobial impact,
(iii) while bacteria develop phage defense mechanisms for their survival, phages continuously adapt to these altered host systems and using phage programming technology will also overcome phage defense mechanisms in bacterial host cell,
(iv) low inherent toxicity, since they consist mostly of nucleic acids and proteins,
(v) phages are relatively cheap and easy to isolate and propagate but may become time consuming when considering the development of programmed phage, a highly virulent, broad-spectrum, and non-transducing phage,
(vi) they can generally withstand food processing environmental stresses (including food physiochemical conditions),
(vii) they have proved to have prolonged shelf life,
(viii) they have a high success rate against the target organism with no reports of harmful side effects to humans to date (Stone 2002; Tanji et al. 2004; Sheng et al. 2006; Skurnik et al. 2007; Sherman 2008; Abedon et al. 2011; Viazis et al. 2011; Sillankorva et al. 2012; Jassim and Limoges 2014),
(ix) the abundance and ubiquitous nature of phages against the major pathogenic bacteria are readily discovered and isolated especially from sewage and

waste materials which contain high bacterial concentration, hence lowering the cost of production (Skurnik et al. 2007),

(x) the remarkable stability of phages in food is another valuable property with some phages able to withstand heat up to 60 °C, freezing conditions, high osmotic pressure and also a pH of 4.0 (Whitman and Marshall 1971). They are also considered to have low environmental impacts as they consist of only nucleic acids and proteins with narrow host ranges (Loc-Carrillo and Abedon 2011),

(xi) some phages have the ability to disrupt some of the bacterial biofilms (Azeredo and Sutherland 2008).

The majority of related studies demonstrated that phage biocontrol has the potential to become the main approach in reducing foodborne pathogens in food producing animals and also in fresh and processed foods. However, there are still many questions to be addressed. Market acceptance may be a challenge for the use of phage in a broader basis for biocontrol of foodborne pathogens in foods (Sulakvelidze 2013). If the cost of phage production causes an increase in food prices, it may not be welcome by producers and consumers. Hence, an education campaign about the safety and advantages of phage-based biocontrol could be implemented to educate the public. Some phages are already commercially available (e.g. ListShield™, EcoShield™ and SalmoFresh™) and approved by the FDA as discussed above, thus the need for evidence on the safety of phages to be applied not only in food but also in animals and humans is likely available.

References

Abdulamir AS, Jassim SAA, Abu Bakar F (2014) Novel approach of using a cocktail of designed bacteriophages against gut pathogenic *E. coli* for bacterial load biocontrol. Ann Clin Microbiol Antimicrob 13:39. http://www.ann-clinmicrob.com/content/pdf/s12941-014-0039-z.pdf

Abedon ST (2010) The 'nuts and bolts' of phage therapy. Curr Pharm Biotechnol 11:1

Abedon ST, Thomas-Abedon C (2010) Phage therapy pharmacology. Curr Pharm Biotechnol 11:28–47

Abedon ST, Kuhl SJ, Blasdel BG, Kutter EM (2011) Phage treatment of human infections. Bacteriophage 1:66–85

Adak GK, Long SM, O'Brien SJ (2000) Trends in indigenous foodborne disease and deaths, England and Wales: 1992 to 2000. Gut 51:832–841

Agunos A, Léger D, Carson C (2012) Review of antimicrobial therapy of selected bacterial diseases in broiler chickens in Canada. Can Vet J 53:1289–1300

Ahmed MF, Schulz J, Hartung J (2013) Air samplings in a *Campylobacter jejuni* positive laying hen flock. Ann Agric Environ Med 20(1):16–20

Aldoori AA, Mahdii EF, Abbas AK, Jassim SAA (2015) Bacteriophage biocontrol rescues mice bacteremic of clinically isolated mastitis from dairy cows associated with methicillin-resistant *Staphylococcus aureus*. Adv Microbiol 5:383–403

Allen VM, Weaver H, Ridley AM et al (2008) Sources and spread of thermophilic *Campylobacter* spp. during partial depopulation of broiler chicken flocks. J Food Prot 71:264–270
Allen HK, Levine UY, Looft T et al (2013) Treatment, promotion, commotion: antibiotic alternatives in food-producing animals. Trends Microbiol 21:114–119. doi:10.1016/j.tim.2012.11.001
Atterbury RJ (2009) Bacteriophage biocontrol in animals and meat products. Microbial Biotechnol 2(6):601–612
Atterbury RJ, Connerton PL, Dodd CE et al (2003a) Isolation and characterization of *Campylobacter* bacteriophages from retail poultry. Appl Environ Microbiol 69:4511–4518
Atterbury RJ, Connerton PL, Dodd CE et al (2003b) Application of host-specific bacteriophages to the surface of chicken skin leads to a reduction in recovery of *Campylobacter jejuni*. Appl Environ Microbiol 69:6302–6306
Atterbury RJ, Van Bergen MAP, Ortiz F et al (2007) Bacteriophage therapy to reduce *Salmonella* colonisation of broiler chickens. Appl Environ Microbiol 73(14):4543–4549
Avens JS, Miller BF (1970) Quantifying bacteria on poultry carcass skin. Poult Sci 49:1309–1315
Azeredo J, Sutherland IW (2008) The use of phages for the removal of infectious biofilms. Curr Pharm Biotechnol 9:261–266
Bailey JS (1993) Control of *Salmonella* and *Campylobacter* in poultry production. A summary of work at Russell Research Center. Poult Sci 72(6):1169–1173
Balogh B, Jones JB, Iriarte FB, Momol MT (2010) Phage therapy for plant disease control. Curr Pharm Biotechnol 11:48–57
Bardina C, Spricigo DA, Cortés P, Llagostera M (2012) Significance of the bacteriophage treatment schedule in reducing *Salmonella* colonization of poultry. Appl Environ Microbiol 78 (18):6600–6607. doi:10.1128/AEM.01257-12
Behling RG, Eifert J, Erickson MC et al (2010) Selected pathogens of concern to industrial food processors: infectious, toxigenic, toxico-infectious, selected emerging pathogenic bacteria. In: Kornacki JL (ed) Principles of microbiology troubleshooting in the industrial food processing environment, Chap 2. Springer Science + Business Media, pp 5–61. doi:10.1007/978-1-4419-5518-0_2
Behravesh B (2008) Salmonellosis, in control of communicable diseases manual, 19th edn. American Public Health Association, pp 535–540
Berndtson E, Emanuelson U, Engvall A, Danielsson-Tham ML (1996) A 1-year epidemiological study of *Campylobacters* in 18 Swedish chicken farms. Prev Vet Med 26:167–185
Berrang ME, Northcutt JK, Fletcher DL, Cox NA (2003) Role of dump cage fecal contamination in the transfer of *Campylobacter* to carcasses of previously negative broilers. J Appl Poult Res 12:190–195
Berrang ME, Northcutt JK, Dickens JA (2004) The contribution of airborne contamination to *Campylobacter* counts on defeathered broiler carcasses. J Appl Poult Res 13(1):1–4
Blasco R, Murphy MJ, Sanders MF, Squirrell DJ (1998) Specific assays for bacteria using phage mediated release of adenylate kinase. J Appl Microbiol 84:661–666
Breitbart M, Hewson I, Felts B et al (2003) Metagenomic analyses of an uncultured viral community from human feces. J Bacteriol 185:6220–6223
Bren L (2007) Bacteria-eating virus approved as food additive. FDA Consumer 41:20–22
Bruttin A, Brüssow H (2005) Human volunteers receiving *E. coli* phage T4 orally: a safety test of phage therapy. Antimicrob Agents Chemother 49:2874–2878
Calci KR, Burkhardt W, Watkins WD, Rippey SR (1998) Occurrence of male specific bacteriophage in feral and domestic animal wastes, human feces, and human-associated wastewaters. Appl Environ Microbiol 64:5027–5029
Callewaert L, Walmagh M, Michiels CW, Lavigne R (2011) Food applications of bacterial cell wall hydrolases. Curr Opin Biotechnol 22:164–171
Callicott KA, Friðriksdóttir V, Reiersen J et al (2006) Lack of evidence for vertical transmission of *Campylobacter* spp. in chickens. Appl Environ Microbiol 72:5794–5798
Capita R, Prieto M, Alonso-Calleja C (2004) Sampling methods for microbiological analysis of red meat and poultry carcasses. J Food Prot 67:1303–1308

Carlton RM, Noordman WH, Biswas B et al (2005) Bacteriophage P100 for control of *Listeria monocytogenes* in foods: genome sequence, bioinformatics analyses, oral toxicity study, and application. Regul Toxicol Pharm 43:301–312
Carvalho CM, Gannon BW, Halfhide DE et al (2010) The in vivo efficacy of two administration routes of a phage cocktail to reduce numbers of *Campylobacter coli* and *Campylobacter jejuni* in chickens. BMC Microbiol 10:232
Carvalho CM, Santos SB, Kropinski AM et al (2012) Phages as therapeutic tools to control major foodborne pathogens: *Campylobacter* and *Salmonella*. In: Kurtboke I (ed) Bacteriophages, Chap 10. InTech Europe, Rijeka, pp 180–214
CDC (2005) Outbreaks reported: 1990, 1991; 1992. MMWR 1993
CDC (2007) Multistate outbreak of human *Salmonella* infections associated with frozen pot pies-United States. Morb Mortal Wkly Rep 57(47):1277–1280
CDC (2013a) Surveillance for foodborne disease outbreaks—United States, 1998–2008. In: MMWR surveillance summaries, vol 62(2). http://www.cdc.gov/mmwr/pdf/ss/ss6202.pdf
CDC (2013b) An atlas of *Salmonella* in the United States, 1968–2011. http://www.cdc.gov/Salmonella/pdf/Salmonella-atlas-508c.pdf
Chantarapanont W, Berrang M, Frank JF (2003) Direct microscopic observation and viability determination of *Campylobacter jejuni* on chicken skin. J Food Prot 66:2222–2230
Chantarapanont W, Berrang ME, Frank JF (2004) Direct microscopic observation of viability of *Campylobacter jejuni* on chicken skin treated with selected chemical sanitizing agent. J Food Prot 67:1146–1152
Chhibber S, Kaur S, Kumari S (2008) Therapeutic potential of bacteriophage in treating *Klebsiella pneumoniae* B5055-mediated lobar pneumonia in mice. J Med Microbiol 57:1508–1513
Coffey B, Mills S, Coffey A et al (2010) Phage and their lysins as biocontrol agents for food safety applications. Annu Rev Food Sci Technol 1:449–468
Coffey B, Rivas L, Duffy G et al (2011) Assessment of *Escherichia coli* O157:H7—specific bacteriophages e11/2 and e4/1c in model broth and hide environments. Int J Food Microbiol 147:188–194
Cohen N, Ennaji H, Bouchrif B et al (2007) Comparative study of microbiological quality of raw poultry meat at various seasons and for different slaughtering processes in Casablanca (Morocco). J Appl Poult Res 16:502–508
Coker AO, Isokpehi RD, Thomas BN et al (2002) Human campylobacteriosis in developing countries. Emerg Infect Dis 8:237–244
Coloma J, Cano-Sarabiab M, Oteroa J et al (2015) Liposome-encapsulated bacteriophages for enhanced oral phage therapy against *Salmonella* spp. Appl Environ Microbiol 81(14):4841–4849. doi:10.1128/AEM.00812-15
Conner DE, Davis MA, Zhang L (2001) Poultry-borne pathogens: plant considerations. In: Sams AR (ed) Poultry meat processing, Chap 9. CRC Press, Boca Raton, pp 137–158
Connerton PL, Loc-Carrillo CM, Swift C et al (2004) Longitudinal study of *Campylobacter jejuni* bacteriophages and their hosts from broiler chickens. Appl Environ Microbiol 70:3877–3883
Corry JEL, Atabay HI (2001) Poultry as a source of *Campylobacter* and related organisms. J Appl Microbiol 90:96S–114S
Courchesne NM, Parisien A, Lan CQ (2009) Production and application of bacteriophage and bacteriophage-encoded lysins. Recent Pat Biotechnol 3:37–45
Coward C, Grant AJ, Swift C et al (2006) Phase-variable surface structures are required for infection of *Campylobacter jejuni* by bacteriophages. Appl Environ Microbiol 72:4638–4647
Cox NA, Mercuri AJ, Juven BJ et al (1974a) Evaluation of succinic acid and heat to improve the microbiological quality of poultry meat. J Food Sci 39:985–987
Cox NA, Mercuri AJ, Thomson JE, Gregory DW Jr (1974b) Quality of broiler carcasses as affected by hot water treatments. Poult Sci 53:1566–1571
Cox NA, Bailey JS, Berrang ME (1996) Extent of *Salmonella* contamination in breeder hatcheries. Poult Sci 70:416–418

Cox NA, Richardson LJ, Carson JA et al (2010) Comparison of neck skin excision and whole carcass rinse sampling methods for microbiological evaluation of broiler carcasses before and after immersion chilling. J Food Prot 73:976–980

Croci DL, de Medici D, Scalfaro C et al (2000) Determination of enteroviruses, hepatitis A virus, bacteriophages and *Escherichia coli* in Adriatic Sea mussels. J Appl Microbiol 88(2):293–298

Crutchfield SR, Buzby JC, Roberts T, et al (1997) Economic assessment of food safety regulations: The new approach to meat and poultry inspection. Agric Econ Rep No. 755. http://purl.umn.edu/34009

Dabrowska K, Switała-Jelen K, Opolski A et al (2005) Bacteriophage penetration in vertebrates. J Appl Microbiol 98:7–13

Denyer SP, Jassim SAA, Stewart GSAB (1991) In vivo bioluminescence for studying the adhesion of bacteria. Biofouling 5(1–2):125–132

Denyer SP, Jassim SAA, Rees CED et al (1998) Genetically engineered reporter bacteria for the detection of bacteriophage. United States Patent 5723330. http://www.patentgenius.com/patent/5723330.html

Dini C, Islan GA, de Urraza PJ, Castro GR (2012) Novel biopolymer matrices for microencapsulation of phages: enhanced protection against acidity and protease activity. Macromol Biosci 12:1200–1208. doi:10.1002/mabi.201200109

Dreyfuss MS, Ransom GM, Russell MD et al (2007) Pathogen control in meat and poultry production: implementing the USDA Food Safety and Inspection Service's Hazard Analysis and Critical Control Point System. In: Simjee S (ed) Infectious disease: foodborne diseases, Chap 15. Humana Press Inc., Totowa, NJ, pp 383–404. http://eknygos.lsmuni.lt/springer/655/383-404.pdf

EFSA (2008) Microbiological risk assessment in feeding stuffs for food-producing animals. EFSA J 720:1–84. http://www.efsa.europa.eu/sites/default/files/scientific_output/files/main_documents/biohaz_op_ej720_mra_feedingstuffs_en_v3,3.pdf

EFSA (2009a) The community summary report on trends and sources of zoonoses, zoonotic agents, antimicrobial resistance and foodborne outbreaks in the European Union in 2007. EFSA J 223:111–131

EFSA (2009b) Scientific opinion: the use and mode of action of bacteriophages in food production. Scientific Opinion of the Panel on Biological Hazards (Question No EFSA-Q-2008-400). Endorsed by the BIOHAZ panel for public consultation 22 January 2009. Public consultation 30 January–6 March 2009. 2. EFSA J 1076:1–26. doi:10.2903/j.efsa.2009.1076

EFSA (2011) The European Union summary report on trends and sources of zoonoses, zoonotic agents and food-borne outbreaks in 2009. EFSA J 9(3):2090–2105

EFSA-ECDC report for 2007 (2007) *Salmonella* remains most common cause of food-borne outbreaks. http://www.efsa.europa.eu/en/press/news/zoonoses090506

El-Shibiny A, Connerton PL, Connerton IF (2005) Enumeration and diversity of *Campylobacter*s and bacteriophages isolated during the rearing cycles of free-range and organic chickens. Appl Environ Microbiol 71(3):1259–1266

El-Shibiny A, Scott A, Timms A et al (2009) Application of a group II *Campylobacter* bacteriophage to reduce strains of *Campylobacter jejuni* and *Campylobacter coli* colonizing broiler chickens. J Food Prot 72(4):733–740

Endersen L, Mahony JO, Hill C et al (2014) Phage therapy in the food industry. Annu Rev Food Sci Technol 5:327–349

ESR (2016) Bacteriophage biocontrol technology. https://ninesights.ninesigma.com/offers/-/offers-portlet/viewOffer/89

European Commission (2002) Risk assessment of food borne bacterial pathogens: quantitative methodology relevant for human exposure assessment. Preliminary report. http://ec.europa.eu/food/fs/sc/ssc/out252_en.pdf

FAO (2010) Agribusiness handbook: poultry meat and eggs. Rome, Italy. http://www.eastagri.org/publications/pub_docs/6_Poultry_web.pdf

FAO-WHO (2009) *Salmonella* and *Campylobacter* in chicken meat. Microbiology risk assessment series 19. Meeting report. http://www.who.int/foodsafety/publications/micro/MRA19.pdf
Fenton M, Ross P, McAuliffe O et al (2010) Recombinant bacteriophage lysins as antibacterials. Bioengineered Bugs 1(1):9–16
Fenton M, Keary R, McAuliffe O et al (2013) Bacteriophage-derived peptidase CHAPk eliminates and prevents staphylococcal biofilms. Int J Microbiol 2013:625341
Fischer CR, Yoichi M, Unno H, Tanji Y (2004) The coexistence of *Escherichia coli* serotype O157: H7 and its specific bacteriophage in continuous culture. FEMS Microbiol Lett 241:171–177
Fischetti VA (2008) Bacteriophage lysins as effective antibacterials. Curr Opin Microbiol 11:393–400
Foley SL, Lynne AM, Nayak R (2008) *Salmonella* challenges: prevalence in swine and poultry and potential pathogenicity of such isolates. J Anim Sci 86(E. Suppl.):E149–E162
Friedman CR, Neimann J, Wegener HC, Tauxe RV (2000) Epidemiology of *C. jejuni* infections in the United States and other industrialized nations. In: Nachamkin I, Blaser MJ (eds) *Campylobacter*, 2nd edn. American Society for Microbiology, Washington, DC, pp 121–138
Furuse K, Osawa S, Kawashiro J et al (1983) Bacteriophage distribution in human faeces: continuous survey of healthy subjects and patients with internal and leukaemic diseases. J Gen Virol 64(Pt 9):2039–2043
Gantois I, Ducatelle R, Pasmans F et al (2009) Mechanisms of egg contamination by *Salmonella enteritidis*. FEMS Microbiol Rev 33(4):718–738
Garcia P, Rodriguez L, Rodriguez A, Martinez B (2010) Food biopreservation: promising strategies using bacteriocins, bacteriophages and endolysins. Trends Food Sci Technol 21: 373–382
Gast RK (2008) Paratyphoid infections. In: Saif YM, Fadly AM, Glisson JR et al (eds) Diseases of poultry, 12th edn. Blackwell Publishing, Ames, Iowa, pp 636–665
Gast RK, Guard-Petter J, Holt PS (2002) Characteristics of *Salmonella enteritidis* contamination in eggs after oral, aerosol, and intravenous inoculation of laying hens. Avian Dis 46:629–635
Georgsson F, Þorkelssonb AE, Geirsdóttira M, et al (2006) The influence of freezing and duration of storage on Campylobacter and indicator bacteria in broiler carcasses. Food Microbiol 23(7):677–683
Gill JJ, Hyman P (2010) Phage choice, isolation and preparation for phage therapy. Curr Pharm Biotechnol 11:2–14
Goode D, Allen VM, Barrow PA (2003) Reduction of experimental *Salmonella* and *Campylobacter* contamination of chicken skin by application of lytic bacteriophages. Appl Environ Microbiol 69(8):5032–5036
Goodridge LD (2010) Designing phage therapeutics. Curr Pharm Biotechnol 11:1527
Goodridge L, Abedon ST (2003) Bacteriophage biocontrol and bioprocessing: application of phage therapy to industry. SIM News 53(6):254–262
Grawjewski BA, Kusek JW, Gelfand HM (1985) Development of a bacteriophage typing scheme for *Campylobacter jejuni* and *Campylobacter coli*. Epidemiol Infect 104:403–414
Greer GG (2005) Bacteriophage control of foodborne bacteria. J Food Prot 68:1102–1111
Greer GG, Dilts BD (2002) Control of *Brochothrix thermosphacta* spoilage of pork adipose tissue using bacteriophages. J Food Prot 65:861–863
Guard-Petter J (2001) The chicken, the egg and *Salmonella enteritidis*. Environ Microbiol 3:421–430
Hagens S, Loessner MJ (2010) Bacteriophage for biocontrol of foodborne pathogens: calculations and considerations. Curr Pharm Biotechnol 11:58–68
Hald B, Wedderkopp A, Madsen M (2000) Thermophilic *Campylobacter* spp. in Danish broiler production: a cross-sectional survey and a retrospective analysis of risk factors for occurrence in broiler flocks. Avian Pathol 29:123–131
Hale CR, Scallan E, Cronquist AB et al (2012) Estimates of enteric illness attributable to contact with animals and their environments in the United States. Clin Infect Dis 54(Suppl. 5): S472–S479

Hammerl JA, Jäckel C, Alter T et al (2014) Reduction of *Campylobacter jejuni* in broiler chicken by successive application of group II and group III phages. PLoS ONE 9(12):e114785

Hannah JF, Cason JA, Richardson JR et al (2011) Effect of stomaching on numbers of bacteria recovered from chicken skin. Poult Sci 90:491–493

Hansson I, Ederoth M, Andersson L et al (2005) Transmission of *Campylobacter* spp. to chickens during transport to slaughter. J Appl Microbiol 99(5):1149–1157

Hansson I, Pudas N, Harbom B, et al (2010) Within-flock variations of Campylobacter loads in caeca and on carcasses from broilers. Int J Food Microbiol 141:51–55

Havelaar AH (1987) Virus, bacteriophages and water purification. Vet Q 9(4):356–360

Havelaar AH, Furuse K, Hogeboom WM (1986) Bacteriophages and indicator bacteria in human and animal faeces. J Appl Bacteriol 60:255–262

Havelaar AH, Mangen MJ, de Koeijer AA et al (2007) Effectiveness and efficiency of controlling *Campylobacter* on broiler chicken meat. Risk Anal 27:831–844

Herman L, Heyndrickx M, Grijspeerdt K et al (2003) Routes for *Campylobacter* contamination of poultry meat: epidemiological study from hatchery to slaughterhouse. Epidemiol Infect 131:1169–1180

Hertwig S, Hammerl JA, Appel B, Alter T (2013) Post-harvest application of lytic bacteriophages for biocontrol of foodborne pathogens and spoilage bacteria. Berl Munch Tierarztl Wochenschr 126:357–369

Hibma AM, Jassim SAA, Griffiths MW (1996) In vivo bioluminescence to detect the attachment of L-forms of *Listeria monocytogenes* to food and clinical contact surfaces. Int J Food Microbiol 33:157–167

Hibma AM, Jassim SAA, Griffiths M (1997) Infection and removal of L-forms of *Listeria monocytogenes* with bred bacteriophage. Int J Food Microbiol 34:197–207

Higgins JP, Higgins SE, Guenther KL et al (2005) Use of a specific bacteriophage treatment to reduce *Salmonella* in poultry products. Poult Sci 84(7):1141–1145

Hinton A, Cason JA (2008) Bacterial flora of processed broiler chicken skin after successive washings in mixtures of potassium hydroxide and lauric acid. J Food Prot 71:1707–1713

Hodgson K (2013) Bacteriophage therapy. Microbiol Aust 34:28–31

Hoffmann S, Batz MB, Morris JG Jr (2012) Annual cost of illness and quality-adjusted life year losses in the United States due to 14 foodborne pathogens. J Food Prot 75:1292–1302

Holah JT, Taylor JH, Dawson DJ, Hall KE (2002) Biocide use in the food industry and the disinfectant resistance of persistent strains of *Listeria monocytogenes* and *Escherichia coli*. J Appl Microbiol 92:111S–120S

Hsu FC, Shieh YSC, Sobsey MD (2002) Enteric bacteriophages as potential fecal indicators in ground beef and poultry meat. J Food Prot 65(1):93–99

Huang H (1999) Evaluation of culture enrichment for use with *Salmonella* detection in immunoassay. Int J Food Microbiol 51(2–3):85–94

Hue O, Allain V, Laisney MJ et al (2011) *Campylobacter* contamination of broiler caeca and carcasses at the slaughterhouse and correlation with *Salmonella* contamination. Food Microbiol 28(5):862–868

Humphrey TJ, Mead GC, Rowe B (1988) Poultry meat as a source of human salmonellosis in England and Wales. Epidemiol Infect 100:175–184

Humphrey TJ, Jorgensen F, Mattick KL (2000) Fit to eat? Food scares and safe food production. Microbiol Today 27:10–12

Humphrey TJ, Martin KW, Slader J, Durham K (2001) *Campylobacter* spp. in the kitchen: spread and persistence. J Appl Microbiol 90:115S–120S

Humphrey T, Brien OS, Madsen M (2007) *Campylobacter* as zoonotic pathogens: a food production perspective. Int J Food Microbiol 117:237–257

ICMSF (1998) Poultry and poultry products. In: Microorganisms in foods, microbial ecology of food commodities, vol 6. Blackie Academic and Professional, London, pp 75–129

Izat AL, Gardner FA, Denton JH, Golan FA (1988) Incidence and level of *Campylobacter jejuni* in broiler processing. Poult Sci 67:1568–1572

James C, Goksoy EO, Corry JEL, James SJ (2000) Surface pasteurisation of poultry meat using steam at atmospheric pressure. J Food Eng 45:111–117
James C, James SJ, Hannay N et al (2007) Decontamination of poultry carcasses using steam or hot water in combination with rapid cooling, chilling or freezing of carcass surfaces. Int J Food Microbiol 114(2):195–203
Janež N, Loc-Carrillo C (2013) Use of phages to control *Campylobacter* spp. J Microbiol Meth 95(1):68–75
Jassim SAA, Griffiths MW (2007) Evaluation of a rapid microbial detection method via phage lytic amplification assay coupled with Live/Dead fluorochromic stains. Lett Appl Microbiol 44:673–678
Jassim SAA, Limoges RG (2013) Impact of external forces on cyanophage-host interactions in aquatic ecosystems. World J Microbiol Biotechnol 29(10):1751–1762
Jassim SAA, Limoges RG (2014) Natural solution to antibiotic resistance: bacteriophages 'The Living Drugs'. World J Microbiol Biotechnol 30(8):2153–2170
Jassim SAA, Ellison A, Denyer SP, Stewart GSAB (1990) In vivo bioluminescence: a cellular reporter for research and industry. J Biolumin Chemilumin 5:115–122
Jassim SAA, Camprubi S, Tomas JM et al (1993) In vivo bioluminescence for studying bacterial adhesion and in vitro phagocytosis. In: Szalay AA, Kricka LJ, Stanley P (eds) Bioluminescence and chemiluminescence. Wiley, New York, pp 491–495
Jassim SAA, Denyer SP, Stewart GSAB (1995) Selective virus culture. International Patent Application, No. WO9523848. http://patentscope.wipo.int/search/en/WO1995023848
Jassim SAA, Abdulamir AS, Abu Bakar F (2010) Methods for bacteriophage design. WIPO Patent Application WO2010/064044 A1 http://www.sumobrain.com/patents/wipo/Methods-bacteriophage-design/WO2010064044A1.pdf
Jassim SAA, Abdulamir AS, Abu Bakar F (2011) Phage-based limulus amoebocyte lysate assay for rapid detection of bacteria. WO2011/098820A1
Jassim SAA, Abdulamir AS, Abu Bakar F (2012) Novel phage-based bio-processing of pathogenic *Escherichia coli* and its biofilms. World J Microbiol Biotechnol 28(1):47–60
Javed MA, Ackermann HW, Azeredo J et al (2014) A suggested classification for two groups of *Campylobacter myoviruses*. Arch Virol 159:181–190
Joerger RD (2003) Alternatives to antibiotics: bacteriocins, antimicrobial peptides and bacteriophages. Poult Sci 82:640–647
Jończyk E, Kłak M, Międzybrodzki R, Górski A (2011) The influence of external factors on bacteriophages-review. Folia Microbiol (Praha) 56:191–200. doi:10.1007/s12223-011-0039-8
Jones JB, Vallad GE, Iriarte FB et al (2012) Considerations for using bacteriophages for plant disease control. Bacteriophage 2(4):208–214
Jorgensen F, Bailey R, Williams S et al (2002) Prevalance and numbers of *Salmonella* and *Campylobacter* spp. on raw, whole chickens in relation to sampling methods. Int J Food Microbiol 76:151–164
Joshua GW, Guthrie-Irons C, Karlyshev AV, Wren BW (2006) Biofilm formation in *Campylobacter jejuni*. Microbiology 152:387–396
Kang HW, Kim JW, Jung TS, Woo GJ (2013) wksl3, a new biocontrol agent for *Salmonella enterica* serovars *enteritidis* and typhimurium in foods: characterization, application, sequence analysis, and oral acute toxicity study. Appl Environ Microbiol 79(6):1956–1968
Kapperued G, Espeland G, Wahl E et al (2003) Factors associated with increased and decreased risk of *Campylobacter* infection: a prospective case-control study in Norway. Am J Epidemiol 158:234–242
Kennedy JE, Wei CI, Oblinger JL (1986) Distribution of coliphages in various foods. J Food Prot 49(12):944–951
Khakhria R, Lior H (1992) Extended phage-typing scheme for *Campylobacter jejuni* and *Campylobacter coli*. Epidemiol Infect 108:403–414
Kim KH, Lee GY, Jang JC et al (2013) Evaluation of Anti-SE bacteriophage as feed additives to prevent *Salmonella enteritidis* (SE) in broiler. Asian-Aust J Anim Sci 26(3):386–393

Kittler S, Fischer S, Abdulmawjood A et al (2013) Effect of bacteriophage application on *Campylobacter jejuni* loads in commercial broiler flocks. Appl Environ Microbiol 79: 7525–7533

Kutter E, De Vos D, Gvasalia G et al (2010) Phage therapy in clinical practice: treatment of human infections. Curr Pharm Biotechnol 11:69–86

Lee MD, Newell DG (2006) *Campylobacter* in poultry: filling an ecological niche. Avian Dis 50:1–9. doi:10.1637/7474-111605r.1

Levin BR, Bull JJ (2004) Population and evolutionary dynamics of phage therapy. Nat Rev Microbiol 2:166–173

Lewis K (2008) Multidrug tolerance of biofilms and persister cells. Curr Top Microbiol Immunol 322:07–131

Li Y, Yang H, Swem BL (2002) Effect of high-temperature inside–outside spray on survival of *Campylobacter jejuni* attached to prechill chicken carcasses. Poult Sci 81:1371–1377

Lillard HS (1986) Role of fimbriae and flagella in the attachment of *Salmonella typhimurium* to poultry skin. J Food Sci 51:54–56

Lillard HS (1988) Comparison of sampling methods and implications for bacterial decontamination of poultry carcasses by rinsing. J Food Prot 51:405–408

Lillard HS (1989) Factors affecting the persistence of *Salmonella* during the processing of poultry. J Food Prot 52:829–832

Lim T-H, Lee D-H, Lee Y-N et al (2011) Efficacy of bacteriophage therapy on horizontal transmission of *Salmonella gallinarum* on commercial layer chickens. Avian Dis 55(3): 435–438

Loc-Carrillo C, Abedon ST (2011) Pros and cons of phage therapy. Bacteriophage 1:111–114

Loc-Carrillo C, Atterbury RJ, El-Shibiny A et al (2005) Bacteriophage therapy to reduce *Campylobacter jejuni* colonization of broiler chickens. Appl Environ Microbiol 71(11): 6554–6563

Loessner MJ, Rees CE, Stewart GS, Scherer S (1996) Construction of luciferase reporter bacteriophage A511: luxAB for rapid and sensitive detection of viable Listeria cells. Appl Environ Microbiol 62:1133–1140

Ly-Chatain HM (2014) The factors affecting effectiveness of treatment in phages therapy. Front Microbiol 5:1–7

Ma Y, Pacan JC, Wang Q et al (2008) Microencapsulation of bacteriophage Felix O1 into chitosanalginate microspheres for oral delivery. Appl Environ Microbiol 74:4799–4805

Ma Y, Pacan JC, Wang Q et al (2012) Enhanced alginate microspheres as means of oral delivery of bacteriophage for reducing *Staphylococcus aureus* intestinal carriage. Food Hydrocoll 26:434–440

Mahony J, McAuliffe O, Ross RP, van Sinderen D (2011) Bacteriophages as biocontrol agents of food pathogens. Curr Opin Biotechnol 22:157–163

Man SM (2011) The clinical importance of emerging *Campylobacter* species. Nat Rev Gastroenterol Hepatol 8:669–685

Maukonen J, Mättö J, Wirtanen G et al (2003) Methodologies for the characterization of microbes in industrial environments: a review. J Ind Microbiol Biotechnol 30(6):327–356

McMeekin TA, Thomas CJ, McCall D (1979) Scanning electron microscopy of microorganisms on chicken skin. J Appl Bacteriol 46:195–200

McMeekin TA, Thomas CJ, Pennington PI (1984) Contamination and decontamination of poultry carcass neck tissue. J Food Saf 6:79–88

Mizoguchi K, Morita M, Fischer CY et al (2003) Coevolution of bacteriophage PP01 and *Escherichia coli* O157:H7 in continuous culture. Appl Environ Microbiol 69:170–176

Moore JE, Corcoran D, Dooley JS et al (2005) *Campylobacter*. Vet Res 36:351–382

Morrison GJ, Fleet GH (1985) Reduction of *Salmonella* on chicken carcasses by immersion treatments. J Food Prot 48:939–943

Nannapaneni R, Soni KA (2015) Use of bacteriophage to remove biofilms of *Listeria monocytogenes* and other foodborne bacterial pathogens in the food environment. In:

Pometto III AL, Demirci A (eds) Biofilms in the food environment, 2nd edn, Chap 5. IFT Press, Wiley Blackwell, Oxford, UK, pp 131–144
Nayak R, Kenney PB, Keswani J, Ritz C (2003) Isolation and characterisation of *Salmonella* in a turkey production facility. Br Poult Sci 44:192–202
Nelson DC, Schmelcher M, Rodriguez-Rubio L et al (2012) Endolysins as antimicrobials. Adv Virus Res 83:299–365
Nesbitt A, Ravel A, Murray R et al (2011) Integrated surveillance and potential sources of *Salmonella enteritidis* in human cases in Canada from 2003 to 2009. Epidemiol Infect 140:1757–1773
Newell DG (2002) The ecology of *Campylobacter jejuni* in avian and human hosts and in the environment. Int J Infect Dis 6:3516–3521
Newell DG, Fearnley C (2003) Sources of *Campylobacter* colonization in broiler chickens. Appl Environ Microbiol 69:4343–4351
Newell D, Wagenaar J (2000) Poultry infections and their control at the farm level. In: *Campylobacter*, 2nd edn. American Society of Microbiology Press, Washington, DC, pp 497–510
Newell DG, Elvers KT, Dopfer D et al (2011) Biosecurity-based interventions and strategies to reduce *Campylobacter* sp. on poultry farms. Appl Environ Microbiol 77(24):8605–8614
Notermans S, Kampelmacher EH (1974) Attachment of some bacterial strains to the skin of broiler chickens. Br Poult Sci 15:573–585
Notermans S, Kampelmacher EH (1975) Heat destruction of some bacterial strains attached to broiler skin. Br Poult Sci 16:351–361
Notermans S, Kampelmacher EH, Van Schothorst M (1975) Studies on sampling methods used in the control of hygiene in poultry processing. J Appl Microbiol 39:55–61
Nyati KK, Prasad KN (2014) Campylobacteriosis and water: an overview. In: Singh PP, Sharma V (eds) Water and health, Chap 8. Springer, Berlin, pp 119–132
O'Flaherty S, Ross RP, Coffey A (2009) Bacteriophage and their lysins for elimination of infectious bacteria. FEMS Microbiol Rev 33:801–819
Oda M, Morita M, Unno H, Tanji Y (2004) Rapid detection of *Escherichia coli* O157:H7 by using green fluorescent protein-labeled PP01 bacteriophage. Appl Environ Microbiol 70(1):527–534
Olsen KN, Lund M, Skov J et al (2009) Detection of *Campylobacter* bacteria in air samples for continuous real-time monitoring of *Campylobacter* colonization in broiler flocks. Appl Environ Microbiol 75(7):2074–2078
Panisello PJ, Rooney R, Quantick PC, Stanwell-Smith R (2000) Application of foodborne disease outbreak data in the development and maintenance of HACCP systems. Int J Food Microbiol 59:221–234
Parisien A, Allain B, Zhang J et al (2008) Novel alternatives to antibiotics: bacteriophages, bacterial cell wall hydrolases, and antimicrobial peptides. J Appl Microbiol 104:1–13
Parka H, Hung Y-C, Brackettb RE (2002) Antimicrobial effect of electrolyzed water for inactivating Campylobacter jejuni during poultry washing. Int J Food Microbiol 72(1–2):77–83
Park SC, Nakai T (2003) Bacteriophage control of *Pseudomonas plecoglossicida* infection in ayu Plecoglossus *altivelis*. Dis Aquat Org 53:33–39
Park SC, Shimamura I, Fukunaga M et al (2000) Isolation of bacteriophages specific to a fish pathogen, *Pseudomonas plecoglossicida*, as a candidate for disease control. Appl Environ Microbiol 66:1416–1422
Perales I, Audicana A (1989) The role of hens' eggs in outbreaks of salmonellosis in north Spain. Int J Food Microbiol 8:175–180
Pickering LK (2006) American Academy of Pediatrics, *Salmonella* infections. Red Book, Report of the Committee on Infectious Diseases, pp 581–584
Purnell G, Mattick K, Humphrey T (2004) The use of 'hot wash' treatments to reduce the number of pathogenic and spoilage bacteria on raw retail poultry. J Food Eng 62:29–36
Rees JC, Voorhees KJ (2005) Simultaneous detection of two bacterial pathogens using bacteriophage amplification coupled with matrix-assisted laser desorption /ionization time-of-flight mass spectrometry. Rapid Commun Mass Spectrom 19:2757–2761

Ridley AM, Allen VM, Sharma M et al (2008) Real-time PCR approach for detection of environmental sources of *Campylobacter* strains colonizing broiler flocks. Appl Environ Microbiol 74:2492–2504

Ridley A, Morris V, Gittins J et al (2011a) Potential sources of *Campylobacter* infection on chicken farms: contamination and control of broiler-harvesting equipment, vehicles and personnel. J Appl Microbiol 111(1):233–244

Ridley AM, Morris VK, Cawthraw SA et al (2011b) Longitudinal molecular epidemiological study of thermophilic *Campylobacters* on one conventional broiler chicken farm. Appl Environ Microbiol 77:98–107

Rindhe SN, Zanjad PN, Doifode VK et al (2008) Assessment of microbial contamination of chicken products sold in Parbhani city. Vet World 1(7):208–210

Riska PF, Su Y, Bardarov S et al (1999) Rapid film-based determination of antibiotic susceptibilities of *Mycobacterium tuberculosis* strains by using a luciferase reporter phage and the Bronx Box. J Clin Microbiol 37(4):1144–1149

Rocourt J, Moy G, Vierk K, Schlundt J (2003) The present state of foodborne disease in OECD Countries. WHO, Geneva. http://www.who.int/foodsafety/publications/foodborne_disease/oecd_fbd.pdf

Rosenquist H, Nielsen NL, Sommer HM et al (2003) Quantitative risk assessment of human Campylobacteriosis associated with thermophilic *Campylobacter* species in chickens. Int J Food Microbiol 83(1):87–103

Rosenquist H, Sommer HM, Nielsen NL, Christensen BB (2006) The effect of slaughter operations on the contamination of chicken carcasses with thermotolerant *Campylobacter*. Int J Food Microbiol 108:226–232

Russell SM (2009) *Salmonella* intervention strategies and testing methods. The poultry site. http://www.thepoultrysite.com/articles/1357/Salmonella-intervention-strategies-and-testing-methods/

Sahin O, Kassem II, Shen Z et al (2015) *Campylobacter* in poultry: ecology and potential interventions. Avian Dis 59(2):185–200

Salama SM, Bolton FJ, Hutchinson DN (1989) Improved method for the isolation of *Campylobacter jejuni* and *Campylobacter coli* bacteriophages. Lett Appl Microbiol 8:5–7

Schmelcher M, Loessner MJ (2016) Bacteriophage endolysins: applications for food safety. Curr Opin Biotechnol 37:76–87 doi:10.1016/j.copbio.2015.10.005

Seal BS, Lillehoj HS, Donovan, Gay CG (2013) Alternatives to antibiotics: a symposium on the challenges and solutions for animal production. Anim Health Res Rev 1–10. doi:10.1017/S1466252313000030

Shakeeba W, Hanifi-Moghaddam P, Coleman R et al (2010) Orally administered P22phage tailspike protein reduces *Salmonella* colonization in chickens: prospects of novel therapy against bacterial infections. PLoS ONE 5:e13904

Sheng H, Knecht HJ, Kudva IT, Hovde CJ (2006) Application of bacteriophages to control intestinal *Escherichia coli* O157:H7 levels in ruminants. Appl Environ Microbiol 72: 5359–5366

Sherman M (2008) Bacteriophages: beyond antibiotics. US Pharmacist 33:46–51

Sillankorva S, Neubauer P, Azeredo J (2008) Isolation and characterization of a T7-like lytic phage for *Pseudomonas fluorescens*. BMC Biotechnol 8:1–11

Sillankorva SM, Oliveira H, Azeredo J (2012) Bacteriophages and their role in food safety. Int J Microbiol. Article ID 863945. doi:10.1155/2012/863945

Simmons M, Fletcher DI, Cason JA, Berrang ME (2003) Recovery of *Salmonella* from retail broiler by a whole-carcass enrichment procedure. J Food Prot 66:446–450

Sklar IB, Joerger RD (2001) Attempts to utilize bacteriophage to combat *Salmonella enterica* serovar *enteritidis* infection in chickens. J Food Saf 21(1):15–29

Skurnik M, Pajunen M, Kiljunen S (2007) Biotechnological challenges of phage therapy. Biotechnol Lett 29:995–1003

Smith HW, Huggins MB (1982) Successful treatment of experimental *Escherichia coli* infections in mice using phage: its general superiority over antibiotics. J Gen Microbiol 128:307–318

Smith HW, Huggins MB, Shaw KM (1987) The control of experimental *Escherichia coli* diarrhoea in calves by means of bacteriophages. J Gen Microbiol 133:1111–1126

Sobel J, Griffin PM, Slutsker L et al (2002) Investigation of multistate foodborne disease outbreaks. Public Health Rep 117:8–19

Son JS, Jun SY, Kim EB et al (2010) Complete genome sequence of a newly isolated lytic bacteriophage, EFAP-1 of *Enterococcus faecalis*, and antibacterial activity of its endolysin EFAL-1. J Appl Microbiol 108:1769–1779

Sorensen MC, van Alphen LB, Harboe A et al (2011) Bacteriophage F336 recognizes the capsular phosphoramidate modification of *Campylobacter jejuni* NCTC11168. J Bacteriol 193: 6742–6749

Stafford RJ, Schluter P, Kirk M et al (2007) A multi-centre prospective case-control study of *Campylobacter* infection in persons aged 5 years and older in Australia. Epidemiol Infect 135 (6):978–988

Stanford K, McAllister TA, Niu YD et al (2010) Oral delivery systems for encapsulated bacteriophages targeted at *Escherichia coli* O157:H7 in feedlot cattle. J Food Prot 73(7): 1304–1312

Stewart GSAB, Jassim SAA, Denyer SP (1993) Engineering microbial bioluminescence and biosensor applications. In: Rapley R, Walker MR (eds) Molecular diagnostics, Chap 27. Blackwell Scientific Publications, Oxford, pp 403–423

Stone R (2002) Bacteriophage therapy. Stalin's forgotten cure. Science 298:728–731

Suárez VB, Quiberoni A, Binetti AG, Reinheimer JA (2002) Thermophilic lactic acid bacteria phages isolated from Argentinian dairy industries. J Food Prot 65(10):1597–1604

Sulakvelidze A (2011) Safety by nature: potential bacteriophage applications. Microbe 6(3): 122–126

Sulakvelidze A (2013) Using lytic bacteriophages to eliminate or significantly reduce contamination of food by foodborne bacterial pathogens. J Sci Food Agric 93:3137–3146

Sulakvelidze A, Alavidze Z, Morris JG Jr (2001) Bacteriophage therapy. Antimicrob Agents Chemother 45:649–659

Suresh T, Hatha AAM, Screenivasa D (2006) Prevalence and antimicrobial resistance of *Salmonella enteritidis* and other *Salmonella* in the eggs and egg-storing trays from retails markets of Coimbatore, South India. Food Microbiol 23(3):294–299

Taitt CR, Shubin YS, Angel R (2004) Detection of *Salmonella enterica* Serovar *typhimurium* by using a rapid, array-based immunosensor. Appl Environ Microbiol 70(1):152–158

Tam CC, Higgins CD, Neal KR et al (2009) Chicken consumption and use of acid-suppressing medications as risk factors for *Campylobacter enteritis*, England. Emerg Infect Dis 15: 1402–1408

Tanji Y, Shimada T, Yoichi M et al (2004) Toward rational control of *Escherichia coli* O157:H7 by a phage cocktail. Appl Microbiol Biotechnol 64:270–274

Tauxe RV, Doyle MP, Kuchenmüller T et al (2010) Evolving public health approaches to the global challenge of foodborne infection. Int J Food Microbiol 139:16–28

Tessari ENC, Kanashiro AMI, Stoppa GFZ et al (2012) Important aspects of *Salmonella* in the poultry industry and in public health. In: Mahmoud BSM (ed) *Salmonella*—a dangerous foodborne pathogen. Publisher InTech, pp 181–206 http://cdn.intechopen.com/pdfs/26428/InTech-Important_aspects_of_Salmonella_in_the_poultry_industry_and_in_public_health.pdf

Thomson JE, Cox NA, Whitehead WK, Mercuri AJ (1974) Effect of hot spray washing on broiler carcass quality. Poult Sci 53:946–952

Thong KL, NgeoW Y, Altwegg M et al (1995) Molecular analysis of *Salmonella enteritidis* by pulsed-field gel electrophoresis and ribotyping. J Clin Microbiol 33(5):1070–1074

Todd EC (1989) Preliminary estimates of costs of foodborne disease in the United States. J Food Prot 52:595–601

Trachoo N, Frank JF, Stern NJ (2002) Survival of *Campylobacter jejuni* in biofilms isolated from chicken houses. J Food Prot 65:1110–1116

Ulitzur S, Kuhn J (1987) Introduction of lux genes into bacteria, a new approach for specific determination of bacteria and their antibiotics susceptibility. In: Scholmerich J, Andreesen R,

Kapp A et al (eds) Bioluminescence and chemiluminescence: new perspectives. Wiley, Chichester, pp 463–472

Ulitzur S, Kuhn J (1989) Detection and/or identification of microorganisms in a test sample using bioluminescence or other exogenous genetically introduced marker. US patent 4,861,70

US FDA/CFSAN (2006) Agency Response Letter, GRAS Notice No. 000198. http://www.fda.gov/Food/IngredientsPackagingLabeling/GRAS/NoticeInventory/ucm154675.htm

US FDA/CFSAN (2007) Agency Response Letter, GRAS Notice No. 000218. http://www.fda.gov/Food/IngredientsPackagingLabeling/GRAS/NoticeInventory/ucm153865.htm

USDA (2011) Food Safety and Inspection Service http://usdasearch.usda.gov/search?utf8=%E2%9C%93&affiliate=fsis&query=phage&commit.x=0&commit.y=0&commit=Search

USDA (2013) Guidance for shell egg cleaners and sanitizers. Last modified 30 Sept 2013. http://www.fsis.usda.gov/wps/portal/fsis/topics/regulatory-compliance/labeling/nonfood-compounds/shell-egg-cleaners-sanitizers/guidance-shell-egg-cleaners-sanitizers

USDA-FSIS (2014) Pre-harvest management controls and intervention options for reducing Shiga toxin-producing *Escherichia coli* shedding in cattle: an overview of current research. http://www.fsis.usda.gov/wps/wcm/connect/d5314cc7-1ef7-4586-bca2-f2ed86d9532f/Reducing-Ecoli-Shedding-in-Cattle.pdf?MOD=AJPERES

Van de Giessen AW, Tilburg JJ, Ritmeester WS, Van der Plas J (1998) Reduction of *Campylobacter* infections in broiler flocks by application of hygiene measures. Epidemiol Infect 121:57–66

Velge P, Cloeckeart A, Barrow P (2005) Emergence of *Salmonella* epidemics: the problem related to *Salmonella enterica* serotype *enteritidis* and multiple antibiotic resistance in other major serotypes. Vet Res 36:267–288

Verheust C, Pauwels K, Mahillon J et al (2010) Contained use of bacteriophages: risk assessment and biosafety recommendations. Appl Biosaf 15(1):32–44

Viazis S, Akhtar M, Feirtag J, Diez-Gonzalez F (2011) Reduction of *Escherichia coli* O157:H7 viability on hard surfaces by treatment with a bacteriophage mixture. Int J Food Microbiol 145:37–42

Wagenaar JA, Van Bergen MA, Mueller MA et al (2005) Phage therapy reduces *Campylobacter jejuni* colonization in broilers. Vet Microbiol 109(3–4):275–283

Wagenaar JA, Mevius DJ, Havelaar AH (2006) Campylobacter in primary animal production and control strategies to reduce the burden of human campylobacteriosis. Rev sci tech Off int Epiz 25(2):581–594

Walker A, Jassim SAA, Holah JT et al (1992) Bioluminescent *Listeria monocytogenes* provide a rapid assay for measuring biocide efficacy. FEMS Microbiol Lett 70:251–256

Waseh S, Hanifi-Moghaddam P, Coleman R et al (2010) Orally administered P22 phage tailspike protein reduces *Salmonella* colonization in chickens: prospects of a novel therapy against bacterial infections. PLoS ONE 5:e13904

Whichard JM, Sriranganathan N, Pierson FW (2003) Suppression of *Salmonella* growth by wild-type and large-plaque variants of bacteriophage Felix O1 in liquid culture and on chicken frankfurters. J Food Prot 66(2):220–225

Whitman PA, Marshall RT (1971) Characterization of two psychrophilic *Pseudomonas* bacteriophages isolated from ground beef. Appl Microbiol 22:463–468

WHO (2001) Biomarkers in risk assessment: validity and validation. Environmental health criteria —EHC 222, Geneva, Switzerland

WHO (2013) The global view of Campylobacteriosis. Report of expert consultation. Utrecht, Netherlands, 9–11 July 2012

Wilson DJ, Gabriel E, Leatherbarrow AJ et al (2008) Tracing the source of Campylobacteriosis. PLoS Genet 4(9):e1000203

Withey S, Cartmell E, Avery LM, Stephenson T (2005) Bacteriophages-potential for application in wastewater treatment processes. Sci Total Environ 339(1–3):1–18

Woolhouse M, Ward M, van Bunnik B, Farrar J (2015) Antimicrobial resistance in humans, livestock and the wider environment. Phil Trans R Soc B 370:20140083. doi:10.1098/rstb.2014.0083

Yang H, Li Y, Johnson MG (2001) Survival and death of *Salmonella typhimurium* and *Campylobacter jejuni* in processing water and on chicken skin during poultry scalding and chilling. J Food Prot 64:770–776
Yemini M, Levi Y, Yagil E, Rishpon J (2007) Specific electrochemical phage sensing for *Bacillus cereus* and *Mycobacterium smegmatis*. Bioelectrochemistry 70:180–184
Young R (1992) Bacteriophage lysis: mechanism and regulation. Microbiol Rev 56:430–481
Zhang L, Jeong JY, Janardhanan KK et al (2011) Microbiological quality of water immersion-chilled and air-chilled broilers. J Food Prot 74:1531–1535
Zimmer M, Vukov N, Scherer S, Loessner MJ (2002) The murein hydrolase of the bacteriophage 3626 dual lysis system is active against all tested *Clostridium perfringens* trains. Appl Environ Microbiol 68:5311–5317

Chapter 4
Control, Prevention and Rapid Detection of Methicillin-Resistant *Staphylococcus aureus*

Abstract Methicillin-resistant *Staphylococcus aureus* (MRSA) infections are a particular problem in hospitals. Wild phages can be reprogrammed for several clinical isolates of MRSA. These phages can then be utilized in a safe fogging biocontrol system, using specialized dispensers, along with another non toxic liquid variation as a surface disinfectant in healthcare or animal settings for the decontamination of MRSA and other antibiotic resistant bacteria. This phage-based biocontrol technology is non-toxic, odourless, inexpensive and effective in controlling and reducing the risk of contracting a hospital-acquired MRSA infection. By using phages as a biocontrol in medical facilities both staff and patients can feel safe from MRSA. This chapter will demonstrate the benefits of phage-based MRSA decontamination coupled with several rapid bacterial detection protocols enabling the development of inexpensive diagnostic kits and safe decontamination methods. There is also an urgent need for a real-time, accurate and inexpensive diagnostic kit that can be used in situ with no need for skilled technicians while also allowing patients to test for some diseases or conditions at home. These tests are cost-effective, quick, and confidential. The aim of this chapter is to provide a comprehensive overview of some innovative phage biocontrol methodologies and diagnostic kits that can be used in healthcare settings, while encouraging further research on the use of phages diagnostically and preventatively in daily decontamination practices.

Keywords Bacteriophage · Biocontrol · Diagnostic · Endolysins enzymes · Epidemiology · Farm animals · Food · Hospitals · Infections · Livestock · Methicillin-resistant *Staphylococcus aureus* · Methicillin-resistant coagulase-negative staphylococci · Poultry · Staphylococci

Contents

S.A.A. Jassim and R.G. Limoges, *Bacteriophages: Practical Applications for Nature's Biocontrol*, DOI 10.1007/978-3-319-54051-1_4

4.1 Introduction

Staphylococci were first studied by Alexander Ogston in 1881. Currently the *Staphylococcus* genus includes at least 45 species and 21 subspecies which are classified in two groups based on their ability to produce the enzyme coagulase: coagulase-positive staphylococci (CoPS) and coagulase-negative staphylococci (CoNS) (Bergeron et al. 2011). Thirty-eight different species are found within the CoNS and only seven exceptions are CoPS, being *S. aureus*, *S. delphini*, *S. hyicus*, *S. intermedius*, *S. lutrae*, *S. pseudintermedius*, *S. schleiferi* subsp. *Coagulans*.

CoNS occur widely in nature and are part of the normal microflora of the skin and mucosal surfaces of the upper respiratory tract of both man and animals (Malik et al. 2005). Colonizing CoNS pathogens have been reported to be responsible for infections in humans, particularly in immunocompromised hosts and neonates (Jain et al. 2004; Ternes et al. 2013; May et al. 2014). *S. haemolyticus* and *S. epidermidis* are the most common bacteria associated with neonatal healthcare-associated infections (Ternes et al. 2013). This is due to their ability to colonize human and animal skin, particularly *S. epidermidis* to form protective biofilms (Von Eiff et al. 2006). This is probably due to the intimate association between potential hosts (dogs and cats have become an integral part of modern society in the developed world). This contact can enhance staphylococcal dispersal which are easily spread by skin-to-skin contact, through aerosols from sneezing and coughing and also through saliva (Malik et al. 2005). They are also able to produce enterotoxins in food such as *S. carnosus*, *S. equorum*, *S. piscifermentans*, and *S. xylosus* (Verasa et al. 2008; Zell et al. 2008).

Presently the *S. aureus* bacteria remains a versatile and dangerous pathogen in humans. Staphylococcal infections have increased steadily, originally causing little change in overall mortality (Lowy 1998). *S. aureus* is a Gram-positive bacterium frequently found in the nose, respiratory tract and on the skin causing several infections such as abscesses, sinusitis and food poisoning. The bacterium can cause a range of illnesses, from minor skin infections such as boils, folliculitis, pimples, impetigo, carbuncles, cellulitis, scalded skin syndrome and abscesses, to life-threatening diseases such as toxic shock syndrome, colonization of the intestinal tracts, pneumonia, meningitis, endocarditis, bacteremia, osteomyelitis and sepsis (Jassim 1989) (see Fig. 4.1). They are capable of producing several virulence factors such as enzymes (coagulase, hyaluronidase, deoxyribonuclease, lipase, staphylokinase, and β-lactamase), several exotoxins, enterotoxins (A to E, G to I, R to T) and haemolysins (α, β, γ, and δ), in addition to leucocidin and protein A (Jassim 1989; Argudín et al. 2010; Hennekinne et al. 2012).

The most significant human pathogen of *S. aureus* is the methicillin-resistant *S. aureus* (MRSA) which is usually acquired by exposure within infected hospitals and other healthcare facilities, causing a variety of serious healthcare-associated infections. Since the introduction of β-lactamase-stable antimicrobial drugs into clinical use, these strains have emerged worldwide as important nosocomial pathogens. Their prevalence in the community is also increasing substantially. About one-quarter of healthy people carry one or more strains asymptomatically at any given time and infections are commonly endogenous, being caused by the patient's own colonizing strain. Major sources of MRSA infections are the patient's own floras (endogenous infection), those acquired from another person (cross-infection), or those from substances recently contaminated by a human source (environmental infection). The report of Emerging Infectious Diseases in

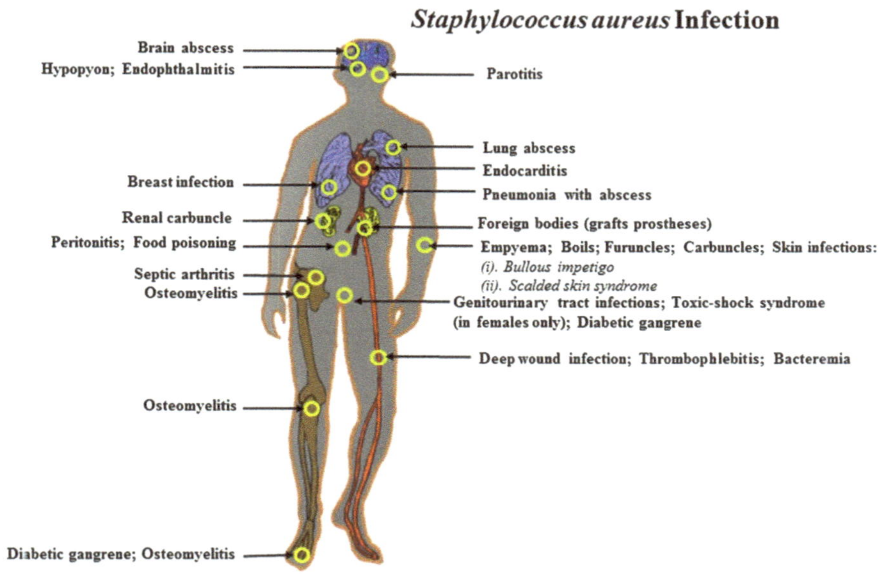

Fig. 4.1 A diagram of *Staphylococcus aureus* infections in humans

2007, a publication of the Centers for Disease Control and Prevention (CDC), estimated the number of MRSA infections in hospitals doubled nationwide, from approximately 127,000 in 1999 to 278,000 in 2005, while at the same time annual deaths increased from 11,000 to more than 17,000 (Klein et al. 2007). Another study led by the CDC and published in the October 17, 2007 issue of the Journal of the American Medical Association, estimated that MRSA was responsible for 94,360 serious infections and associated with 18,650 hospital stay-related deaths in the US in 2005 (CDC 2007; Klevens et al. 2007).

Hospital-associated or acquired (HA) *S. aureus* are infections that are contracted in medical facilities such as hospitals or nursing homes. Since these infections are resistant to numerous antibiotics, MRSA has created significant challenges to health care facilities for over 40 years (Doyle et al. 2012). Since many *S. aureus* strains are resistant to penicillin, penicillinase-resistant β-lactams and many other antibiotics, they represent a major threat to the public-health system (Palavecino 2014). MRSA was first discovered in the UK in 1961 (Jevons 1961), thus this event represents the birth of MRSA. It made its first major appearance in the US in 1981 (Klevens et al. 2007) and since then it's often referred to in the media as a "superbug". Traditionally, MRSA infections have been limited to hospitals and predisposed individuals, such as the elderly, immunocompromised patients and those undergoing surgery. More recently however, MRSA infections have occurred outside of the hospital setting (Palavecino 2014). These infections are called community-associated, (CA)-MRSA, with infections that are transmitted through close personal contact with an infected person or through direct contact with an infected wound. CA-MRSA infections were first reported in the Northern US 1980s, where they caused fatal infections in otherwise healthy children and adults (MDH 2004; Wang et al. 2007; Palavecino 2014). It has become the most frequent cause of skin and soft tissue infections presenting to emergency departments in the US (Moran et al. 2006). Since that time, CA-MRSA has become a global problem, with the most serious epidemic seen in the US (Wang et al. 2007; Diep and Otto 2008). Infections, accidents or illness that result in trips to the emergency room can sometimes result in necrotizing fasciitis or even death in otherwise healthy adults outside of healthcare settings (Wang et al. 2007).

What makes CA-MRSA strains more successful in causing human disease as compared with their HA-MRSA counterparts is that they secrete a new class of staphylococcal peptides that have a remarkable ability to recruit, activate and subsequently lyse human neutrophils, thus eliminating the main cellular defense against *S. aureus* infection (Wang et al. 2007). Furthermore, it was found that these peptides are produced in high concentrations by standard CA-MRSA strains and contribute significantly to the strains' ability to cause disease in animal models of infection (Wang et al. 2007). Moreover, recently the CA-MRSA ST93 clone is becoming dominant in Australia and is clinically highly virulent. The data of Chua et al. (2014) demonstrates that hyper expression of α-haemolysin mediates enhanced virulence in ST93 CA-MRSA, and provides additional control of exotoxin production, in particular α-haemolysin, mediated by regulatory systems other than the gene regulator and have the potential to fine-tune virulence in CA-MRSA. In general, methicillin resistance rates of *S. aureus* vary considerably between countries. In the US the figure is reported as 0.8–1.2% of the general population although in HIV-positive individuals it has been found to be as high as 10–17% (Peters et al. 2013). In the European Union (EU), the figure in long-term facilities in Germany was (1%), Belgium (2–43%), Spain (16%), UK (5–23%), France (38%) (Kock et al. 2010). In German hospitals more than 20% of clinical *S. aureus* isolates are found to be methicillin resistant (Stürenburg 2009). Despite this variation, in the majority of cases, the clonal structure of MRSA isolates from nursing home residents was closely related to that found among patients in neighbouring acute care hospitals (Kock et al. 2010).

According to recent estimates, MRSA infections cause more deaths annually in the US than HIV/AIDS (Klevens et al. 2007). This is alarming because methicillin resistance in *S. aureus* not only means limited effectiveness of antibiotic treatment, but also leads to prolonged hospital stay and higher morbidity and mortality rates (Cosgrove et al. 2003, 2005). This problem is not unique to one country; the British National Audit Office estimated that the incidence of nosocomial infections in the EU ranges from 4 to 10% of all hospital admissions. As of early 2005, the number of deaths in the UK attributed to MRSA has been estimated by various sources to lie in the area of 3000 per year (Johnson et al. 2005) while staphylococci bacteria account for almost half of all UK hospital infections. In addition, it was found that MRSA bacteremia have a higher attributable mortality than methicillin-susceptible *S. aureus* (MSSA) bacteremia (Blot et al. 2002), which further complicates the whole problem of MRSA infections. Another study showed that patients with MRSA infection had three times the length of hospital stay (14.3 vs. 4.5 days), incurred three times the total cost ($48,824 vs. $14,141), and experienced five times the risk of in-hospital death (11.2% vs. 2.3%) than patients without this infection (Noskin et al. 2005).

The main mode of MRSA transmission is from MRSA-colonized or infected patients to others by direct contact, through the touch of healthcare workers (HCWs) (Blok et al. 2003; Nour et al. 2007; Kumar et al. 2011; Morioka et al. 2014). This is due to HCWs including students who are the interface between the hospital and the community and may serve as agents of cross transmission of HA-MRSA and CA-MRSA (Fadeyi et al. 2010). Prolonged hospital stay and improper hand washing were found to be the major risk factors for MRSA colonization (Bhaskaran et al. 2014). CA-MRSA infection from human sources seems to be on the rise in America with about 80,500 MRSA infections in 2005, compared to about 111,300 in 2011. This has not been mirrored in the EU but an increase in infection from farm animals has been detected (Seaman 2013). Therefore, MRSA became a major public health problem, primarily related to health care but no longer confined to intensive care units, acute care hospitals, or any health care institution (Klevens et al. 2007). MRSA infections constitute an important and still evolving public health challenge for European countries (Köck et al. 2010).

Worldwide, MRSA affects all communities including diverse populations, ethnicity and race, sports teams, prison inmates and child care attendees. It usually involves skin disease and can also cause severe, sometimes fatal invasive disease (Klevens et al. 2007). The main factors perpetuating and aggravating the problem of MRSA infections worldwide is that MRSA is difficult to control and its survival is promoted by the increasing use of antibiotics (Schentag et al. 1998). The Society for Healthcare Epidemiology of America (SHEA) in a careful analysis of potential interventions, did not quote any specific example of successful general control using antibiotic policy (Shlaes et al. 1997). Reduced use of antibiotics in the past has coincided with elimination of certain clones' resistant to the drug e.g. the reduced use of tetracyclines in the 1970s was associated with reductions in tetracycline-resistant MRSA in Denmark and Birmingham (Knudsen and Rosdahl 1991). In general, the rate at which *S. aureus* can develop or acquire resistance to new antibiotics seems to be higher than the rate at which new antibiotics are discovered and developed. Many MRSA isolates are multiple resistant and are susceptible only to glycopeptide antibiotics such as vancomycin and investigational drugs. MRSA isolates that also have decreased susceptibility to glycopeptides (glycopeptide intermediately susceptible *S. aureus*) reported in recent years, are a cause of great public health concern (Enright et al. 2002).

4.2 The Occurrence of MRSA in Animals

S. aureus is also known to colonize a variety of animal species including cattle, pigs, poultry, dogs, cats, rabbits, horses, and exotic species (Weese 2010) and the bacterium are most commonly reported as a cause of mastitis in cattle and sheep, chondronecrosis with osteomyelitis in chickens (McNamee and Smyth 2000), and infection of farmed rabbits (Viana et al. 2007). Most reports characterizing animal-associated *S. aureus* have demonstrated that strains affecting animals are distinct from those infecting humans, suggesting that specific host-dependent pathogenic factors may have evolved independently in these species, which only rarely cross species boundaries (Viana et al. 2007; Shepheard et al. 2013). Methicillin-resistant CoNS from healthy dogs in Nsukka, Nigeria was recently reported (Chah et al. 2014). The authors concluded the existence of multidrug resistant (MDR) strains of MDRCoNS, particularly *S. sciuri* subspecies rodentium. Since dogs are in close contact with their owners, the risk of transmission of MDR staphylococci between animals and humans, as well as the possibility of transfer of mecA and other resistance genes from the CoNS to human pathogenic *S. aureus* are serious public health concerns. Up to 90% of *S. epidermidis* strains are now resistant to methicillin, (May et al. 2014).

4.2.1 Household Pets

Colonization of people and their pets with indistinguishable MSSA strains is not uncommon (Malik et al. 2005; Hanselman et al. 2009; Chah et al. 2014), indicating the potential for transmission of MRSA between people and pets in a household (Morris et al. 2006; Boost et al. 2008). MRSA wound infections, surgical site infections, pyoderma, otitis, and urinary tract infections are most common but various other opportunistic infections have been reported (Baptiste et al. 2005; Leonard et al. 2006; Morris et al. 2006; Vitale et al. 2006; Weese et al. 2006; Griffeth et al. 2008; Tomlin et al. 1999). There have been reports of MRSA infections in dogs caused by sequence type (ST1) 398, a clone associated with food animals and an emerging cause of human CA-MRSA in some regions (Witte et al. 2007; Nienhoff et al. 2009).

4.2.2 Livestock

The emergence in the last decade of a livestock-associated MRSA (LA-MRSA) clone which also has the capacity to cause zoonotic infections in humans has raised important questions regarding its origin and its potential to cause human epidemics. It was first detected in swine (Doyle et al. 2012). Food animal infections appear to have taken a different course from that in companion animals. Rather than emerging from human sources, MRSA seems to have evolved independently in one or more food animal species, with subsequent dissemination and interspecies transmission to humans (see below 4.4 MRSA Transmission Between Animals, Humans and in Hospitals).

From the initial report of MRSA in a dairy cow in 1972 (Devriese et al. 1972), to more recent reports (Lee 2003, 2006; Kwon et al. 2005; Juhász-Kaszanyitzky et al. 2007; Moon et al. 2007), the incidence of MRSA mastitis and the prevalence of methicillin resistance among bovine *S. aureus* isolates appears to be quite low (Lee 2003; Kwon et al. 2005). Thus MRSA does not appear to be a common or important bovine mastitis pathogen at this time. This is quite surprising given the number of years since the first identification of MRSA in cattle and the close contact of humans with the udders of dairy cattle. The appearance of MRSA in cattle varies considerably between countries, for example a Dutch study found 28% of veal calves colonized with MRSA (Graveland et al. 2008), but in Switzerland the pathogen was isolated from only 1.3% of calves and 0.4% of adult cows (Huber et al. 2009) and in Canada from none of 491 Canadian feedlot cattle (Weese et al. 2009a). However, the resulting prevalence data are quite variable between farms, from 0.4 to 28%, and comparisons of the data are inadvisable because of differing study populations (ages, cattle production systems) and methods.

4.2.3 Pigs

Pig farming is a possible source of MRSA. Voss et al. (2005) demonstrated transmission of MRSA between a pig and a human, between family members (pig farmers and their families), and between a nurse and patient in the hospital. The unexpected high frequency of MRSA among the group of regional pig farmers (>760× higher than in the general Dutch population) indicates that their profession might put them at risk for MRSA colonization. Furthermore, they found that the strain with spa-type t108 appears to be more prevalent and widespread, given that the strain spread from animal to human, between family members, between patient and nurse, and among pig farmers from different regions. They proposed if these observations hold true in other regions that pig farming poses a significant risk factor for MRSA carriage in humans that warrants screening wherever pig farmers or their family members are admitted to a hospital. This hypothesis is supported by the observation of Seaman (2013) who reported that residents who lived closest to farms using pig manure were at 38% increased risk of MRSA infections, compared to those who lived farthest away. The MRSA isolates with ST398 type were found prevalent among pigs in many countries; in the Netherlands (Huijsdens et al. 2006; de Neeling et al. 2007; van Duijkeren et al. 2008), Canada (Khanna et al. 2007), Germany (Witte et al. 2007), Singapore (Sergio et al. 2007), Belgium (Dewaele et al. 2008) and the US (Smith et al. 2008a). Nevertheless, the data are quite variable between farms, ranging from nil to 100%, and comparisons of the data are inadvisable because of differing study populations (ages, pig production systems) and methods.

It is worth noting that ST398 strains have predominated in reports of MRSA in pigs internationally (de Neeling et al. 2007; Khanna et al. 2007; Sergio et al. 2007; Dewaele et al. 2008; Smith et al. 2008b; van Duijkeren et al. 2008, 2010; Hasman et al. 2010), therefore it is plausible that this strain emerged in pigs and was subsequently disseminated to other species. Furthermore, an unrelated clone, ST9, the source and implications of which are unknown, was also reported in pig farms in China (Cui et al. 2009; Wagenaar et al. 2009). Risk factors for MRSA colonization of pigs have not been adequately investigated but it has been suggested that MRSA strains are transmitted between farms through the purchase of colonized pigs (van Duijkeren et al. 2008). The appearance of ST398 was implicated because tetracycline resistance was found typical in ST398 and this has led to the assumption that the emergence of this strain resulted from heavy tetracycline use in pig production (de Neeling et al. 2007). This assumption is perfectly consistent with results of Price et al. (2012) who strongly suggest that LA-MRSA CC398 which originated in humans as MSSA, were commonly present in pigs but not present at all in the other reservoirs tested (Hasman et al. 2010). The MSSA later spread to livestock via rapid radiation, where it subsequently acquired tetracycline and methicillin resistance (Price et al. 2012). Hence, the rapid and wider prevalence of clinical MRSA infections in pigs of exudative dermatitis, urinary tract infection, and mastitis-metritis-agalactia syndrome (Schwarz et al. 2007; van Duijkeren et al. 2007) was found after tetracycline administration (van Duijkeren et al. 2008). Furthermore, antibiotics have been in widespread use on pig farms to improve the farms' hygiene which has contributed to the emergence and dissemination of MRSA, this is discussed in greater detail in Chap. 2. From the above reports it appears that pig farming poses a significant risk factor for human MRSA exposure and that humans also contributed to spreading the organism (van Duijkeren et al. 2010), hence the need for effective decontamination on pig farms.

4.2.4 Poultry

Poultry is not immune to MRSA which have also been isolated from healthy and sick chickens (Lee 2003, 2006). Furthermore, the MRSA strain ST398 has also been isolated from healthy chickens (Nemati et al. 2008; Persoons et al. 2009), and from infected poultry (Hasman et al. 2010). Surprisingly, almost all of the poultry isolates (96%) belonged to CC5 (spa types t002 and t306), which is also known to be commonly found among human blood isolates and subsequent pulsed-field gel electrophoresis (PFGE) analysis identified indistinguishable PFGE patterns among a poultry isolate and selected human isolates (Hasman et al. 2010).

4.3 MRSA in Food

MRSA has been considered primarily a health-care associated pathogen, causing invasive diseases in which multidrug resistance poses a substantial challenge to successful treatment. However, food has been implicated as a source of contamination in one outbreak of blood and wound infections in hospitalized immunocompromised patients. Furthermore, MRSA are well known for causing pyoderma and other soft tissue infections via cuts, wounds and tissue abrasions. Thus MRSA can colonize the skin, nasopharyngeal cavities and other sites of both humans and animals, possibly without evidence of infection. This leads to the conclusion that the potential exists for its transmission to humans. Since MRSA is an important nosocomial pathogen it is certainly plausible that surface contamination of food could lead to colonization if people touch their noses after contaminating their hands. This was clearly evidenced when a contaminated worker peeled a banana that was implicated as a source of an outbreak in a hospital in the Netherlands (Kluytmans et al. 1995).

MRSA has also been identified in retail meat worldwide, involving a range of meat products and with prevalence ranging from 0.4 to 25% (de Boer et al. 2009; Pu et al. 2009; Lim et al. 2010; Weese et al. 2010; Bhargava et al. 2011). For example, in Spain MRSA was detected in 1.6% of food samples (from pork, chicken, rabbit, veal and wild boar), two strains from pork and veal corresponded to ST398-SCC*mec*V (*spa* types t011 and t1197, respectively), two strains from chicken and rabbit were typed as ST125-SCC*mec*IVa-t067, and the strain from a wild boar was ST217-SCC*mec*IVa-t032 (Lozano et al. 2009). In Taiwan during 2004–2006, MRSA-contaminated retail meat was recovered from 8.8, 11.3, and 4.3% of pork carcass samples, respectively, collected at 53 slaughterhouses (Lin et al. 2009). A Canadian study identified MRSA in 6.3% of ground pork and 5.6% of ground beef which had up to 20 cfu per g^{-1}; most quantifiable samples had fewer than 10^2 cfu g^{-1} and the highest had 3590 cfu g^{-1} (Weese et al. 2009b). MRSA in retail meat in Detroit, Michigan were found in raw beef (20.5%), chicken (25.0%), turkey (24.6%) (Bhargava et al. 2011), whilst in Iowa it was found in pork meat (1.2%) (Hanson et al. 2011). Recently it was reported that multidrug resistant *S. aureus* was more prevalent in the pork isolates, followed by beef, then beef livers (Abdalrahman et al. 2015). The latter authors showed that the molecular typing of a subset of the recovered isolates was highly diverse, where spa typing was more discriminatory than PFGE. They concluded that the alarmingly high incidence of *S. aureus* in retail beef livers in this study should raise awareness about the food safety of such meat products. In general, pork meat had the highest contamination rate in the US and Canada (Pu et al. 2009; Weese et al. 2010), as did beef in South Korea (Lim et al. 2010) and poultry in the Netherlands (de Boer et al. 2009). A new MRSA clone related to pig and cattle farming detected in raw meat in the Netherlands was multilocus sequence type 398, the clone related to farming (van Loo et al. 2007). The study in South Korea observed MRSA from chicken, which demonstrated sequence type 692 by multilocus sequence typing, a type distinct from that isolated in beef and pork (Lim et al. 2010).

In general, from all the above reports, despite sample size variations, these studies suggested that MRSA contamination in different meat categories can vary by location and that molecular distinction may exist among MRSA isolates in meat of different origin. On the other hand, although MRSA prevalence in raw food is low, the risk of its transmission through the food chain cannot be disregarded, especially in uncooked meat. Yet *S. aureus* is one of the top five pathogens contributing to foodborne illnesses, causing an estimated quarter million cases every year in the US (Abdalrahman et al. 2015). Indeed, foodborne disease outbreaks caused by MRSA have been repeatedly proposed (Takesue et al. 1991; Inamatsu et al. 1992; Kodama et al. 1997; Gravet et al. 1999; Jones et al. 2002; Boyce and Havill 2005; Marler 2010; Okuyama and Yoshida 2012; Wendlandt et al. 2013; Ogawa et al. 2014). Moreover, contaminated foods were suspected as being a vehicle for MRSA and may also constitute a health risk for food handlers. Carcasses obtained from animals colonized by MRSA may become contaminated during slaughter and foods can also become contaminated during processing by food handlers colonized by MRSA.

From the literature review it appears that the absence of a detectable enterotoxin gene among ST398 LA-MRSA strains from a wide host range is reassuring regarding their foodborne pathogenic potential (Hallin et al. 2011). Nevertheless, the increases both in MRSA food contamination and in human incidence of CA-MRSA and MRSA ST398 infections, point to a direct link between the two, but no objective data are available to support or refute such a hypothesis. Noteworthy,

LA-MRSA strains represent a homogeneous lineage distinct from co-local HA- and CA-MRSA strains, especially in its accessory genome content, characterized by a lack of human-associated virulence and adhesion determinants (Hallin et al. 2011). Likewise, LA-MRSA and CA-MRSA and even HA-MRSA can be present in or on food intended for human consumption, but they concluded on the basis of the published literature that this does not equate to MRSA being considered a food-borne pathogen (Wendlandt et al. 2013). Several studies suggest that colonization of the intestinal tracts of hospitalized patients may have important clinical implications (Gravet et al. 1999; Ray et al. 2003; Boyce and Havill 2005). Staphylococcal food-borne intoxication, caused by heat-stable enterotoxins produced by *S. aureus*, causes over 240,000 cases of food-borne illness in the US annually (Doyle et al. 2012; Kadariya et al. 2014). To date, more than 20 staphylococcal enterotoxins (SEs) have been described: SEA to SElV (Hennekinne et al. 2012).

The widespread and rapid growth in CA-MRSA and LA-MRSA has raised the question as to whether MRSA is indeed a food-borne pathogen (Wendlandt et al. 2013). Furthermore, LA-MRSA are a major cause of human infection and thus the need for close surveillance (Köck et al. 2013), since there is increasing evidence of MRSA in food animals (Vanderhaeghen et al. 2010), which may also produce these enterotoxins (Doyle et al. 2012; Hennekinne et al. 2012). Classical staphylococcal 'MSSA' produced food poisoning enterotoxins caused by ingestion of preformed enterotoxins (Pereira et al. 2009; Marler 2010). Several reports indicated that MRSA isolates can possess various enterotoxins A, AB, D, C, G and I which caused staphylococcal food poisoning (Takesue et al. 1991; Inamatsu et al. 1992; Kodama et al. 1997; Gravet et al. 1999; Jones et al. 2002; Boyce and Havill 2005; Marler 2010; Okuyama and Yoshida 2012; Ogawa et al. 2014). Recently, Abdalrahman et al. (2015) found that the prevalence of enterotoxin genes such as SEG, SHE, and SEI and the toxic shock syndrome gene tst was higher in the pork isolates than in the beef ones, whilst the haemolysin genes, particularly β haemolysin, were more prevalent in isolates from beef livers.

It appears food may be a concern as a vehicle of MRSA extra intestinal colonization and infection (Bhalla et al. 2007), since staphylococcal enterotoxins (SEs) can be produced at the following extreme conditions at 10–45 °C, pH 4–9.6, water activity 0.85–0.99, NaCl 0–10%, redox potential < -100 to $>+200$ mV, and atmosphere anaerobic–aerobic (Jassim 1989; Hennekinne et al. 2012). Predominantly these conditions are similar with most methods of food preparation. Accordingly, foods may also be contaminated with MRSA during food preparation and may produce SEs, and therefore, it is logical that concerns would emerge about MRSA contamination of food and to assume that colonized animals were the source of contamination. The presence of MRSA ST398 in meat, more commonly found in people, including strains not yet reported in food animals (Lozano et al. 2009; Pu et al. 2009; Weese et al. 2009b; Vanderhaeghen et al. 2010), is raising questions about whether contamination is from food animals, other sources such as people involved in food processing, or both. Therefore, persons working with animals or handling meat products may be at increased risk for antibiotic-resistant infections which could become more common with an increase in food contamination and colonization of food handlers (Doyle et al. 2012).

4.4 MRSA Transmission Between Animals, Humans and in Hospitals

The occurrence of MRSA transmission between animal farms and the community have been studied in detail in order to find correct measures to prevent an epidemic MRSA (EMRSA) in the human community (Hallin et al. 2008). The livestock (pigs, cattle and poultry) are colonized with MRSA and the zoonotic transmission of such MRSA to humans via direct animal contact, environmental contamination or meat are a matter of concern. Furthermore, LA-MRSA mostly belongs to clonal complex (CC) 398 as defined by multilocus sequence typing (Köck et al. 2013). One potential explanation for the increased virulence of these strains is the presence of bacteriophage-associated genes for Panton-Valentine leukocidin (PVL). PVL genes are commonly found in MRSA strains linked to community transmission in most parts of the world but are rarely found in strains classically associated with health care associated transmission (Vandenesch et al. 2003; Labandeira-Rey et al. 2007). Because the PVL bacteriophage and SCCmec integrate into distinct sites on the *S. aureus* chromosome, it is not likely that PVL genes have been transmitted incidentally in association with SCCmec IV or V (Boyle-Vavra and Daum 2007). This lack of physical proximity on the chromosome makes it even more intriguing that these genetic elements are closely associated in diverse strains of MRSA that emerged in community settings worldwide at approximately the same time. Therefore, it is possible to draw the conclusion that this represents human-to-pig transmission. Whereas, LA-MRSA strains may evolve on farms via lysogenic bacteriophages and the use of antibiotics in animal husbandry for prevention of disease within a herd, for treatment of an existing disease outbreak or as feed additives for growth promotion in industrial livestock and poultry (Schelin et al. 2011; Fitzgerald 2012; this is discussed in greater detail in Chap. 2).

LA-MRSA spreads to the general human population in a number of different ways: contact with contaminated meat products (via handling or ingestion); occupational contact (farmers, meat packers, butchers, etc.); potential secondary spread into the larger community from those who are occupationally exposed; entry into and transmission via hospitals or other health care facilities; or spread via environmental routes including air, water, or manure in areas in proximity to live animal farms or crop farms where manure has been used as a fertilizer (Fig. 4.2; this is discussed in greater detail in Chap. 2).

It was found that hospitalized patients with *S. aureus* nares and/or stool carriage frequently had *S. aureus* on their skin and on nearby environmental surfaces. *S. aureus* intestinal colonization was associated with increased frequency of

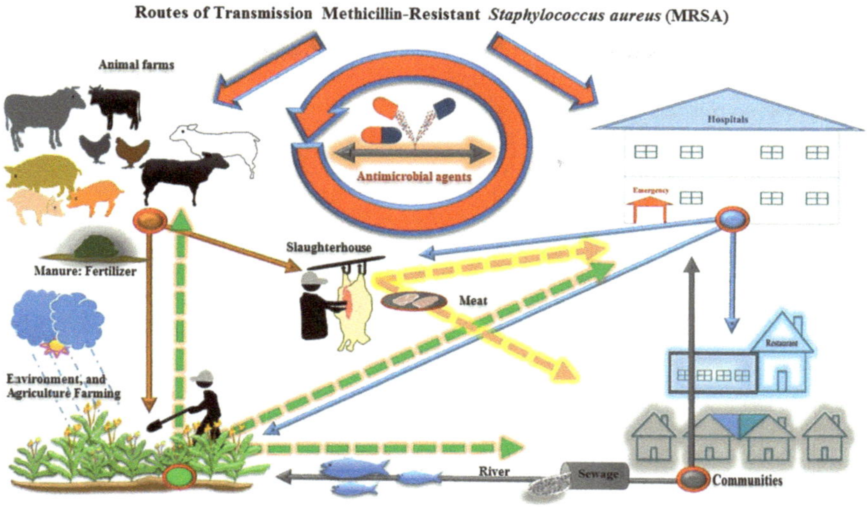

Fig. 4.2 Diagram of MRSA transmission between farm animals, the environment and humans

positive skin cultures, which could potentially facilitate staphylococcal infections and nosocomial transmission (Bhalla et al. 2007). Furthermore, it was found that intensive care and liver transplant unit patients with both rectal and nares MRSA colonization had significantly higher rates of *S. aureus* infection than did patients with nares carriage alone by 40.4% versus 18.2% (Squier et al. 2002). Moreover, it was reported MRSA is increasingly recovered from nursing home residents with established risk factors and among healthy individuals with no recognizable risk factors, and it seems clear that CA-MRSA strains are epidemiologically and clonally unrelated to hospital-acquired strains (Palavecino 2004). Therefore, there is clear evidence that MRSA can be transmitted between humans in hospitals and between humans and animals and vice versa and between animals on farms, yet the frequency, direction, and overall relevance along with the roles of different animal species in the transmission of MRSA to humans are not well established. It seems that close contact between humans and various animal species creates the potential for exposure of animals to "human" organisms, and extensive antimicrobial use in animals presumably facilitates the emergence and spread of antimicrobial resistance (AMR) with the mechanisms, risk factors, and other information about the emergence of AMR in animals (this is discussed in greater detail in Chap. 2).

In 2003 Frank predicted that all health care facilities, regardless of location or complexity of care rendered, will experience sporadic or endemic MRSA over the next few years. Since the discovery of MRSA in the early 2000s, LA-MRSA multilocus sequence type 398 (ST398) strains have been rapidly emerging globally among production animals such as cattle, pigs and poultry as well as among humans (Hasman et al. 2010) and the cause of human infections, is most often associated with livestock exposure (Price et al. 2012). LA-MRSA has emerged as a pressing concern in some countries (Köck et al. 2013). The ST398 strains were first identified in France (Armand-Lefevre et al. 2005) and in the same year in the Netherlands as a common component of the skin microbiome of pigs and pig farmers (Voss et al. 2005). Since then it was isolated in the EU, North America, and Asia from pigs, veal calves, turkeys, and chickens (Graveland et al. 2011a). Severe infections in pig farmers and their family members from LA-MRSA ST398 are increasing (Rasigade et al. 2010; Schijffelen et al. 2010). It was noticed that ST398 MRSA strains do not readily transmit between humans and furthermore the carriage rates decrease rapidly in livestock farmers once the infection is removed (Graveland et al. 2011b; van Cleef et al. 2011). We can conclude that LA-MRSA ST398 which is largely associated with pigs (or sometimes with veal calves) has adapted and has a low level of epidemicity for humans. Although the strains of MSSA CC398 were commonly present in pigs (Hasman et al. 2010), several publications have reported the occurrence of MSSA ST398 infections in humans without recent exposure to pig farms (McCarthy et al. 2011; van der Mee-Marquet et al. 2011), implying that subtypes of ST398 exist which are circulating among human populations. Overall it is clear that ST398 rates are rising and are to a large degree associated with animal contact, as colonization rates have been particularly high in various groups with livestock contact, such as veterinarians, farmers, and people living on farms (Weese 2010; Garcia-Graells et al. 2012).

In conclusion it is clear that MRSA can reciprocally affect human and animal populations. The first documented case of direct transmission of MRSA between cows and humans was reported (Juhász-Kaszanyitzky et al. 2007). MRSA isolated from cows with subclinical mastitis and from a person who worked with these animals were indistinguishable by phenotyping and genotyping methods and were of a low frequency spa type. Clearly the contact with pigs or cattle is a risk factor for ST398 or nontypable MRSA colonization (Weese 2010). The latter author indicated not surprisingly, working in barns with colonized pigs is a risk factor and more intensive and longer contact is probably an important factor, as evidenced by transient MRSA carriage in people who visit farms to collect samples versus higher rates of colonization in workers with more intensive or longer contact. As MRSA becomes (or has become) established in different animal and human populations, elimination of all risk will be impossible (Weese 2010). Therefore, MRSA in animals require careful control methods to better eliminate the emergence and dissemination of MRSA in different species and to interspecies transmission to humans and vice versa. This approach can reduce the implications of MRSA in both animals and humans while avoiding an excessive response that might unnecessarily harm human interactions with pets, food animals, and food. Bacteriophage can be developed and used as biocontrol measures to reduce the impact on animal health and welfare, human health, agriculture and human-animal interaction. A safe and non-toxic bacteriophage mixture (nature's biocontrol) can be used to disinfect health care facilities, workers, animals and their environs, replacing less effective and more expensive and often toxic chemical cleaners.

4.5 Epidemiology of MRSA in Hospitals

For several decades, MRSA was almost exclusively a health care associated pathogen and was not considered a significant threat to healthy persons in the general community. Currently MRSA is one of the most prevalent micro-organisms involved with healthcare-associated infections. It is usually confined to hospitals and in particular to vulnerable or debilitated patients. Environmental contamination with a MDR bacterium was the major predictive factor in HCW contamination during patient care, suggesting that efforts to decontaminate patient rooms during their stay may decrease transmission (Morgan et al. 2012). As with ordinary strains of *S. aureus*, some patients harbor MRSA on their skin or nose without harm (such patients are said to be colonized). However, these patients may develop infections if the MRSA spreads to other parts of the body. Patients with breaks in their skin due to wounds or in dwelling catheters which allow MRSA to enter the body, and those with certain types of deficiency in their immune system are at increased risk of developing an infection. Individuals colonized with MRSA may also serve as a 'reservoir' of MRSA that may spread to other patients, for example, if hospital staff attending to a colonized or infected patient become contaminated or colonized with MRSA themselves and spread the bacteria to other patients with whom they subsequently have contact. It was estimated around 20–40% of hospital-acquired infections have been attributed to cross-infection via the hands of HCW, who have become contaminated from direct contact with the patient or indirectly by touching contaminated environmental surfaces (Weber et al. 2010; Falagas et al. 2011).

The transmission of MDR bacteria, namely, MRSA to HCW clothing was common for all MDR bacteria, emphasizing the importance of improving compliance with contact precautions (Morgan et al. 2012). In addition, MRSA may spread via contaminated equipment or through the environment. Some strains of MRSA are particularly successful at spreading between patients and may also spread between hospitals, presumably when colonized patients or staff move from one hospital to another (Hallin et al. 2008), and to their homes and eventually to the whole community (Fig. 4.2). These strains are known as an EMRSA (Hallin et al. 2008). As previously reported MRSA can survive in extra intestinal colonization and infection (Bhalla et al. 2007) and they can adapt to harsh environmental conditions, including the presence of antimicrobial agents and yet they can still produce toxins (Jassim 1989; Hennekinne et al. 2012). Some strains can grow and produce toxins in the presence of only 4 amino acids (cysteine, leucine, proline and valine) (Jassim et al. 1989) and survive the process of phagocytosis and intracellular killing by polymorphonuclear leukocytes (Jassim 1989).

In general, Boyce et al. (1997) concluded that inanimate surfaces near affected patients commonly become contaminated with MRSA and that the frequency of contamination is affected by the body site at which patients are colonized or infected. That personnel may contaminate their gloves (or possibly their hands) by touching such surfaces suggests that contaminated environmental surfaces may serve as a reservoir of MRSA in hospitals. In this regard transfer of the pathogens in hospital occurs via surfaces, and this depends on two important factors:

(1) the pathogens able to survive on dry surfaces for a long period, and
(2) the role of patients and healthcare staff carrying a sufficiently high level of pathogens to spread and contaminate the hospital surfaces (Boyce 2007).

It was demonstrated that 36% of surfaces sampled in the rooms of patients with MRSA in a wound or urine were contaminated, compared with 6% of surfaces in the rooms of patients infected with MRSA (Boyce et al. 1997). In general the most frequently contaminated surfaces are floors, doorknobs, bed linens, bed-frame lockers, the patient's gown, over bed tables, blood pressure cuffs, wound dressing, infusion pumps, catheters or drain, television remote control devices, various plastic items in the hospital, artificial airways, side rails, linens, pagers, cell phones, mattresses, gowns/uniforms, white coats, shirts, ties of doctors, HCW gloves, bedside tables, toilet seats in rooms previously occupied by an infected patient, hospital dust, and mops used for cleaning (Boyce et al. 1997; Talon 1999; Frank 2003; Boyce 2007; Dancer 2009; Lopez et al. 2009; Page et al. 2009; Berendt et al. 2011; Morgan et al. 2012; Zollfrank et al. 2012). Moreover, ventilation grills were found to be sources of MRSA outbreaks in hospitals (Cotterill et al. 1996). MRSA are able to survive or persist on surfaces in hospitals for months and can thereby be a continuous source of transmission if no regular preventive surface disinfection is performed (Kramer et al. 2006; Boyce 2007). Therefore, the adhesion of MRSA to a surface followed by biofilm formation will occur in less than 24 h, whereas the biofilm of MRSA will create a critical microbiological problem for healthcare services in the hospital (Joshi et al. 2010).

4.6 Decontamination of the Hospital Environment

Many MRSA infections occur in hospitals and healthcare facilities, with a higher incidence rate in nursing homes or long-term care facilities. Rates of MRSA infection are also increased in hospitalised patients who are treated with quinolones. From the above it is clear that the healthcare provider-to-patient transfer is common, especially when healthcare providers move from patient to patient without performing necessary hand-washing techniques between patients (Muto et al. 2003; Tacconelli et al. 2008). Patients with MRSA frequently contaminate environmental surfaces in their immediate vicinity. This organism can remain viable in the environment for weeks or months. Cleaning practices are often suboptimal, where daily cleaning of surfaces near patients is often performed poorly. Additionally, terminal cleaning of rooms after patient discharge is often inadequate. Carling et al. (2006) found that only 47% of surfaces targeted for terminal cleaning had been cleaned.

Obviously, contaminated environmental surfaces can contribute to transmission of pathogens by serving as a source from which HCWs contaminate their hands or gloves, contaminated medical equipment that comes into direct contact with the patient can also serve as a source of transmission (Boyce et al. 1997; Bhalla et al. 2004; Hayden et al. 2008). It has been reported that patients admitted to a room formerly occupied by a patient with MRSA are at an increased risk of acquiring the organism, suggesting that terminal cleaning of rooms is inadequate; patients acquire the organism directly from contaminated surfaces from HCWs who contaminate their hands in the room (Martinez et al. 2003; Huang et al. 2006; Drees et al. 2008). Therefore, to reduce environmental contamination pathogens transmission, a prospective 9-month study was carried-out at medical intensive care units in admission and daily screening of patients. The results suggested that environmental samples and HCW hands should be cultured twice weekly. The study also recommended including a baseline period, education, monitoring and feedback for housekeepers, wash-out period with no specific intervention and multimodal hand hygiene intervention (Hayden et al. 2006). The Centres for Disease Control's, Healthcare Infection Control Practices Advisory Committee (HICPAC), SHEA, Infectious Diseases Society of America (IDSA) and Healthcare-Associated Infections Task Force have all recommended that healthcare facilities must pay greater attention to cleaning and disinfection of equipment and the environment (Siegel et al. 2007; Yokoe et al. 2008).

In general, it appears that housekeepers and nursing staff often do not agree on who should clean what, furthermore, housekeepers do not always understand which detergent or disinfectant to use, what concentration should be used or how often to change cleaning cloths or mop heads. Other contributing factors are demands for fast room 'turnaround times' and staff shortages or frequent turnover of personnel (Calfee et al. 2008). In this regard, several strategies to disinfect both clinical equipment and the environment are available, often involving the use of antimicrobial chemicals in hospitals (Abreu et al. 2013). For instance, a fluorescent dye marker system used to test high-touch objects in patient rooms that were marked with invisible fluorescent solution. After terminal thorough cleaning, the target surfaces were evaluated using a portable UV light to see if the marker had been wiped off (Carling et al. 2006, 2008; Goodman et al. 2008). ATP bioluminescence methods used for years to monitor adequacy of cleaning procedures in beverage and food processing industries (Jassim et al. 1990; Simpson et al. 1990; Denyer et al. 1991; Walker et al. 1992; McIntyre et al. 1996) have also been introduced in hospital hygiene monitoring systems (Griffith et al. 2000; Malik et al. 2003; Whiteley et al. 2012).

In conclusion, it appears that improving cleaning and disinfection practices in hospitals requires developing detailed protocols, educating housekeepers, monitoring cleaning and providing feedback to housekeepers. New technology has been introduced such as liquid disinfectants, hydrogen peroxide vapor (McAnoy 2006), mobile UV light systems (Carling et al. 2006, 2008; Goodman et al. 2008), and other new area decontamination systems still warrant further evaluation to determine their effectiveness and impact on transmission of healthcare-associated pathogens. Nevertheless, issues to consider when evaluating liquid disinfectants or area decontamination systems are still needed to address the following:

i. Effectiveness against bacteria, including MRSA,
ii. Ease of application,
iii. Speed of action or room turn-around time,
iv. Materials compatibility,
v. Adverse effects on housekeepers, patients, environment,
vi. Impact on MRSA transmission,
vii. Cost.

In general, infection can be spread via inadequately decontaminated items. Requirements for chemical disinfection are mainly limited to cleaning areas of contaminated spillage; decontamination of certain heat-labile instruments e.g. fiberscopes, and disinfection of the skin. In Japan the use of alcohol-based hand rub and gloves by clinical staff to control and prevent MRSA-positive patients was more common in 2011 than in 2000 or 2003 (Morioka et al. 2014). Chemical disinfectants can fail if not selected and used properly. Therefore, environmental contamination with MDR bacterium is the major predictive factor in HCW contamination during patient care, suggesting that efforts to decontaminate patient rooms during their stay may decrease transmission (Morgan et al. 2012). Accordingly, new control measures, rather than antibiotics, are needed to contain the increasing hazard of MRSA, especially when it is known that antibiotics are also part of the problem. For example, glycopeptides, cephalosporins and in particular quinolones are associated with an increased risk of colonisation of MRSA. Moreover, reducing the use of antibiotic classes which promote MRSA colonization, especially fluoroquinolones is recommended in current guidelines (Tacconelli et al. 2008). On the other hand, most antimicrobial agents fail to kill bacterial biofilm (Olson et al. 2002; Toté et al. 2009; Jassim et al. 2012; Abdulamir et al. 2015). The concentration of biocides required to kill MRSA biofilm is approximately 100–1000-times greater than that required to kill planktonic MRSA (Otter et al. 2014). Currently, there are few prevention techniques to control biofilm formation without causing side effects. The use of bio-solutions has been suggested such as enzymes, bacteriophages, interspecies interactions and antimicrobial molecules from microbial origin (Simões et al. 2010).

Predominately, the current control measures in hospitals and on farms for MRSA have proven to be inadequate, leaving a need for new rapid control methods to curb MRSA infections in situ (Aldoori et al. 2015). Clearly there is a need for safe, widely accepted, and innovative measures for the control and therapy of MRSA since worldwide, an estimated 2 billion people carry some form of *S. aureus*; of these, up to 53 million (2.7% of carriers) are thought to carry MRSA (Graham et al. 2006). Therefore, novel approaches for reducing such antibiotic selective pressure on MRSA must be taken urgently. One promising new proposal for the control of MRSA uses bacteriophages. The following in this chapter we describe fogging and dispensers designed to enlist phage against MRSA.

4.7 Bacteriophages: Nature's MRSA Control Agents

Bacteriophages 'Phages' are viruses that can attack bacteria including MRSA. An overview of current knowledge on staphylococcal infections and phages able to kill *Staphylococcus*, including experimental studies and available data on their clinical use were reported (Kaźmierczak et al. 2014). Regulatory approval of the safety and efficacy of phages in human therapy, animal biocontrol (prophylaxis) and in non-clinical settings has been reported (Jassim and Limoges 2014). The lytic cycle or 'virulent phages' fit in the class of 'natural antimicrobial controlling agents' and are arguably the most abundant biological entities on the planet, while also considered safe for humans and animals (Jassim and Limoges 2014; USDA-FSIS 2014; Abdulamir et al. 2015; Aldoori et al. 2015). These phages are being exploited in various areas of biotechnology, including rapid bacterial detection, food bio-processing and removal of bacterial biofilm (Jassim and Limoges 2014). Phages can also be used in normal cleaning products where they can lie dormant for weeks and only awake to devour the MRSA when it arrives. This approach differs drastically from past efforts to kill the bacterial pathogens with antibiotics, which have proven counter-productive (Jassim and Limoges 2014).

A study showed that phages can be used effectively as a bio-processer (Jassim et al. 2012), on the carcasses post-harvest (USDA-FSIS 2014), to deliver a photosensitizer, or light-activated antimicrobial agent, to a target MRSA, resulting in enhanced and selective killing of the target organism (Embleton et al. 2005; Hope et al. 2009). Such attributes are desirable in photodynamic therapy for infectious diseases. In another study, a lytic phage (phage K) was assessed in vitro for its ability to lyse clinically isolated MRSA strains, *S. aureus* with heterogeneous vancomycin resistance and teicoplanin-resistant strains from hospitals and bovine infections (O'Flaherty et al. 2005). In the same study, phage-enriched wash solution resulted in a 100-fold reduction in staphylococcal numbers on human skin by comparison with numbers remaining after washing in phage-free solution (O'Flaherty et al. 2005). In mice artificially infected with MRSA a cow's mastitis isolates, it was found that wild highly lytic phage isolated from milk samples of cows with mastitis were able to rescue 100% of the mice from a median lethal dose (LD_{50}) or (1×10^8 cfu ml^{-1} per mouse) achieved when the phage:bacteria ratio was 100:1. Even when treatment was delayed for 6 h post lethal infection, to the point where all mice were moribund, 80% of them were rescued by a single injection of this phage preparation (Aldoori et al. 2015). Based on the current results, a comprehensive study is needed to guide further research on the MRSA phage as a biocontrol for MRSA mastitis in dairy cows to replace or reduce the use of antibiotics in animal husbandry and to prevent horizontal and vertical transmission of MRSA in agriculture animals. Horizontal gene transfer has been reported to occur between *S. aureus* and *S. intermedius* and *S. aureus* and CoNS (Malik et al. 2005). A single wild breed of phage was able to infect and lyse both cells of CoPS and CoNS (Sabah A.A. Jassim, unpublished data). In this regard, the phages can be fed to animals via drinking water, oral injection and vegetable capsules and it was found that a cocktail of designed phages feeding for 20 days via vegetable capsules yielded the highest reduction of fecal *Escherichia coli* 4.62 $\log_{10}$ (Abdulamir et al. 2014), and the *E. coli* biofilms were reduced by >3 $\log_{10}$ (Jassim et al. 2012).

Staphylococcal biofilms are a major concern in both clinical and food settings because they are an important source of contamination. The ability of staphylococci phage K and modified derivatives to prevent biofilm formation and to reduce established biofilm density of *S. aureus* was investigated (Kelly et al. 2012). Results suggested that these biofilms were eliminated in a time-dependent manner, with biofilm biomass reduction significantly greater after 72 h than after 24–48 h. Using the phage cocktail resulted in the complete inhibition of biofilm formation over a 48-h period with no appearance of phage resistance. Recently, Jassim and his research team Abdulamir et al. (2015) carried-out an experiment to mimic the sterilization conditions in hospitals where the sterilizing agents dry up after 30 min. In this regard 10^{12} pfu (plaque-forming unit) ml^{-1} phages, as anti-MRSA, were applied on the formed biofilms in the tissue culture plates with and without chemical adjuvant substances then kept, without lid, stationary up to 24 h at room temperature. Sub-lethal concentrations of chemical adjuvant together with applied phages were used in order to increase the permeability of the matrix of the biofilm as well as to increase the exposure of phage binding receptors on the cell walls of MSSA and MRSA bacteria. A sub-lethal dose, 0.08%, of benzethonium chloride was found to be the best in disintegrating polysaccharide intercellular adhesion (PIA) type biofilms when mixed with applied phages. For proteinacious fibronectin-binding protein A (FnBPA) type biofilms, a sub-lethal dose 0.06% of ethanol was found to be the best in disintegrating the proteinacious matrix of FnBPA type biofilms when mixed with applied phages. However, both above doses were of no effect on biofilms or phages when benzethonium chloride and ethanol were used alone, respectively. It was found that the most resistant type of biofilms to phage-based control was FnBPA in MRSA where 50% of biofilms were reduced but not eradicated completely. They concluded that PIA-disturbing agent and protein (namely FnBPA) denatured alcohol can increase the efficiency of phages in accessing host cell walls and lysing them, which in turn lead to much more efficient MRSA and MSSA biofilm treatment and prevention.

Another approach explored the potential of phage endolysins (peptidoglycan hydrolases), where 'phage lysins' have been the focus of research into combatting planktonic (free floating) and sessile (anchored) microorganisms that form biofilms (Sass and Bierbaum 2007; Fischetti 2008; Fenton et al. 2010, 2013; Son et al. 2010; Gutiérrez et al. 2014; Bai et al. 2016; Schmelcher and Loessner 2016). Biofilms are collections of microorganisms and the extracellular polymers they secrete, attach to either inert or living substrata. These cell wall peptidoglycan hydrolases possess a number of advantages over conventional antibiotics including, rapid lytic activity against bacterial cells, low probability of developing bacterial resistance, and significantly lower chance of disrupting commensal microflora, due to the enzymes' specificity (Fenton et al. 2013). The phage-derived peptidase, cysteine- and histidine-dependent amidohydrolase/peptidase ($CHAP_K$) as a biocidal agent for the rapid disruption of biofilm-forming staphylococci, commonly associated with bovine mastitis was reported (Fenton et al. 2013). It was found that the purified $CHAP_K$ applied to biofilms of *S. aureus* completely eliminated the staphylococcal biofilms within 4 h. In addition, $CHAP_K$ was able to prevent biofilm formation by this strain. The $CHAP_K$ lysin also reduced *S. aureus* in a skin decolonization model. The potential of $CHAP_K$ as a biocidal agent for prevention and treatment of biofilm-associated staphylococcal infections or as a decontaminating agent in the food and healthcare sectors have been suggested. Regarding the planktonic cell of MRSA, the purified phage cloned lysin encoded by the novel *S. aureus* phage, phi MR11, designated MV-L has also rapidly and completely lysed cells of MRSA and vancomycin-intermediate *S. aureus* strains in growing conditions (Rashel et al. 2007). In vivo study based an ointment on the potential for an engineered staphylococcus-specific phage lysin (ClyS) as active ingredient in a mouse model of skin infection was reported (Pastagia et al. 2011). It was found that the therapeutic effect for both MRSA and methicillin-susceptible *S. aureus* was significantly higher than that of mupirocin, which is commonly used for the treatment of skin infections.

The emergence of MRSA has prompted interest in alternatives to conventional drugs. One possible option is to use phage as antimicrobial agents. Built upon this background, this chapter proposes to use phage biocontrol in hospital settings to improve and customize care and biocontrol to the individual needs of patients and HCWs for MRSA carriage and to identify risk factors for MRSA colonization using an MRSA bio alarm. We also suggest new applications for the best use of phage as a biocontrol to target pathogens causing HA-MRSA and the relevant procedures to safely control phage applications in healthcare and veterinary settings.

4.8 Using Phage for Decontamination of MRSA in Healthcare Settings

4.8.1 Phage Hunting

Corresponding phages of many MRSA are easily discovered, often from hospital sewage, walls, floors, from cows on the farm (Abdulamir et al. 2015; Aldoori et al. 2015), or other waste materials that contain high bacterial concentrations. Interestingly, human anatomical niches seem to represent natural reservoirs of phages even for potential pathogenic bacteria such as *E. coli* (Nakamura et al. 2009), *E. faecalis* (Bachrach et al. 2003) or MRSA (Ma et al. 2011). Evidence exists that the ratio of natural phages to bacteria seems to have an impact on human health and might be unbalanced during infection (Viertel et al. 2014). Natural phage diversity and phage-host interactions within the human and the environment at large are reviewed (Jassim and Limoges 2014) leading to new concepts for treating or preventing endogenous infections.

4.8.2 Phage Formulation

Phages can be formulated and developed as with certain antibiotics (Alisky et al. 1998; Goodridge 2010; Kutter et al. 2010; Rastogi et al. 2016). They have been formulated in different forms, as liquids, creams, impregnated into solids, capsule, tablets (Carlton 1999; Krylov 2001; Górski et al. 2007; Kutateladze and Adamia 2010; Kutter et al. 2010; Abdulamir et al. 2014) to suit various phage applications and routes of administration. Different phages furthermore can be mixed as cocktails to broaden their properties, typically resulting in a collectively greater antibacterial spectrum of activity (Jassim et al. 2012; Abdulamir et al. 2015).

4.8.3 Phage Decontamination

4.8.3.1 Dry Fog Phage System (DFPS)

From the above it is clear that the most contaminated surfaces in hospitals are the floors, doorknobs, bed linens, bed-frame lockers, the patient's gown, over bed tables, blood pressure cuffs, wound dressing, television remote control devices, side rails, linens, pagers, cell phones, mattresses, gowns/uniforms, white coats, shirts, ties of doctors, HCW gloves, bedside tables, toilet seats in rooms previously occupied by an infected patient and mops. It is well known that there is not available an in situ decontamination method that is able to kill bacterial pathogens on all the above surfaces collectively in one-step to obtain rapid (around 40 min) decontamination of MRSA. The DFPS is proposed as a natural bio-disinfectant. Phage technology can be used safely in healthcare settings. The DFPS and other applications are also discussed in greater detail in Chap. 3. In brief, the idea is to develop a phage fogging system. An MRSA super phage (this discussed in greater detail in Chap. 1) that will infect and subsequently lyse the MRSA strains (within 40 min) wherever present including, but not limited to, the air, walls, floors, medical equipment, on human skin, in the nose, on animals' bodies, etc., where it will selectively reduce or eliminate the target MRSA from the healthcare facility (Fig. 4.3).

The DFPS works by using a special spray nozzle that produces a dry fog of ultra-fine (1–10 μm) droplets of the λ-buffer containing phages via compressed air and the aqueous λ-buffer to produce the fog. The outlet port or stainless steel nozzle is 2 μ diameter with a pin and open diffusion cone is used to evaporate the solution into the air (Fig. 4.3). The optimal pressure is found and adjusted with regard to the room size and air characteristics including humidity and temperature (5–30 °C).

In general, the λ-buffer droplets resemble the size of most bacterial species where they are around 0.5–1.0 μm. Since each milliliter (ml) of the λ-buffer contains 10^9 pfu, one ml equals 1000 microliter (μl) this means that each μl contains 10^6 pfu

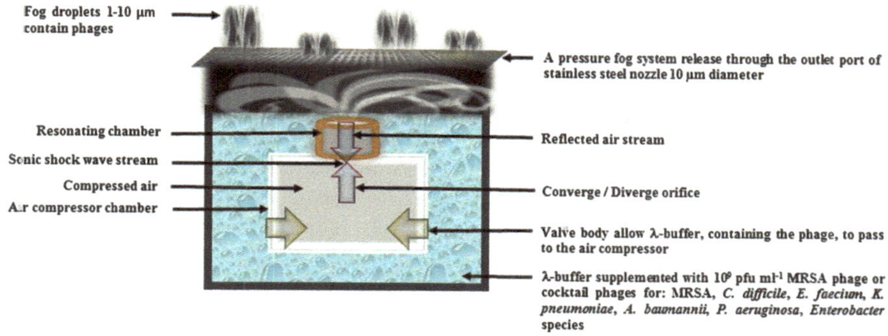

Fig. 4.3 DFPS for rapid deployment through ventilation to disinfect healthcare environments

(phage particles). Therefore, each 1.0 µl is equivalent to 1.0 µm and that means each 1.0 µm of the λ-buffer fog droplets carries around 10^6 pfu of phage particles. An average MRSA is about 0.5–1.0 µm in diameter. So if one λ-buffer droplet 1.0 µm carries 10^6 pfu of phage, which attaches to a similar size MRSA, this will represent a phage-to-bacteria ratio of 10^6:1. This is a massive phage concentration to infect 1 single bacterial cell. This range of phage concentration is ideal for anti-bacterial and for anti-biofilm applications (Jassim et al. 2012; Abdulamir et al. 2014; Aldoori et al. 2015).

It is noteworthy, that these tiny droplets will also attach to dust particles which may contain airborne bacteria causing the dust to be slightly wet and to agglomerate, becoming heavy enough to drop from the air and fall back into the concentration of phage which will kill the bacteria in the dust. This system requires only an aqueous λ-buffer supplemented with suitable phages to target the bacteria in the setting. Also it works with low power consumption, requiring no chemical disinfectants or significant wetting of the room (always less than 1/2% by weight, typically no more than 0.1% moisture addition). A moderate pressure fog system is without odour which will not change climatic conditions in the hospital, house, rooms and mall centre, or any area or industry which needs to control pathogenic bacteria. The system will be under temperature and humidity sensor control as well as a phage control-timer coupled with external signal and guarantees a perfect control of the MRSA or any pathogenic bacteria.

The DFPS decontamination depends on the phage density, aggressiveness and replication speed. We recommend using a more sophisticated phage programming technology, as discussed in Chap. 1, to produce finite specific phages that rapidly kill their bacterial host and continue to replicate until no host is encountered. In this case, the phages persist for some time even after no bacteria are present. The phage density should be not $<10^9$ pfu per ml in λ-buffer. The phages can be resistant to unfavourable physical and chemical factors, such as low and high temperatures, pH, salinity, and ions (Jończyk et al. 2011). If MRSA is present on surfaces or in the air, the phages will rapidly infect the bacterium with a lytic phage binding to host cell within 5–10 min (Jassim et al. 2010a, b), followed by the stage that the phage undergoes a latent period of 25–40 min. The latent period encompasses the process of nucleic acid translocation into host cytoplasm, progeny virion assembly, and finally, the burst, with release of newly formed phage progeny (Jassim et al. 2010a, b; Kaur et al. 2012). The minimal burst size is around 73 phage particles per cycle and the maximal burst size can be as much as 336 phage particles per cycle (Jassim et al. 2010a, b). After this period, the bacteria enter the lytic state and become unable to reproduce. In this case more phages will be produced from the lysed bacteria. However, the space of the site remains occupied by bacterial cells until phage spreading occurs. When the phage population at a given site is updated, it spreads to the nearest-neighbour sites. The transport mechanism is similar to the bacterial case; however, phages always die after some time once empty sites are reached, namely, no bacteria are present.

The phage DFPS can kill many bacterial species by supplying the DFPS with a cocktail of multiple phage species for example phages for MRSA, *Clostridium difficile*, *Enterococcus faecium* (vancomycin-resistant enterococci), *Klebsiella pneumoniae*, *Acinetobacter baumannii*, *Pseudomonas aeruginosa*, *Enterobacter* species, or any other phages for medically important bacterial pathogens that are present. In this regard, the infected bacteria can no longer be invaded by other phages. Bacteria at any given site are always sacrificed to only one of the phage species and are not split between infections of distinct predators. In Fig. 4.3 diagram illustrating the DFPS for decontamination of MRSA patient's room surfaces during their stay and accordingly will decrease transmission of MRSA by selectively reducing or eliminating target MRSA from the healthcare facility.

4.8.3.2 Phage-Based Hand Rubs (PBHR)

MRSA survives well in a hospital environment, and this raises a big concern for healthcare, since patients, HCWs and visitors can contaminate their hands by touching contaminated surfaces and objects. Antimicrobial soap and hand disinfection with alcohol-based hand rubs (ABHRs) are used for routine decontamination of hands in hospitals. It was found that ABHRs are superior to disinfect MRSA than antimicrobial soap and to prevent nosocomial infections in healthcare (Bessonneau et al. 2010; IHI 2014). In this regard it is now preferred for routine decontamination of hands for reducing MRSA infections in healthcare settings (Kramer and Kampf 2007; WHO 2009, IHI 2014). ABHRs contain on average 70% by weight of one or more alcohols, including ethanol, n-propanol and isopropanol (Bessonneau et al. 2010). During the hand rubbing procedure, users are exposed to these alcohols not only through dermal contact, but also via inhalation, due to the physical and chemical properties of alcohols volatilizing from alcoholic solutions or gels into the air (Bessonneau et al. 2010).

The studies of Bessonneau et al. (2010) show that using an average of 30 hand rubbings of ABHRs per healthcare professional per day, a healthcare worker may be exposed to as much as 5500 mg m^{-3} per work shift, five times above the recommended occupational time weighted average limit. Ethanol ingestion is well known to increase risks of several diseases affecting the pancreas, liver cirrhosis, cardiovascular disorders, fetal alcohol syndrome, cancer, neurological disorders, eye, nose and throat irritation in addition to the possible health effects related to unintentional alcoholization (via inhalation and dermal contact) (Bessonneau et al. 2010). Overuse can lead to severe hand dermatitis, dryness, and irritation, itching, and even cracking and bleeding (WHO 2009). On the other hand, there have been some fire incidents reported in German hospitals related to the use of ABHRs (Kramer and Kampf 2007). It is worth mentioning, ABHRs do not provide any ongoing disease protection since alcohol will be evaporated from hands after a few seconds from the application. Thus if the HCWs touch for example doorknobs of MRSA patients' room obviously they will re-contaminate their hand.

The ability of phages to recognize precisely their hosts, rendered them favourable as environmentally friendly disinfectant agents that can be used and stay active as long as the MRSA is present on surfaces having no side-effects for replacement of ABHRs and other chemical disinfectants in healthcare settings. Evidence suggested that phages can also be used as a hand wash to disinfect for MRSA. Ten MRSA infected human hands were soaked in a solution containing 10^8 pfu ml^{-1} of a single phage K, revealing 2 log_{10} reduction, suggesting that MRSA can be controlled using a specific phage (O'Flaherty et al. 2005). This is the first model in situ hand wash studies using a MRSA phage-enriched wash solution resulting in a 100-fold reduction in staphylococcal numbers on human skin by comparison with numbers remaining after washing in phage-free solution. Infusion of the phage into a nonimmunogenic bismuth-based cream resulted in strong anti-Staphylococcus activity from the cream on plates and in broth (O'Flaherty et al. 2005).

Hospital staff, visitors and patients can take positive steps to reduce the risk of acquiring MRSA infection by applying λ-buffer gel supplemented with reprogrammed MRSA phages at titre (10^9 pfu per ml) in dispensers (Fig. 4.4). Placing phage dispensers in convenient locations, for example patient rooms, visitor areas, lobbies, near elevators, and so on, alongside or replacing chemical sanitizers. The following procedure is used for dispensers with MRSA phage: Wet hands with phage MRSA phage-based hand rub. It can be applied in water form on hair and face (especially around the nose). The phages will adhere invisibly to the skin and continue killing MRSA after the water dries for up to 24 h after use. This means that MRSA phage dispenser's, unlike conventional chemical antibacterial agents such as ABHRs, chlorhexidine gluconate, or soaps, will keep killing MRSA providing peace of mind knowing that people have taken steps to reduce the risk of infection.

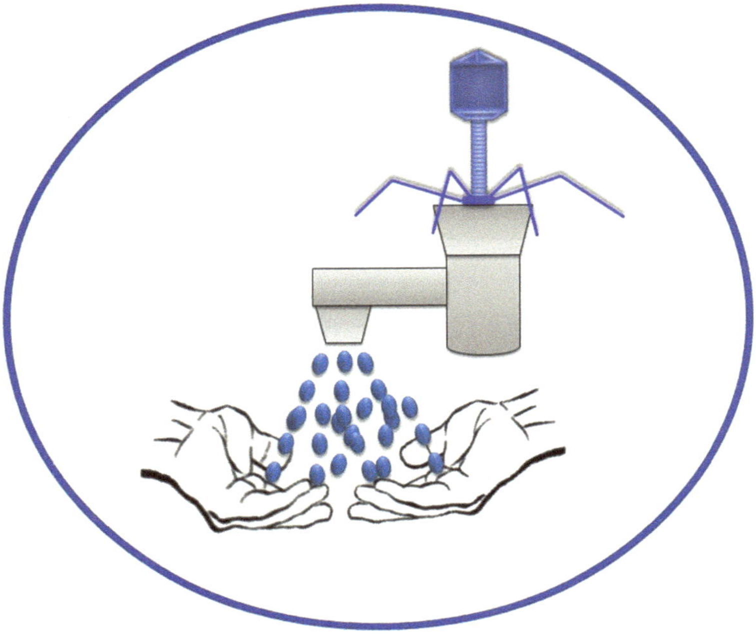

Fig. 4.4 Phage-based hand rubs: an environmentally friendly sanitizer

4.9 Rapid Diagnostic for MRSA

Over the last two decades' major steps have been taken to increase the diagnostic potential of the microbiology laboratory, mainly by implementing methods for DNA and RNA amplification. PCR is the most well-developed molecular technique to date, and has a wide range of clinical applications, including specific or broad-spectrum pathogen detection, evaluation of emerging novel infections, surveillance, early detection of biohazard threat agents, and antimicrobial resistance profiling. On the other hand, the high cost and relatively high operator skill requirement for these tests are obstacles to their widespread routine use. Another important limitation of PCR is that it does not differentiate between live bacteria and dead remnants and in general it is still too complicated for many clinical or field applications (Aellen et al. 2006). A new PCR was developed that can discriminate between live and dead cells by combining ethidium bromide monoazide (EMA) staining with real-time PCR. The detection of viable bacteria demonstrates this method to effectively bind DNA from dead cells. It was found that the optimized PCR method could detect viable bacterial cells of *L. monocytogenes* and *E. coli* in samples without interference from dead cells (Soejima et al. 2008; Wang et al. 2009). This led to the development of other methods of bacterial detection able to discriminate between live and dead bacteria cells by using simple culture-based albeit rapid detection methods, such as the chromogenic agar, which contains media substrates that change color in the presence of *S. aureus*. Selectivity for MRSA is achieved by incorporating antibiotics into the agar. Use of such agar allows identification of MRSA from primary isolation plates within 48–72 h, obviating the need for additional subcultures or biochemical tests (Malhotra-Kumar et al. 2008).

The implementation of PCR tests reduces the time to detect MRSA carriers from 48–72 h down to 2–5 h. Nevertheless, specificity is sometimes impaired due to false-positive PCR signals occurring in mixed flora specimens. In order to rule out any false-positive PCR results, a culture screen (48–72 h) must always be carried out simultaneously (Stürenburg 2009). This reinforces the need for an active, simple, rapid and accurate detection method for carriage of MRSA to facilitate early contact isolation and even treatment, thus preventing spread in the hospital and reducing costs. There are several MRSA diagnostic methods reported in great detail comparing the effectiveness and cost of each method (Stürenburg 2009). There is currently no single stand-alone rapid bacteria test assay that can uncover only live cells of MRSA strains qualitatively and quantitatively. In addition, a need for even more rapid and inexpensive diagnostic tests is being felt, highlighted by the mortality associated with MRSA infections, the ever-increasing problem of antimicrobial resistance and the lack of new antibiotics in the pharmaceutical pipeline.

Furthermore, advanced technology is needed to improve automation, optimise detection sensitivity and specificity and expand our capacity to detect multiple targets simultaneously. The speed with which MRSA carriage is detected has an important role to play, as it is a key component of any effective strategy to prevent the pathogen from spreading (Stürenburg 2009). There is, therefore, a substantial need for rapid diagnostic methods that identify clinical isolates of MRSA, on farms, or on various surfaces. The ideal rapid diagnostic method should provide the following characteristics and parameters: qualitative and quantitative test, rapid <60 min, reliable and easy to use, inexpensive, with no need for highly skilled technicians that is easy to carry to endemic locations. Recognizing this, there is a need to develop rapid diagnostic interventions on MRSA colonization and infection rates in hospitals and to apply cutting edge technological advances to the development of rapid diagnostic assays for MRSA detection using phages. For over two decades Jassim and coauthors have used phages for rapid bacterial detection of different bacterial strains and species and the results have demonstrated that phage is a reliable and accurate tool to report the presence of bacteria in a sample as compared to CFU (measure of viable bacterial on agar plates) (Jassim et al. 1996, 2010a, b; McIntyre et al. 1996; Denyer et al. 1998; Stewart et al. 1998; Favrin et al. 2001; 2003, Jassim and Griffiths 2007). Several innovative rapid diagnostic protocols are illustrated below that would decrease the time to detection, thereby reducing the risk of nosocomial transmission and infection, especially in high-risk patients. Fast detection will enable the medical professional to choose an appropriate treatment and swiftly control the MRSA in healthcare settings.

4.9.1 *Phage Endolysins Enzymes*

The use of phage endolysins enzymes 'lytic enzymes' that digest the bacterial peptidoglycan of Gram-positive bacteria, as a novel class of alternative antimicrobials have been discussed in greater detail in Chap. 3. The phage-derived endolysins with high host specificity and host lysis activities has been used for rapid detection of food-borne pathogens (Bai et al. 2016). These systems revealed advantages such as rapid detection of only viable pathogenic cells without interference by food components in a very short reaction time, suggesting that these systems may be suitable for monitoring pathogens in foods (Bai et al. 2016). It was found that endolysin LysK from *S. aureus* phage K has a high host lysis activity with broad host spectrum including general and clinical MRSA (O'Flaherty et al. 2005). This can be a suitable phage candidate to use for rapid MRSA detection in healthcare settings and on animal farms. Figure 4.5 illustrates a new phage lysin-enzymes-based reporter MRSA assay that can detect very low levels of bacteria from human or animal specimen or other medical community samples (e.g., 1 bacterial count in 25 ml or g), and to fairly rapidly (≤ 5 min) give a read-out as to the species/strain of MRSA detected (Fig. 4.6). The concept of the diagnostic test is based on the fact that phage lysin enzymes will instantly lyse their target bacteria within a few minutes and thereby release both MRSA ATP and DNA.

ATP detects the presence of exogenous organisms by the addition of luciferase enzyme and its substrate, such as firefly luciferase and luciferin respectively. Bioluminescence is emitted light from the reaction of the lysed cells with ATP. This light can be readily detected using either a luminometer (Fig. 4.6) or photographic film. Since the background level of light emission is extremely low, we need to use a luminometer with high sensitivity where relatively few numbers of cells can be

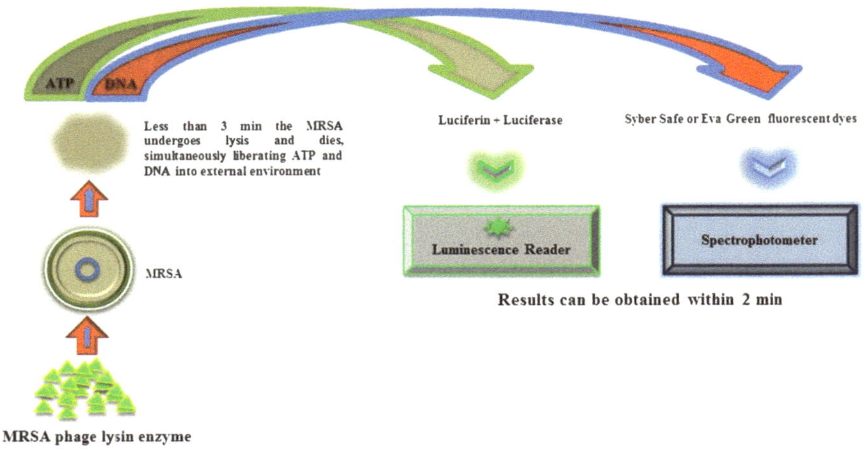

Fig. 4.5 Rapid (5 min) qualitative and quantitative detection of MRSA using phage lysin enzymes

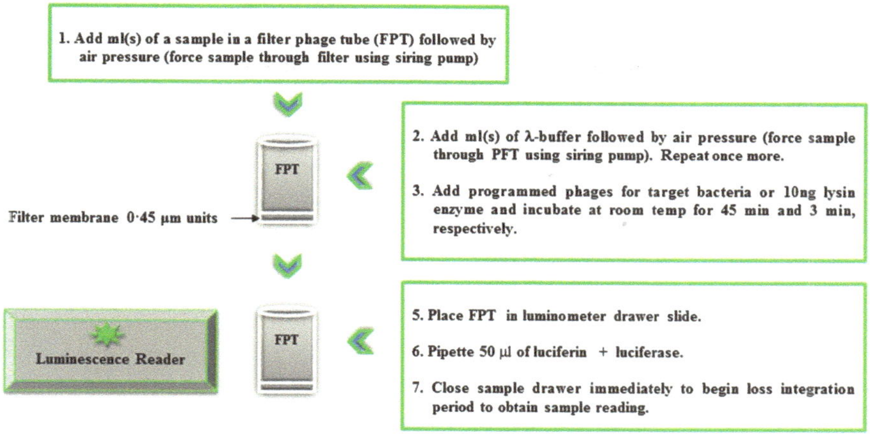

Fig. 4.6 Qualitative and quantitative rapid (45 min or 5 min/real time) MRSA detection via ATP assay

detected in a short period of time ($\leq$5 min). This method holds considerable promise as a novel and rapid quantitative and qualitative bacterial detection, which will have a significant impact in controlling MRSA in hospitals, saving thousands of lives every year. By the same token, instead of using phage exogenous enzymes, the phages can be used to lyse their target bacteria and using the same ATP assay principal and light detection (Fig. 4.6) as discussed in greater detail in Chap. 3. On the other hand, the MRSA phage can be genetically engineered to carry a lux luciferase reporter gene from fireflies (FFlux) (Denyer et al. 1998). When these reporter phages are used to infect cultures of MRSA, the luciferase gene is expressed and (when a luciferin substrate is added) the infected cells emit light. This light can be readily detected within 30–40 min using either a luminometer or photographic film. Since the background level of light emission is extremely low, the sensitivity is high and relatively few numbers of cells can be detected in a short period of time. This phage luciferase reporter holds considerable promise as a novel rapid bacterial detection and also for drug susceptibility tests for MRSA.

4.9.2 MRSA Urinary Tract Infections (UTIs) Detection

In this section we will discuss a rapid detection protocol for bacterial urinary tract infections (UTIs). MRSA UTIs are increasing throughout the world and transmissions acquired in hospitals and care facilities are of particular concern. MRSA UTIs occur in both an endemic and epidemic fashion among patients undergoing urologic surgical procedures (Jones et al. 1999; Bentley et al. 2001; Araki et al. 2002; Mylotte et al. 2002; Pacio et al. 2003; Muder et al. 2006; Gould et al. 2009). MRSA bacteriuria occurs among long-term care patients as well and is significantly associated with urinary catheterization and antibiotic use (Coll et al. 1994). A MRSA UTI cannot be treated with traditional antibiotics and the prevalence of antibiotic resistant bacteria is a growing problem, especially for the elderly living in residential care facilities, as more strains of bacteria adapt to commonly used antibiotics. Efforts to limit the acquisition of MRSA by catheterized patients through appropriate infection-control measures and limitation of unnecessary antibiotic administration are warranted in long-term care facilities (Muder et al. 2006).

Furthermore, *Staphylococcus pseudintermedius* is a commensal and common opportunistic pathogen mainly causing infections of the integumentary system in dogs (Moodley et al. 2014). The CoPS methicillin-resistant *S. pseudintermedius* (MRSP) is an increasingly important cause of opportunistic infections in dogs and cats (Weese et al. 2013), and has emerged as a highly drug-resistant small animal veterinary pathogen (Lehner et al. 2014). Treatment options are often limited because of the high level of antimicrobial resistance (Weese et al. 2013). *S. pseudintermedius* cause UTIs in a dog (Rubin and Gaunt 2011). A dairy cattle herd with UTIs was reported (Yeruham et al. 2006). It was found that *E. coli* was the most frequent cause of UTIs. According to the report of Brownlee (2005) there is a link between a multidrug-resistant UTI strain of *E. coli* isolated from beef cattle and clusters of UTIs in women. If women can get UTIs from the food they eat, it is a major public health concern (Brownlee 2005). The diagnosis of infection in older adults is more challenging, yet early diagnosis and treatment in these patients are imperative because of the higher incidence of morbidity and mortality. MRSA and MRSP require a lengthy test routine that involves the laboratory and the patient population being tested, and for older adults, the process is more challenging. Quantitative urine cultures alone are not sufficient to detect infection in some patient populations unless lower colony counts are considered. Currently two methods are used for diagnosing UTIs:

i. *Urinalysis*:
This is the typical test that looks at physical colour, appearance and clarity and may also include a microscopic evaluation to look for bacteria and/or red and white blood cell counts. The tests may include chemical appearance (urine chemistry): a special stick (dipstick) tests for various substances in the urine. The stick contains little pads of chemicals that change colour when they come in contact with the substances of interest. However, this test alone will not confirm if someone has MRSA, MRSP or any other bacterial infection.

ii. *Urine culture and sensitivity*:
In most cases, the urine sample is obtained from a 'midstream clean catch' urine specimen. To do this, urine is collected in a sterile cup. The cup is then given to the doctor, who sends it to a lab for analysis. Sometimes, urine has to be collected directly from the bladder. To do this, the doctor inserts a sterile tube called a catheter into the bladder. Urine then drains from the bladder into a sterile container to minimize contamination and is transported to the laboratory in a timely fashion to limit independent bacterial growth. Testing, particularly

culturing should be done within 2 h of specimen collection; if not, the sample should be refrigerated. This test is recommended for patients whose characteristics and symptoms suggest complicated UTI or an indication for treatment of bacteriuria.

4.9.3 Phage Based UTI Diagnostic Kits

Phages allow for a rapid specific screen and their use is a more practical approach for reconfirming a bacterial culture. In this regard, we present a simple one step rapid diagnostic for urine screening to detect UTIs, as an alternative to the classic culture method. Nature supplies us with a corresponding phage for every bacterium ever detected. Using phage programming technology (this is discussed in greater detail in Chap. 1) creates a highly selective and sensitive phage master that is able to detect its corresponding bacteria quickly and accurately. This technology can also be adapted to detect virtually any bacteria in any situation in a short time.

4.9.3.1 Smart Media

Envelope contains plastic diagnostic pad, 1 ml and/or 100 µl sterile plastic pipette, 5 ml sterile plastic bottle, gloves, sterile transparent plastic chamber with lid and return envelope (Fig. 4.7).

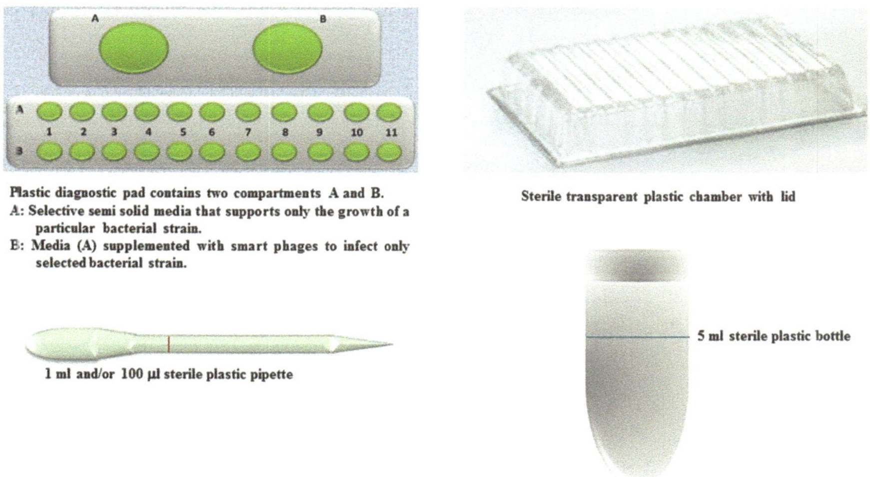

Fig. 4.7 A diagram showing examples of proposed rapid smart media to detect bacterial infections in humans or animals

Urine specimen collection should be done in a manner that minimizes contamination, for example, gloved hand-in/out of the urine flow for a dependent patient. Collect urine sample in the morning. Discard first few ml of urine and collect the rest in a sterile container to the mark (Fig. 4.8). For a resident with an intrauterine device, the catheter should be changed before specimen collection. In cases where we need to detect one bacteria for example MRSA, the below single test kit can be used.

Where we need to detect most common bacteria that are implicated in UTI's the below multiple test kit can be used (Fig. 4.9).

Place the single or multiple kit in a sterile plastic chamber and close the lid. Place the chamber in the envelope and send to a lab (Fig. 4.10).

Insert the kit in return envelope to lab for incubation and reading the results or incubate the kits in a safe place at room temperature (18–20 °C) for 18–24 h then share picture of kit (e.g. through a smart phone) or send to the physician.

Interpretation of results: The test can also show if the patient is infected with more than one bacterial strain. The compartments in row A contain many different selective semi solid media. Each medium will allow the growth of only some particular bacteria, for example in Fig. 4.11 there is bacterial growth in compartments number 1, 2, 6 and 9. Bacteria growing in their own selective media are: *E. coli*, *Klebsiella*, MRSA and *S. marcescens*, respectively. Row B contains the

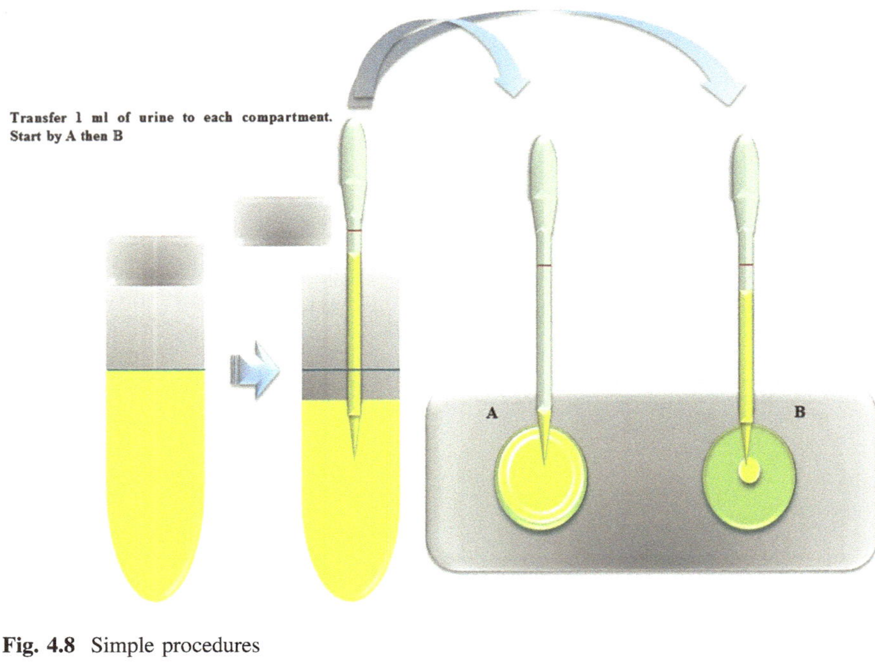

Fig. 4.8 Simple procedures

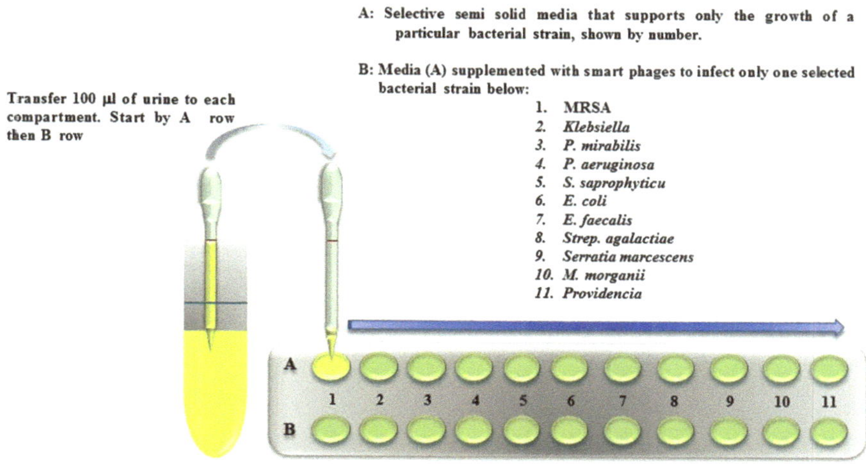

Fig. 4.9 Diagnosis of multiple pathogens

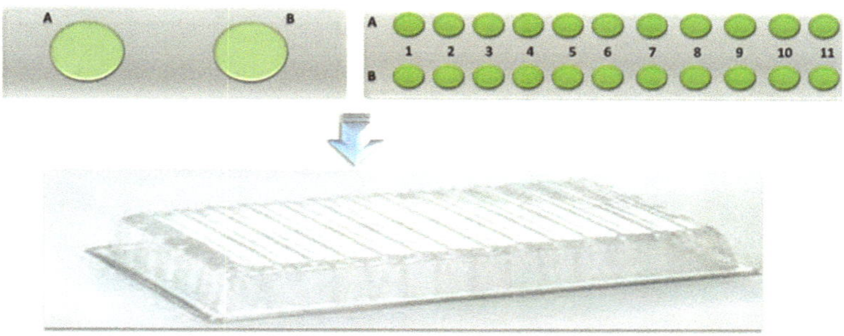

Fig. 4.10 Sterile plastic chamber

same media as in row A compartments but each media is supplemented with smart cocktail phages that will lyse with only their corresponding target bacteria. This test rapidly and precisely diagnoses the bacterial strains and also works as a re-confirmatory test for the results in row A.

This test can also help patients to properly manage the progress of their treatment and if and when bacteria resistant to antibiotics occurs, during or after finishing a course of antibiotics, patients can inform their physician and share the results, possibly even via a smart phone picture, in order to adapt the treatment swiftly to save the patient's life or reduce the cost of longer treatment and prolonged illness.

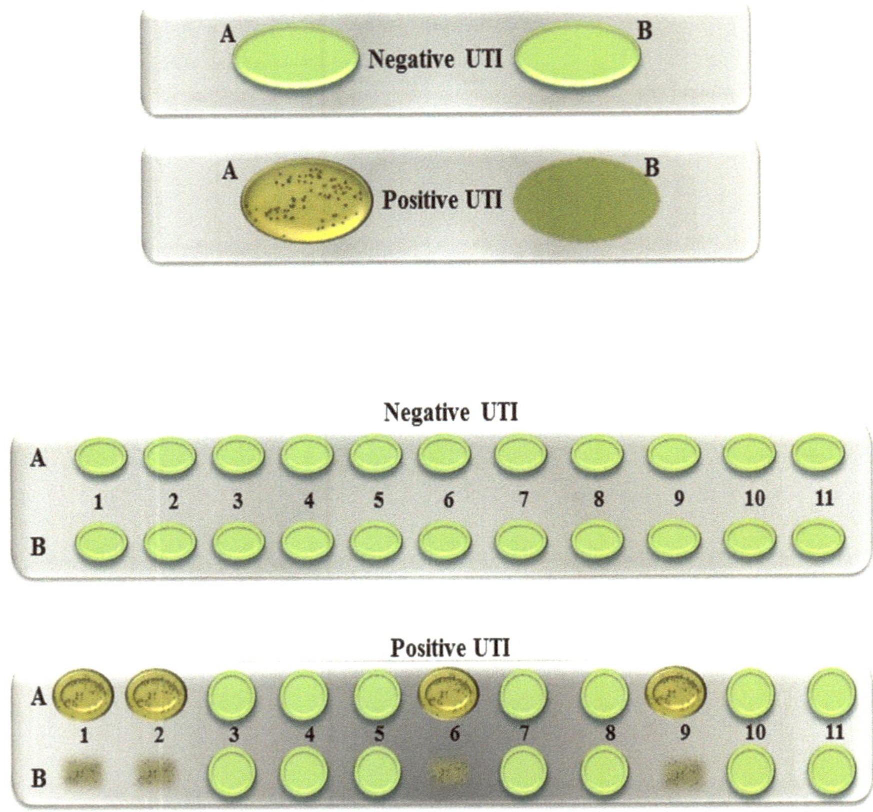

Fig. 4.11 Interpretation of results

4.9.3.2 Phage-Fluorescent Media

Plastic diagnostic pad contains 11 compartments with selective semi solid media that supports only the growth of a particular bacterial strain, supplemented with smart cocktail phages to each infect only one bacterial strain and Syber Safe or Eva Green (Non-hazardous fluorescent dyes) (Fig. 4.12).

Incubate for 1–2 h at room temperature or 37 °C. If the target bacteria are present in the urine sample, the phages will lyse them and bacterial DNA are released. Bacterial DNA stained with Syber Safe or Eva Green stain can be viewed using a blue light transilluminator such as Invitrogen's Safe Imager™ instrument, or a standard UV transilluminator, respectively. If there are no bacteria in the sample of urine, there will not be any DNA fragments present and thus no visualization of DNA. The total number of blue colonies is counted to evaluate the total number of colonies cfu ml^{-1} of the urine sample. Each experiment can be conducted in duplicate, and the average cloning efficiency can be determined.

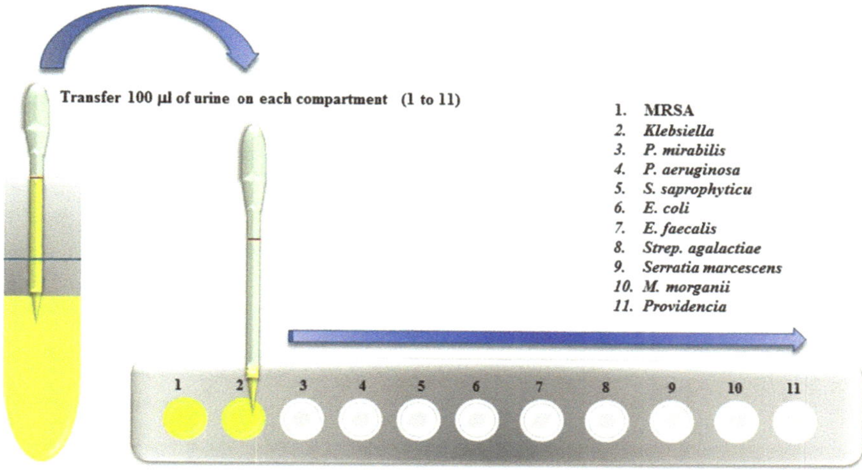

Fig. 4.12 Plastic diagnostic using phage and fluorescent media

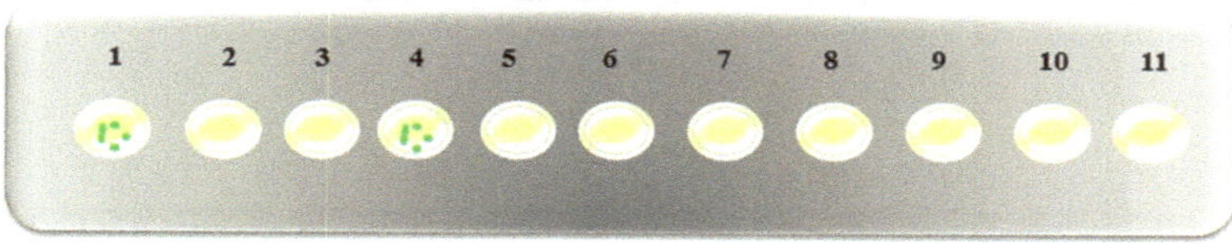

Fig. 4.13 Interpretation of results

Qualitative and quantitative UTIs can be measured. For example, in the below Fig. 4.13 there is bacterial growth in compartments number 1 and 4. Bacteria growing in their own selective media are: *E. coli* and *P. aeruginosa*, respectively. Since the media is supplemented with smart cocktail phages that will lyse only their corresponding target bacteria. This test rapidly and precisely diagnoses the bacterial strains and also works as a re-confirmatory test for the results in classical culture techniques.

4.9.3.3 Phage-Fluorescent Smart Card

The smart card contains programmed phages, which swiftly infect selected UTI bacterial strains, and Syber Safe or Eva Green fluorescent dyes. As illustrated the procedure to conduct the test in Fig. 4.14.

As described above in the phage-fluorescent media kit the phage has been developed that is lethal to for example MRSA and will cause bacterial cell lysis within 45 min. The bacterial DNA in bacterial lysate will react with Syber Safe or

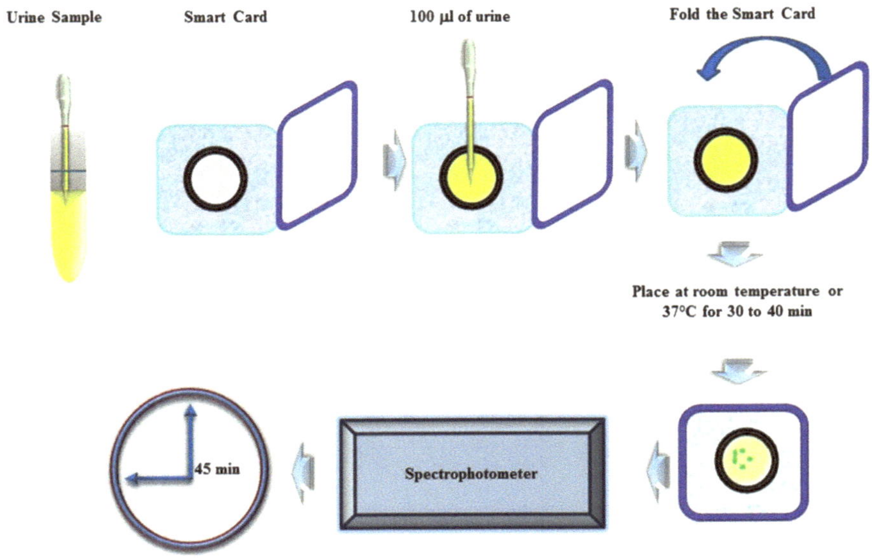

Fig. 4.14 Smart card

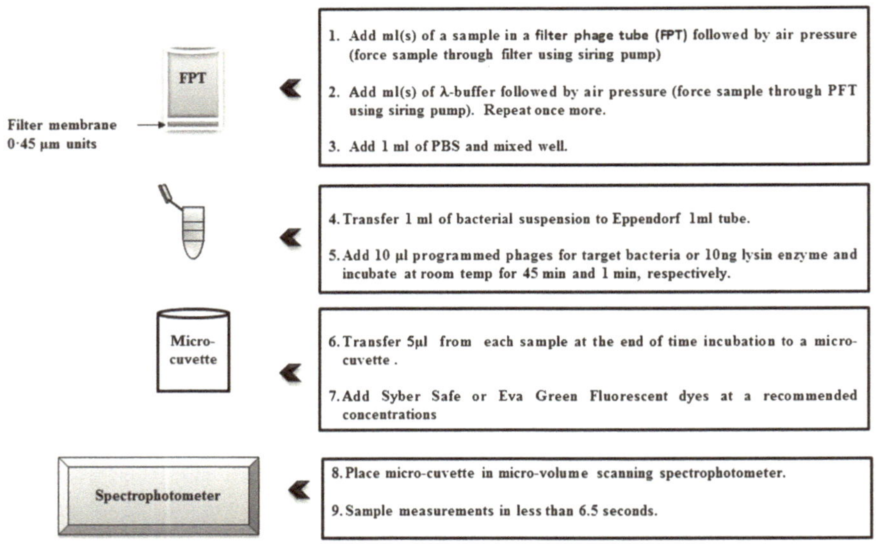

Fig. 4.15 Steps for qualitative and quantitative rapid (45 min or 5 min/real time) MRSA phage molecular diagnostic technique. *Note* This protocol can also be adapted for ATP as described above

Eva Green fluorescent dyes. To visualize the DNA, insert the card in scanning spectrophotometer (to be designed) to measure the blue-light transilluminator from the target bacterial DNA of UTI patient's urine. The density of the fluorescent light can be used to quantify the bacterial count per ml and the smart phage will identify the bacterial strain and species. We could slightly modify the above protocol in Fig. 4.14 by supplementing the card with MRSA phage lysin enzyme (Fig. 4.5) instead of the MRSA phage. In this regard the total bacterial detection from start to finish is around 5 min. This is obviously the most rapid qualitative and quantitative bacterial detection available and can save millions of lives and help to control antibiotic resistance in bacteria that cause common infections (e.g. UTI, pneumonia, bloodstream infections, etc.). In cases where using large volume ≥ 25 ml of sample (urine) the following protocol (Fig. 4.15) can be adopted as follows.

4.10 Conclusions and Perspective

S. aureus is one of the most common etiological factors of hospital- and community acquired infections. All humans are believed to be susceptible to *S. aureus* colonization; however, the intensity of symptoms may vary. In the past 25 years, *S. aureus* infections have increased in number, and the rise in incidence has been accompanied by a rise in antibiotic-resistant strains, in particular MRSA and, more recently, vancomycin-resistant strains. Newer strains of MRSA capable of causing infection in previously healthy persons have spread rapidly in the community. They have now entered and are being transmitted in healthcare settings, with the potential for increased morbidity and mortality from MRSA infections. The genetic backgrounds of these strains vary geographically but are distinct from those of MRSA strains that were established as health-care pathogens in the corresponding regions. This situation suggests that the new strains arose independently in different geographic areas via acquisition of MRSA circulating in the community (Gorwitz 2008).

MRSA can reciprocally affect human and animal populations. Direct transmission of MRSA between animal farms, community and hospitals has been documented. Contact with animal farms and in particular with pigs or cattle is considered a risk factor for ST398 or non-typable MRSA colonization. MRSA is well established in different pets, livestock and human populations, therefore elimination of the risk of contracting MRSA is impossible. MRSA in animals requires careful control methods to eliminate the emergence and dissemination of MRSA. Programmed bacteriophages are easy-to-use natural alternatives to chemical disinfectants which can safely be used in healthcare and farm or veterinary settings to disinfect equipment, walls, floors, cages, slaughter houses and so on. They can target specific species and prevent interspecies transmission.

Phages are naturally occurring viruses that are highly specific for their bacterial hosts which they infect. They can rapidly kill their host, amplifying themselves in the process. Phages are unaffected by antibiotic resistance and are able to disrupt bacterial biofilms. Biofilms are a major line of defense for MRSA, contributing to antibiotic resistance. Phages are able to penetrate MRSA biofilms and replicate locally to high levels, to produce strong local biocontrol effects. Phage technology can be used in many industries as a biocontrol, bioprocess, and bacteria diagnostic or for certain therapies. Wild isolates of phages able to target and kill MRSA strains and/or other staphylococci strains can be evaluated, programmed and compared to other gold standard protocols that are used in hospitals for MRSA disinfection. The phage 1×10^9 pfu ml^{-1} can be adapted into different forms such as a dry fog phage system, phage-based hand rubs, spray or wipes. These can decontaminate certain strains of MRSA, MRSP or any other medically important bacteria thereby preventing bacterial zoonoses or diseases transmitted from different objects and surfaces in healthcare settings, farms and foods. Another advantage of using phages as a biocontrol is that the phage can survive on the skin and continue killing MRSA for up to 24 h after use. This means that unlike conventional chemical antibacterials such as chlorhexidine gluconate, soaps or alcohol gels, phage based dispensers will keep killing MRSA, providing protection and peace of mind. Additionally, phage based systems can be used in animal and food production areas to replace or supplement antibiotics in feed or to kill environmental bacteria which adversely affect animals and/or our food supply.

One challenging problem remains, bacterial resistance against the 'lysogenic phages'. Bacteria can also develop resistance to phages and their lysins, however, when this occurs, unlike with antibiotics, a new phage matched to the resistant bacteria can be rapidly chosen or developed. Alternatively, bacterial resistance to phages can be controlled by inhibiting the resistance development. This can be achieved by using phages or their lysins which are programmed and formulated in a cocktail to be used in MRSA control or treatment. Since the current pace for isolating wild anti-MRSA phages can never keep up with the resistance race, an accelerated approach is needed to provide high throughput of new lytic anti-MRSA phages in a relatively short time. This can be achieved by applying the breakthrough technique of horizontal phage programming technology which does not employ genetic modification.

In conclusion, rapid bacterial detection is crucial to the control of MRSA or other medically important bacterial pathogens in animal related, or in healthcare settings. The disadvantages of PCR include complexity of use, protracted delays in reporting and that it cannot discriminate between dead and live bacteria. Phages and phage endolysin enzymes can be used to rapidly, qualitatively and quantitatively detect and identify bacterial pathogens in samples. Several innovative methods were proposed in this chapter to enable researchers and physicians to use the most effective method of phage-based diagnostic for MRSA with assay time ≤ 120 min in situ. Phage technology is environmentally friendly, can be used in cosmetics, disinfectants, bio-decontamination or elimination of biofilms, while replacing chemical disinfectants and detergents. Rapid MRSA diagnostic kits using phage technology can lead to tremendous innovative medical and pharmaceutical applications.

References

Abdalrahman LS, Wells H, Fakhr MK (2015) *Staphylococcus aureus* is more prevalent in retail beef livers than in pork and other beef cuts. Pathogens 4(2):182–198. doi:10.3390/pathogens4020182

Abdulamir AS, Jassim SAA, Abu Bakar F (2014) Novel approach of using a cocktail of designed bacteriophages against gut pathogenic *E. coli* for bacterial load biocontrol. Ann Clin Microbiol Antimicrob 13:39. doi:10.1186/s12941-014-0039-z

Abdulamir AS, Jassim SAA, Hafidh RR, Bakar FA (2015) The potential of bacteriophage cocktail in eliminating Methicillin-resistant *Staphylococcus aureus* biofilms in terms of different extracellular matrices expressed by PIA, ciaA-D and FnBPA genes. Ann Clin Microbiol Antimicrob 14:49. doi:10.1186/s12941-015-0106-0. http://www.ncbi.nlm.nih.gov/pubmed/26558683

Abreu AC, Tavares RR, Anabela Borges A et al (2013) Current and emergent strategies for disinfection of hospital environments. J Antimicrob Chemother 68(12):2718–2732. doi:10.1093/jac/dkt281

Aellen S, Que Y-A, Guignard B et al (2006) Detection of live and antibiotic-killed bacteria by quantitative real-time PCR of specific fragments of rRNA. Antimicrob Agents Chemother 50 (6):1913–1920. doi:10.1128/AAC.00869-05

Aldoori AA, Mahdii EF, Abbas AK, Jassim SAA (2015) Bacteriophage biocontrol rescues mice bacteremic of clinically isolated mastitis from dairy cows associated with methicillin-resistant *Staphyloccocus aureus*. Adv Microbiol 5:383–403. doi:10.4236/aim.2015.56040

Alisky J, Iczkowski K, Rapoport A, Troitsky N (1998) Bacteriophages show promise as antimicrobial agents. J Infect 36(1):5–15

Araki M, Kariyama R, Monden K et al (2002) Molecular epidemiological studies of *Staphylococcus aureus* in urinary tract infection. J Infect Chemother 8:168–174

Argudín MÁ, Mendoza MC, Rodicio MR (2010) Food poisoning and *Staphylococcus aureus* enterotoxins. Toxins (Basel) 2(7):1751–1773. doi:10.3390/toxins2071751

Armand-Lefevre L, Ruimy R, Andremont A (2005) Clonal comparison of *Staphylococcus aureus* isolates from healthy pig farmers, human controls, and pigs. Emerg Infect Dis 11:711–714

Bachrach G, Leizerovici-Zigmond M, Zlotkin A et al (2003) Bacteriophage isolation from human saliva. Lett Appl Microbiol 36:50–53

Bai J, Kim Y-T, Ryu S, Lee J-H (2016) Biocontrol and rapid detection of food-borne pathogens using bacteriophages and endolysins. Front Microbiol 7:article 474. doi:10.3389/fmicb.2016.00474

Baptiste K, Williams K, Willams N et al (2005) Methicillin-resistant staphylococci in companion animals. Emerg Infect Dis 11(12):1942–1944

Bentley DW, Bradley S, High K et al (2001) Practice guideline for evaluation of fever and infection in long-term care facilities. J Am Geriatr Soc 49:210–222

Berendt AE, Turnbull L, Spady D et al (2011) Three swipes and you're out: how many swipes are needed to decontaminate plastic with disposable wipes? Am J Infect Control 39(5):442–443. doi:10.1016/j.ajic.2010.08.014

Bergeron M, Dauwalder O, Gouy M et al (2011) Species identification of staphylococci by amplification and sequencing of the tuf gene compared to the gap gene and by matrix-assisted laser desorption timeof-flight mass spectrometry. Eur J Clin Microbiol Infect Dis 30:343–354

Bessonneau V, Clément M, Thomas O (2010) Can intensive use of alcohol-based hand rubs lead to passive alcoholization? Int J Environ Res Public Health 7(8):3038–3050. doi:10.3390/ijerph7083038

Bhalla A, Aron DC, Donske CJ (2007) *Staphylococcus aureus* intestinal colonization is associated with increased frequency of *S. aureus* on skin of hospitalized patients. BMC Infect Dis 7:105. doi:10.1186/1471-2334-7-105

Bhalla A, Pultz NJ, Gries DM et al (2004) Acquisition of nosocomial pathogens on hands after contact with environmental surfaces near hospitalized patients. Infect Control Hosp Epidemiol 25:164–167
Bhargava K, Wang X, Donabedian S et al (2011) Methicillin-Resistant *Staphylococcus aureus* in Retail Meat, Detroit, Michigan, USA. Emerg Infect Dis 17(6):1135–1137. doi:10.3201/eid1706.101095
Bhaskaran K, Hemalatha G, Sethumadhavan K (2014) Methicillin resistant *Staphylococcus aureus* carriage amongst students of the medical college. Int J Adv Res Biol Sci 1(7):188–192
Blok HEM, Trolestra A, Kamp-Hopmans TEM et al (2003) Role of health care workers in outbreaks of methicillin resistant *Staphylococcus aureus*: a 10-year evaluation from a Dutch university hospital. Infect Control Hosp Epidemiol 24(9):679–685
Blot S, Vandewoude K, Hoste E, Colardyn F (2002) Outcome and attributable mortality in critically Ill patients with bacteremia involving methicillin-susceptible and methicillin-resistant *Staphylococcus aureus*. Arch Intern Med 162:2229–2235
Boost MV, O'Donoghue MM, James A (2008) Prevalence of *Staphylococcus aureus* carriage among dogs and their owners. Epidemiol Infect 136(7):953–964
Boyce JM (2007) Environmental contamination makes an important contribution to hospital infection. J Hosp Infect 65(2):50–54. doi:10.1016/S0195-6701(07)60015-2
Boyce JM, Havill NL (2005) Nosocomial antibiotic-associated diarrhea associated with enterotoxin-producing strains of methicillin-resistant *Staphylococcus aureus*. Am J Gastroenterol 100(8):1828–1834. doi:10.1111/j.1572-0241.2005.41510.x
Boyce JM, Potter-Bynoe G, Chenevert C et al (1997) Environmental contamination due to methicillin-resistant *Staphylococcus aureus*: possible infection control implications. Infect Control Hosp Epidemiol 18(9):622–627
Boyle-Vavra S, Daum RS (2007) Community acquired methicillin-resistant *Staphylococcus aureus*: the role of Panton-Valentine leukocidin. Lab Invest 87:3–9
Brownlee C (2005) The Beef about UTIs. Food for thought. Biomedicine. Science News. 12 Jan 2005. https://www.sciencenews.org/blog/food-thought/beef-about-utis
Calfee DP, Salgado CD, Classen D et al (2008) Strategies to prevent transmission of methicillin-resistant *Staphylococcus aureus* in acute care hospitals. Infect Control Hosp Epidemiol 29(Suppl 1):S62–S80. doi:10.1086/591061
Carling PC, Briggs JL, Perkins J, Highlander D (2006) Improved cleaning of patient rooms using a new targeting method. Clin Infect Dis 42(3):385–388
Carling PC, Parry MF, Von Beheren SM (2008) Identifying opportunities to enhance environmental cleaning in 23 acute care hospitals. Infect Control Hosp Epidemiol 29:1–7
Carlton RM (1999) Phage therapy: past history and future prospects. Arch Immunol Ther Exp 47 (5):267–274
CDC (2007) MRSA: Methicillin-resistant *Staphylococcus aureus* in healthcare settings. Bulletin 14
Chah KF, Gómez-Sanz E, Nwanta JA et al (2014) Methicillin-resistant coagulase-negative staphylococci from healthy dogs in Nsukka, Nigeria. Braz J Microbiol 45(1):215–220
Chua KY, Monk IR, Lin Y-H et al (2014) Hyperexpression of α-hemolysin explains enhanced virulence of sequence type 93 community-associated methicillin-resistant *Staphylococcus aureus*. BMC Microbiol 14:31. doi:10.1186/1471-2180-14-31
Coll PP, Crabtree BF, O'Connor PJ, Klenzak S (1994) Clinical risk factors for methicillin-resistant *Staphylococcus aureus* bacteriuria in a skilled-care nursing home. Arch Fam Med 3:357–360
Cosgrove SE, Sakoulas G, Perencevich EN et al (2003) Comparison of mortality associated with methicillin-resistant and methicillin-susceptible *Staphylococcus aureus* bacteremia: a meta-analysis. Clin Infect Dis 36:53–59
Cosgrove SE, Qi Y, Kaye KS et al (2005) The impact of methicillin resistance in *Staphylococcus aureus* bacteremia on patient outcomes: mortality, length of stay, and hospital charges. Infect Control Hosp Epidemiol 26:166–174

Cotterill S, Evans R, Fraise AP (1996) An unusual source for an outbreak of methicillin-resistant *Staphylococcus aureus* on an intensive therapy unit. J Hosp Infect 32:207–216
Cui S, Li J, Hu C et al (2009) Isolation and characterization of methicillin-resistant *Staphylococcus aureus* from swine and workers in China. J Antimicrob Chemother 64:680–683
Dancer SJ (2009) The role of environmental cleaning in the control of hospital-acquired infection. J Hosp Infect 73:378–385
de Boer E, Zwartkruis-Nahuis JT, Wit B et al (2009) Prevalence of methicillin-resistant *Staphylococcus aureus* in meat. Int J Food Microbiol 134:52–56. doi:10.1016/j.ijfoodmicro.2008.12.007
de Neeling A, van den Broek M, Spalburg E et al (2007) High prevalence of methicillin resistant *Staphylococcus aureus* in pigs. Vet Microbiol 122:366–372
de San Nour, Gasarira ODM-F, Mendonca RD et al (2007) Controlled evaluation of the IDI-MRSA assay for detection of colonization by methicillin-resistant *Staphylococcus aureus* in diverse musculocutaneous specimens. J Clin Microbiol 45:1098–1101
Denyer SP, Jassim SAA, Stewart GSAB (1991) *In vivo* bioluminescence for studying the adhesion of bacteria. Biofouling 5(issue 1–2):125–132
Denyer SP, Jassim SAA, Fearon PS et al (1998) Genetically engineered reporter bacteria for the detection of bacteriophage. United States Patent 5723330. http://www.patentgenius.com/patent/5723330.html
Devriese LA, Van Damme LR, Famaree L (1972) Methicillin (cloxacillin) resistant *Staphylococcus aureus* strains isolated from bovine mastitis cases. Zentralbl Veterinarmed B 19:598–605
Dewaele I, De Man I, Stael A et al (2008) Methicillin-resistant *Staphylococcus aureus* (MRSA) in Belgian pig farms. In: American Society for Microbiology (ASM) conference on antimicrobial resistance in zoonotic bacteria and foodborne pathogens, Copenhagen
Diep BA, Otto M (2008) The role of virulence determinants in community-associated MRSA pathogenesis. Trends Microbiol 16:361–369
Doyle ME, Hartmann FA, Lee Wong AC (2012) Methicillin-resistant staphylococci: implications for our food supply? Anim Health Res Rev 13(2):157–180. doi:10.1017/S1466252312000187
Drees M, Snydman DR, Schmid CH et al (2008) Prior environmental contamination increases the risk of acquisition of vancomycin-resistant enterococci. Clin Infect Dis 46:678–685
Embleton ML, Nair SP, Heywood W et al (2005) Development of a novel targeting system for lethal photosensitization of antibiotic-resistant strains of *Staphylococcus aureus*. Antimicrob Agents Chemother 49(9):3690–3696
Enright MC, Robinson DA, Randle G et al (2002) The evolutionary history of methicillin-resistant *Staphylococcus aureus* (MRSA). PNAS 99(11):7687–7692. doi:10.1073/pnas.122108599
Fadeyi A, Adeboye MAN, Fowotade A et al (2010) Methicillin Resistant *Staphylococcus aureus* carriage amongst healthcare workers of the critical care units in a Nigerian hospital. Am J Infect Dis 6(1):18–23
Falagas ME, Thomaidis PC, Kotsantis IK et al (2011) Airborne hydrogen peroxide for disinfection of the hospital environment and infection control: a systematic review. J Hosp Infect 78:171–177
Favrin SJ, Jassim SAA, Griffiths MW (2001) Development and optimization of a novel immunomagnetic separation—bacteriophage assay for the detection of *Salmonella enterica* Serovar enteritidis in broth. Appl Environ Microbiol 67(1):217–224. doi:10.1128/AEM.67.1.217-224
Favrin SJ, Jassim SAA, Griffiths MW (2003) Application of a novel immunomagnetic separation-bacteriophage assay for the detection of *Salmonella enteritidis* and *Escherichia coli* O157:H7 in food. Int J Food Microbiol 85(1–2):63–71. doi:10.1016/S0168-1605(02)00483-X
Fenton M, Ross P, McAuliffe O et al (2010) Recombinant bacteriophage lysins as antibacterials. Bioeng Bugs 1(1):9–16
Fenton M, Keary R, McAuliffe O et al (2013) Bacteriophage-derived peptidase $CHAP_K$ eliminates and prevents staphylococcal Biofilms. Int J Microbiol 2013(ArticleI D625341):8. http://dx.doi.org/10.1155/2013/625341

Fischetti VA (2008) Bacteriophage lysins as effective antibacterials. Curr Opin Microbiol 11: 393–400

Fitzgerald JR (2012) Human origin for livestock-associated methicillin-resistant *Staphylococcus aureus*. mBio 3(2):e00082-12. doi:10.1128/mBio.00082-12

Frank U (2003) Prevention and control of methicillin-resistant *Staphylococcus aureus* (MRSA). In: Fluit AC, Schmitz F-J (eds) MRSA current perspectives. Caister Academic Press, England. ch12, p 317–336. ISBN: 0-954264-5-4

Garcia-Graells C, Antoine J, Larsen J et al (2012) Livestock veterinarians at high risk of acquiring methicillin-resistant *Staphylococcus aureus* ST398. Epidemiol Infect 140(03):383–389

Goodman ER, Platt R, Bass R et al (2008) Impact of an environmental cleaning intervention on the presence of methicillin-resistant *Staphylococcus aureus* and vancomycin-resistant enterococci on surfaces in intensive care unit rooms. Infect Control Hosp Epidemiol 29(7):593–599

Goodridge LD (2010) Designing phage therapeutics. Curr Pharm Biotechnol 11(1):15–27. doi:10.2174/138920110790725348

Górski A, Borysowski J, Miedzybrodzki R, Weber-Dabrowska B (2007) Bacteriophages in medicine. In: McGrath S, van Sinderen D (eds) Bacteriophage: genetics and microbiology. Caister Academic Press, Norfolk, UK, pp 125–158

Gorwitz RJ (2008) Understanding the success of methicillin-resistant *Staphylococcus aureus* strains causing epidemic disease in the community. J Infect Dis 197(2):179–182. doi:10.1086/523767

Gould FK, Brindle R, Chadwick PR et al (2009) Guidelines (2008) for the prophylaxis and treatment of methicillin-resistant *Staphylococcus aureus* (MRSA) infections in the United Kingdom. J Antimicrob Chemother. doi:10.1093/jac/dkp065

Graham P, Lin S, Larson E (2006) A U.S. population-based survey of *Staphylococcus aureus* colonization. Ann Intern Med 144:318–325

Graveland H, Wagenaar JA, Broekhuizen-Stins MJ et al (2008) Methicillin-resistant *Staphylococcus aureus* (MRSA) in veal calf farmers and veal calves in the Netherlands. In: American Society for Microbiology (ASM) conference on antimicrobial resistance in zoonotic bacteria and foodborne pathogens, Copenhagen. pp 62–63

Graveland H, Duim B, van Duijkeren E et al (2011a) Livestock-associated methicillin-resistant *Staphylococcus aureus* in animals and humans. Int J Med Microbiol 301:630–634

Graveland H, Wagenaar JA, Bergs K et al (2011b) Persistence of livestock associated MRSA CC398 in humans is dependent on intensity of animal contact. PLoS ONE 6:e16830

Gravet A, Rondeau M, Harf-Monteil C et al (1999) Predominant *Staphylococcus aureus* isolated from antibiotic-associated diarrhea is clinically relevant and produces enterotoxin A and the bicomponent toxin LukE-lukD. J Clin Microbiol 37(12):4012–4019

Griffeth GC, Morris DO, Abraham JL et al (2008) Screening for skin carriage of methicillin-resistant coagulase-positive staphylococci and *Staphylococcus schleiferi* in dogs with healthy and inflamed skin. Vet Dermatol 19(3):142–149. doi:10.1111/j.1365-3164.2008.00663.x

Griffith CJ, Cooper RA, Gilmore J et al (2000) An evaluation of hospital cleaning regimes and standards. J Hosp Infect 45(1):19–28

Gutiérrez D, Ruas-Madiedo P, Martínez B et al (2014) Effective removal of Staphylococcal biofilms by the endolysin LysH5. PLoS ONE 9(9):e107307. doi:10.1371/journal.pone.0107307

Hallin M, Denis O, Deplano A et al (2008) Evolutionary relationships between sporadic and epidemic strains of healthcare-associated methicillin-resistant *Staphylococcus aureus*. Clin Microbiol Infect 14(7):659–669. doi:10.1111/j.1469-0691.2008.02015.x

Hallin M, De Mendonça R, Denis O et al (2011) Diversity of accessory genome of human and livestock-associated ST398 methicillin resistant *Staphylococcus aureus* strains. Infect Genet Evol 11(2):290–299. doi:10.1016/j.meegid.2010.10.021

Hanselman BA, Kruth SA, Rousseau J, Weese JS (2009) Coagulase positive staphylococcal colonization of humans and their household pets. Can Vet J 50:954–958

Hanson BM, Dressler AE, Harper AL et al (2011) Prevalence of *Staphylococcus aureus* and methicillin-resistant *Staphylococcus aureus* (MRSA) on retail meat in Iowa. J Infect Public Health 4(4):169–174. doi:10.1016/j.jiph.2011.06.001

Hasman H, Moodley A, Guardabassi L et al (2010) Spa type distribution in *Staphylococcus aureus* originating from pigs, cattle and poultry. Vet Microbiol 141:326–331

Hayden MK, Bonten MJ, Blom DW et al (2006) Reduction in acquisition of vancomycin-resistant Enterococcus after enforcement of routine environmental cleaning measures. Clin Infect Dis 42:1552–1560

Hayden MK, Blom DW, Lyle EA (2008) Risk of hand or glove contamination after contact with patients colonized with vancomycin-resistant enterococcus or the colonized patients' environment. Infect Control Hosp Epidemiol 29:149–154

Hennekinne J-A, De Buyser M-L, Dragacci S (2012) *Staphylococcus aureus* and its food poisoning toxins: characterization and outbreak investigation. FEMS Microbiol Rev 36 (4):815–836. doi:10.1111/j.1574-6976.2011.00311.x

Hope CK, Packer S, Wilson M, Nair SP (2009) The inability of a bacteriophage to infect *Staphylococcus aureus* does not prevent it from specifically delivering a photosensitizer to the bacterium enabling its lethal photosensitization. J Antimicrob Chemother 64(1):59–61. doi:10.1093/jac/dkp157

Huang SS, Datta R, Platt R (2006) Risk of acquiring antibiotic-resistant bacteria from prior room occupants. Arch Intern Med 166(18):1945–1951. doi:10.1001/archinte.166.18.1945

Huber H, Koller S, Giezendanner N et al (2009) Methicillin-resistant *Staphylococcus aureus* (MRSA) in livestock animals and foods of animal origin in Switzerland. In: American Society for Microbiology (ASM)-ESCMID conference on methicillin-resistant Staphylococci in animals, London

Huijsdens X, van Dijke B, Spalburg E et al (2006) Community acquired MRSA and pig-farming. Ann Clin Microbiol Antimicrob 5:26. doi:10.1186/1476-0711-5-26

Inamatsu T, Ooshima H, Masuda Y (1992) Clinical spectrum of antibiotic associated enterocolitis due to methicillin resistant *Staphylococcus aureus*. Nihon Rinsho 50(5):1087–1092

Institute for Healthcare Improvement (2014) Reducing MRSA infections: staying one step ahead. http://www.ihi.org/resources/Pages/ImprovementStories/ReducingMRSAInfectionsStayingOneStepAhead.aspx

Jain A, Agarwal J, Bansal S (2004) Prevalence of methicillin-resistant, coagulase-negative staphylococci in neonatal intensive care units: findings from a tertiary care hospital in India. J Med Microbiol 53:941–944. doi:10.1099/jmm.0.45565-0

Jassim SAA (1989) Aspects of staphylococcal growth, haemolysis and phagocytosis. Ph.D. Thesis. University of Loughborough, UK. ASIN: B001A1RT34

Jassim SAA, Griffiths MW (2007) Evaluation of a rapid microbial detection method via phage lytic amplification assay coupled with Live/Dead fluorochromic stains. Lett Appl Microbiol 44 (6):673–678. doi:10.1111/j.1472-765X.2007.02115.x

Jassim SAA, Limoges RG (2014) Natural solution to antibiotic resistance: bacteriophages 'The Living Drugs'. World J Microbiol Biotechnol 30(8):2153–2170. doi:10.1007/s11274-014-1655-7

Jassim SAA, Salt WG, Stretton RJ (1989) *In vitro* studies of haemolysis by some staphylococci grown in chemically defined media. J Appl Bacteriol 67(5):511–520

Jassim SAA, Ellison A, Denyer SP, Stewart GSAB (1990) *In vivo* bioluminescence: a cellular reporter for research and industry. J Biolumin Chemilumin 5(2):115–122

Jassim SAA, Akoush S, Griffiths MW (1996) Rapid detection using thermal change to monitor infection by host specific bacteriophage. IUFOST Meeting Food Associated Pathogens Uppsala, Sweden, May 1996

Jassim SAA, Abdulamir AS, Abu Bakar F (2010a) Phage-based limulus amoebocyte lysate assay for rapid detection of bacteria. WO2011/098820A1. http://www.lens.org/images/patent/WO/2011098820/A1/WO_2011_098820_A1.pdf

Jassim SAA, Abdulamir AS, Abu Bakar F (2010b) Methods for bacteriophage design. WIPO Patent Application WO2010/064044 A1. http://www.sumobrain.com/patents/wipo/Methods-bacteriophage-design/WO2010064044A1.pdf

Jassim SAA, Abdulamir AS, Abu Bakar F (2012) Novel phage-based bio-processing of pathogenic *Escherichia coli* and its biofilms. World J Microbiol Biotechnol 28(1):47–60. doi:10.1007/s11274-011-0791-6

Jevons MP (1961) Celbenin: resistant Staphylococci. Br Med J 1(5219):124–125

Jończyk E, Kłak M, Międzybrodzki R, Górski A (2011) The influence of external factors on bacteriophages—review. Folia Microbiol 56(3):191–200. doi:10.1007/s12223-011-0039-8

Jones JW, Carter A, Ewings P, O'Boyle PJ (1999) An MRSA outbreak in a urology ward and its association with Nd:YAG coagulation laser treatment of the prostate. J Hosp Infect 41:39–44

Jones TF, Kellum ME, Porter SS et al (2002) An outbreak of community-acquired foodborne illness caused by methicillin-resistant *Staphylococcus aureus*. Emerg Infect Dis 8:82–84

Johnson AP, Pearson A, Duckworth G (2005) Surveillance and epidemiology of MRSA bacteraemia in the UK. J Antimicrob Chemother 56:455–462

Joshi SG, Paff M, Friedman G et al (2010) Control of methicillin-resistant *Staphylococcus aureus* in planktonic form and biofilms: a biocidal efficacy study of nonthermal dielectric-barrier discharge plasma. Am J Infect Control 38:293–301

Juhász-Kaszanyitzky E, Jánosi S, Somogyi P et al (2007) MRSA transmission between cows and humans. Emerg Infect Dis 13(4):630–632

Kadariya J, Smith TC, Thapaliya D (2014) *Staphylococcus aureus* and staphylococcal food-borne disease: an ongoing challenge in public health. BioMed Res Int 2014:827965. doi:10.1155/2014/827965

Kaur S, Harjai K, Chhibber S (2012) Methicillin-resistant *Staphylococcus aureus* phage plaque size enhancement using sublethal concentrations of antibiotics. Appl Environ Microbiol 78 (23):8227–8233. doi:10.1128/AEM.02371-12

Kaźmierczak Z, Górski A, Dąbrowska K (2014) Facing antibiotic resistance: *Staphylococcus aureus* phages as a medical tool. Viruses 6:2551–2570. doi:10.3390/v6072551

Kelly D, McAuliffe O, Ross RP, Coffey A (2012) Prevention of *Staphylococcus aureus* biofilm formation and reduction in established biofilm density using combination of phage K and modified derivatives. Lett Appl Microbiol 54(4):286–291

Khanna T, Friendship R, Dewey C, Weese JS (2007) Methicillin resistant *Staphylococcus aureus* colonization in pigs and pig farmers. Vet Microbiol 128:298–303

Klein E, Smith DL, Laxminarayan R (2007) Hospitalizations and deaths caused by methicillin-resistant *Staphylococcus aureus*, United States, 1999–2005. Emerg Infect Dis 13:1840–1846

Klevens RM, Morrison MA, Nadle J et al (2007) Invasive methicillin-resistant *Staphylococcus aureus* infections in the United States. JAMA 298:1763–1771

Kluytmans J, van Leeuwen W, Goessens W et al (1995) Food-initiated outbreak of methicillin-resistant *Staphylococcus aureus* analyzed by pheno- and genotyping. J Clin Microbiol 33:1121–1128

Knudsen AM, Rosdahl VT (1991) The decline of methicillin resistance among Danish *Staphylococcus aureus* strains. Infect Control Hosp Epidemiol 12:83–88

Köck R, Becker K, Cookson B et al (2010) Methicillin-resistant *Staphylococcus aureus* (MRSA): burden of disease and control challenges in Europe. Euro Surveill 15(41):19688. http://www.eurosurveillance.org/ViewArticle.aspx?ArticleId=19688A

Köck R, Schaumburg F, Mellmann A et al (2013) Livestock-associated methicillin-resistant *Staphylococcus aureus* (MRSA) as causes of human infection and colonization in Germany. PLoS ONE 8(2):e55040. doi:10.1371/journal.pone.0055040

Kodama T, Santo T, Yokoyama T et al (1997) Postoperative enteritis caused by methicillin-resistant *Staphylococcus aureus*. Surg Today 27(9):816–825. doi:10.1007/BF02385272

Kramer A, Kampf G (2007) Hand rub-associated fire incidents during 25038 hospital-years in Germany. Infect Control Hosp Epidemiol 28(6):745–746

Kramer A, Schwebke I, Kampf G (2006) How long do nosocomial pathogens persist on inanimate surfaces? A systematic review. BMC Infect Dis 6:130. doi:10.1186/1471-2334-6-130
Krylov VN (2001) Phagotherapy in terms of bacteriophage genetics: hopes, perspectives, safety, limitations. Genetika 37(7):869–887
Kumar P, Shukla I, Varshney S (2011) Nasal screening of healthcare workers for nasal carriage of coagulase positive MRSA and prevalence of nasal colonization with *Staphylococcus aureus*. Biol Med 27(1):62–64
Kutateladze M, Adamia R (2010) Bacteriophages as potential new therapeutics to replace or supplement antibiotics. Trends Biotechnol 28(12):591–595
Kutter E, De Vos D, Gvasalia G et al (2010) Phage therapy in clinical practice: treatment of human infections. Curr Pharm Biotechnol 11(1):69–86
Kwon NH, Park KT, Moon JS et al (2005) Staphylococcal cassette chromosome MEC (SCCMEC) characterization and molecular analysis for methicillin-resistant *Staphylococcus aureus* and novel SCCMEC subtype IVG isolated from bovine milk in Korea. J Antimicrob Chemother 56:624–632
Labandeira-Rey M, Couzon F, Boisset S et al (2007) *Staphylococcus aureus* Panton-Valentine leukocidin causes necrotizing pneumonia. Science 315(5815):1130–1135. doi:10.1126/science.1137165
Lee J (2003) Methicillin (oxacillin)-resistant *Staphylococcus aureus* strains isolated from major food animals and their potential transmission to humans. Appl Environ Microbiol 69:6489–6494
Lee J (2006) Occurrence of methicillin-resistant *Staphylococcus aureus* strains from cattle and chicken, and analyses of their mecA, mecR1 and mecI genes. Vet Microbiol 114:155–159
Lehner G, Linek M, Bond R et al (2014) Case-control risk factor study of methicillin-resistant *Staphylococcus pseudintermedius* (MRSP) infection in dogs and cats in Germany. Vet Microbiol 168(1):154–160. doi:10.1016/j.vetmic.2013.10.023
Leonard F, Abbott Y, Rossney A et al (2006) Methicillin-resistant *Staphylococcus aureus* isolated from a veterinary surgeon and five dogs in one practice. Vet Rec 158(5):155–159
Lim SK, Nam HM, Park HJ et al (2010) Prevalence and characterization of methicillin-resistant *Staphylococcus aureus* in raw meat in Korea. J Microbiol Biotechnol 20:775–778
Lin J, Yeh KS, Liu HT, Lin JH (2009) *Staphylococcus aureus* isolated from pork and chicken carcasses in Taiwan: prevalence and antimicrobial susceptibility. J Food Prot 72:608–611
Lopez P-J, Ron O, Parthasarathy P et al (2009) Bacterial counts from hospital doctors' ties are higher than those from shirts. Am J Infect Control 37:79–80
Lowy FD (1998) *Staphylococcus aureus* infections. N Engl J Med 339:520–532. doi:10.1056/NEJM199808203390806
Lozano C, López M, Gómez-Sanz E et al (2009) Detection of methicillin-resistant *Staphylococcus aureus* ST398 in food samples of animal origin in Spain. J Antimicrob Chemother 64:1325–1326
Ma XX, Sun DD, Wang S et al (2011) Nasal carriage of methicillin-resistant *Staphylococcus aureus* among preclinical medical students: epidemiologic and molecular characteristics of methicillin-resistant *S. aureus* clones. Diagn Microbiol Infect Dis 70:22–30
Malhotra-Kumar S, Haccuria K, Michiels M et al (2008) Current trends in rapid diagnostics for methicillin-resistant *Staphylococcus aureus* and glycopeptide-resistant enterococcus species. J Clin Microbiol 46(5):1577–1587. doi:10.1128/JCM.00326-08
Malik RE, Cooper RA, Griffith CJ (2003) Use of audit tools to evaluate the efficacy of cleaning systems in hospitals. Am J Infect Control 31(3):181–187
Malik S, Peng H, Barton MD (2005) Antibiotic resistance in staphylococci associated with cats and dogs. J Appl Microbiol 99(6):1283–1293
Marler B (2010) About MRSA (Methicillin-resistant *Staphylococcus aureus*). Food Poison J. http://www.foodpoisonjournal.com/food-poisoning-information/about-mrsa-methicillinresistant-staphylococcus-aureus/
Martinez JA, Ruthazer R, Hansjosten K et al (2003) Role of environmental contamination as a risk factor for acquisition of vancomycin-resistant Enterococci in patients treated in a medical intensive care unit. Arch Intern Med 163(16):1905–1912

May L, Klein EY, Rothman RE, Laxminarayanc R (2014) Trends in antibiotic resistance in coagulase-negative staphylococci in the United States, 1999 to 2012. Antimicrob Agents Chemother 58(3):1404–1409

McAnoy AM (2006) Vaporous decontamination methods: potential uses and research priorities for chemical and biological contamination control. Report number: DSTO-GD-0465. Publisher Human Protection and Performance Division, DSTO Defence Science and Technology Organisation, Commonwealth of Australia. http://dspace.dsto.defence.gov.au/dspace/handle/1947/3415

McCarthy AJ, Witney AA, Gould KA et al (2011) The distribution of mobile genetic elements (MGEs) in MRSA CC398 is associated with both host and country. Genome Biol Evol 3:1164–1174. doi:10.1093/gbe/evr092

McIntyre L, Jassim SAA, Griffiths MW (1996) Development of a bacteriophage-mediated ATP bioluminescence detection system for *Listeria monocytogenes*. Presented at 83rd Annual Meeting of IAMFES, Seattle, WA, June 30–July 3, 1996, p 70

McNamee PT, Smyth JA (2000) Bacterial chondronecrosis with osteomyelitis (femoral head necrosis) of broiler chickens. Avian Pathol 29:253–270. doi:10.1080/03079450075004 7243

MDH (Minnesota Department of Health) (2004) Community-associated methicillin resistant Staphylococcus aureus in Minnesota. Dis Control News 32:61–72

Moodley A, Damborg P, Nielsen SS (2014) Antimicrobial resistance in methicillin susceptible and methicillin resistant *Staphylococcus pseudintermedius* of canine origin: literature review from 1980 to 2013. Vet Microbiol 171(3–4):337–341. doi:10.1016/j.vetmic.2014.02.008

Moon JS, Lee AR, Kang HM et al (2007) Phenotypic and genetic antibiogram of methicillin resistant staphylococci isolated from bovine mastitis in Korea. J Dairy Sci 90(3):1176–1185

Moran GJ, Krishnadasan A, Gorwitz RJ et al (2006) Methicillin-resistant *S. aureus* infections among patients in the emergency department. N Engl J Med 355(7):666–674

Morgan DJ, Rogawski E, Thom KA et al (2012) Transfer of multidrug-resistant bacteria to healthcare workers' gloves and gowns after patient contact increases with environmental contamination. Crit Care Med 40(4):1045–1051. doi:10.1097/CCM.0b013e31823bc7c8

Morioka I, Takahashi N, Kitajima H (2014) Prevalence of MRSA colonization in Japanese neonatal care unit patients in 2011. Pediatr Int 56(2):211–214. doi:10.1111/ped.12232

Morris D, Mauldin E, O'Shea K et al (2006) Clinical, microbiological, and molecular characterization of methicillin-resistant *Staphylococcus aureus* infections of cats. Am J Vet Res 67(8):1421–1425

Muder RR, Brennen C, Rihs JD et al (2006) Isolation of *Staphylococcus aureus* from the urinary tract: association of isolation with symptomatic urinary tract infection and subsequent staphylococcal bacteremia. Clin Infect Dis 42(1):46–50. doi:10.1086/498518

Muto CA, Jernigan JA, Ostrowsky BE et al (2003) SHEA guideline for preventing nosocomial transmission of multidrug-resistant strains of *Staphylococcus aureus* and enterococcus. Infect Control Hosp Epidemiol 24(5):362–386

Mylotte JM, Tayara A, Goodnough S (2002) Epidemiology of bloodstream infection in nursing home residents: evaluation in a large cohort from multiple homes. Clin Infect Dis 35:1484–1490

Nakamura S, Yang CS, Sakon N et al (2009) Direct metagenomic detection of viral pathogens in nasal and fecal specimens using an unbiased high-throughput sequencing approach. PLoS ONE 4:e4219

Nemati M, Hermans K, Lipinska U et al (2008) Antimicrobial resistance of old and recent *Staphylococcus aureus* isolates from poultry: first detection of livestock-associated methicillin-resistant strain ST398. Antimicrob Agents Chemother 52:3817–3819

Nienhoff U, Kadlec K, Chaberny IF et al (2009) Transmission of methicillin-resistant *Staphylococcus aureus* strains between humans and dogs: two case reports. J Antimicrob Chemother 64(3):660–662

Ogawa Y, Saraya T, Koide T et al (2014) Methicillin-resistant *Staphylococcus aureus* enterocolitis sequentially complicated with septic arthritis: a case report and review of the literature. BMC Res Notes 7:21. doi:10.1186/1756-0500-7-21

Okuyama Y, Yoshida N (2012) Staphylococcal food poisoning and MRSA enterocolitis. Nihon Rinsho 70(8):1362–1365

Noskin GA, Rubin RJ, Schentag JJ et al (2005) The burden of *Staphylococcus aureus* infections on hospitals in the United States: an analysis of the 2000 and 2001 nationwide inpatient sample database. Arch Intern Med 165:1756–1761

O'Flaherty S, Ross RP, Meaney W et al (2005) Potential of the polyvalent anti-Staphylococcus bacteriophage K for control of antibiotic-resistant staphylococci from hospitals. App environ microbiol 71:1836–1842. doi:10.1128/AEM.71.4.1836-1842.2005

Olson M, Ceri H, Morck D et al (2002) Biofilm bacteria: formation and comparative susceptibility to antibiotics. Can J Vet Res 66:86–92

Otter JA, Vickery K, Walker JT et al (2014) Surface-attached cells, biofilms and biocide susceptibility: implications for hospital cleaning and disinfection. J Hosp Infect. doi:10.1016/j.jhin.2014.09.008

Pacio GA, Visintainer P, Maguire G et al (2003) Natural history of colonization with vancomycin-resistant enterococci, methicillin-resistant *Staphylococcus aureus*, and resistant gram-negative bacilli among ling-term-care facility residents. Infect Control Hosp Epidemiol 24:246–250

Page K, Wilson M, Parkin IP (2009) Antimicrobial surfaces and their potential in reducing the role of the inanimate environment in the incidence of hospital-acquired infections. J Mater Chem 19:3819–3831. doi:10.1039/B818698G

Palavecino E (2004) Community-acquired methicillin-resistant *Staphylococcus aureus* infections. Clin Lab Med 24(2):403–418

Palavecino EL (2014) Clinical, epidemiologic, and laboratory aspects of methicillin-resistant *Staphylococcus aureus* infections. Meth mol boil 1085:1–24

Pastagia M, Euler C, Chahales P et al (2011) A novel chimeric lysin shows superiority to mupirocin for skin decolonization of methicillin-resistant and -sensitive *Staphylococcus aureus* strains. Antimicrob Agents Chemother 55:738–744

Pereira V, Lopes C, Castro A et al (2009) Characterization for enterotoxin production, virulence factors, and antibiotic susceptibility of *Staphylococcus aureus* isolates from various foods in Portugal. Food Microbiol 26(3):278–282

Persoons D, Van Hoorebeke S, Hermans K et al (2009) Methicillin-resistant *Staphylococcus aureus* in poultry. Emerging Infect Dis 15:452–453

Peters PJ, Brooks JT, McAllister SK et al (2013) Methicillin-resistant *Staphylococcus aureus* colonization of the groin and risk for clinical infection among HIV-infected adults. Emerg Infect Dis 19(4):623–629. doi:10.3201/eid1904.121353

Price LB, Steggerb M, Hasman H et al (2012) *Staphylococcus aureus* CC398: host adaptation and emergence of methicillin resistance in livestock. mBio 3(1):e00305-11

Pu S, Han F, Ge B (2009) Isolation and characterization of methicillin-resistant *Staphylococcus aureus* strains from Louisiana retail meats. Appl Environ Microbiol 75:265–267. doi:10.1128/AEM.01110-08

Rashel M, Uchiyama J, Ujihara T et al (2007) Efficient elimination of multidrug-resistant *Staphylococcus aureus* by cloned lysin derived from bacteriophage phi MR11. J Infect Dis 196 (8):1237–1247

Rasigade JP, Laurent F, Hubert P et al (2010) Lethal necrotizing pneumonia caused by an ST398 *Staphylococcus aureus* strain. Emerg Infect Dis 16(8):1330. doi:10.3201/eid1608.100317

Rastogi V, Pragya, Verma N et al (2016) An overview on bacteriophages: a natural nanostructured antibacterial agent. Curr Drug Deliv 13(8) (E-pub ahead of print)

Ray AJ, Pultz NJ, Bhalla A et al (2003) Coexistence of vancomycin-resistant enterococci and *Staphylococcus aureus* in the intestinal tracts of hospitalized patients. Clin Infect Dis 37:875–881. doi:10.1086/377451

Rubin JE, Gaunt MC (2011) Urinary tract infection caused by methicillin-resistant *Staphylococcus pseudintermedius* in a dog. Can Vet J 52(2):162–164

Sass P, Bierbaum G (2007) Lytic activity of recombinant bacteriophage 11 and 12 endolysins on whole cells and biofilms of *Staphylococcus aureus*. Appl Environ Microbiol 73(1):347–352

Schmelcher M, Loessner MJ (2016) Bacteriophage endolysins: applications for food safety. Curr Opin Biotechnol 37:76–87. doi:10.1016/j.copbio.2015.10.005

Seaman A (2013) Hospital-acquired MRSA infection rates falling: CDC, Reuters. http://www.reuters.com/article/us-hospital-mrsa-idUSBRE98F0X920130916

Sergio D, Koh T, Hsu L et al (2007) Investigation of methicillin-resistant *Staphylococcus aureus* in pigs used for research. J Med Microbiol 56:1107–1109

Schelin J, Wallin-Carlquist N, Thorup Cohn M et al (2011) The formation of *Staphylococcus aureus* enterotoxin in food environments and advances in risk assessment. Virulence 2(6):580–592. doi:10.4161/viru.2.6.18122

Schentag JJ, Hyatt J, Carr J (1998) Genesis of methicillin-resistant *Staphylococcus aureus* (MRSA). How treatment of MRSA infections has selected for vancomycin-resistant *Enterococcus faecium* and the importance of antibiotic management and infection control. Clin Infect Dis 26:1204–1214

Schijffelen MJ, Boel CH, van Strijp JA, Fluit AC (2010) Whole genome analysis of a livestock-associated methicillin-resistant *Staphylococcus aureus* ST398 isolate from a case of human endocarditis. BMC Genom 11:376. doi:10.1186/1471-2164-11-376

Schwarz S, Kadlec K, Strommenger B (2007) Methicillin-resistant *Staphylococcus aureus* and staphylococcus pseudintermedius detected in the BFT-germvet monitoring programme 2004-2006 in Germany. J Antimicrob Chemother 61:282–285

Shepheard MA, Fleming VM, Connor TR et al (2013) Historical zoonoses and other changes in host tropism of *Staphylococcus aureus*, identified by phylogenetic analysis of a population dataset. PLoS ONE 8:e62369. doi:10.1371/journal.pone.0062369

Shlaes DM, Gerding DN, John JF (1997) Society for healthcare epidemiology of America and infectious diseases society of America joint committee on the prevention of antimicrobial resistance: guidelines for the prevention of antimicrobial resistance in hospitals. Infect Control Hosp Epidemiol 18:275–2791

Siegel JD, Rhinehart E, Jackson M et al (2007) Management of multidrug-resistant organisms in health care settings, 2006. Am J Infect Control 35(10 Suppl 2):S165–S193

Simões M, Simões LC, Vieira MJ (2010) A review of current and emergent biofilm control strategies. Food Sci Technol 43(4):573–583

Simpson WJ, Fernandez JL, Hammond JRM et al (1990) A Highly sensitive assay for adenosine triphosphate employing an improved firefly luciferase reagent. Lett Appl Microbiol 11(4):208–210. doi:10.1111/j.1472-765X.1990.tb00162.x

Smith TC, Male MJ, Harper AL et al (2008a) Isolation of methicillin-resistant *Staphylococcus aureus* (MRSA) from swine in the midwestern United States. In: International conference on emerging infectious diseases, Atlanta

Smith TC, Male MJ, Harper AL et al (2008b) Methicillin-resistant *Staphylococcus aureus* (MRSA) strain ST398 is present in Midwestern US swine and swine workers. PLoS ONE 4: e4258

Soejima T, Iida K, Qin T et al (2008) Method to detect only live bacteria during PCR amplification. J Clin Microbiol 46(7):2305–2313. doi:10.1128/JCM.02171-07

Son JS, Lee SJ, Jun SY et al (2010) Antibacterial and biofilm removal activity of a podoviridae *Staphylococcus aureus* bacteriophage SAP-2 and a derived recombinant cell-wall-degrading enzyme. Appl Microbiol Biotechnol 86(5):1439–1449

Squier C, Rihs JD, Risa KJ et al (2002) *Staphylococcus aureus* rectal carriage and its association with infections in patients in a surgical intensive care unit and a liver transplant unit. Infect Control Hosp Epidemiol 23(9):495–501. doi:10.1086/502095

Stewart GSAB, Jassim SAA, Denyer SP et al (1998) The specific and sensitive detection of bacterial pathogens within 4 h using bacteriophage amplification. J Appl Bacteriol 84(5):777–783. doi:10.1046/j.1365-2672.1998.00408.x

Stürenburg E (2009) Rapid detection of methicillin-resistant *Staphylococcus aureus* directly from clinical samples: methods, effectiveness and cost considerations. Ger Med Sci 7:Doc06. doi:10.3205/000065

Tacconelli E, De Angelis G, Cataldo MA et al (2008) Does antibiotic exposure increase the risk of methicillin-resistant *Staphylococcus aureus* (MRSA) isolation? A systematic review and meta-analysis. J Antimicrob Chemother 61:26–38

Takesue Y, Yokoyama T, Kodama T et al (1991) Toxin involvement in methicillin-resistant *Staphylococcus aureus* enteritis in gastroenterological surgery. Gastroenterol Jpn 26(6): 716–720

Talon D (1999) The role of the hospital environment in the epidemiology of multi-resistant bacteria. J Hosp Infect 43:13–17

Ternes YM, Lamaro-Cardoso J, André MCP et al (2013) Molecular epidemiology of coagulase-negative Staphylococcus carriage in neonates admitted to an intensive care unit in Brazil. BMC Infect Dis 13:572. doi:10.1186/1471-2334-13-572

Tomlin J, Pead MJ, Lloyd DH et al (1999) Methicillin-resistant *Staphylococcus aureus* infections in 11 dogs. Vet Rec 144(3):60–64. doi:10.1136/vr.144.3.60

Toté K, Berghe DV, Deschacht M et al (2009) Inhibitory efficacy of various antibiotics on matrix and viable mass of *Staphylococcus aureus* and *Pseudomonas aeruginosa* biofilms. Int J Antimicrob Agents 33:525–531

USDA-FSIS (2014) Pre-harvest management controls and intervention options for reducing Shiga toxin-producing *Escherichia coli* shedding in cattle: an overview of current research. http://www.fsis.usda.gov/wps/wcm/connect/d5314cc7-1ef7-4586-bca2-f2ed86d9532f/Reducing-Ecoli-Shedding-in-Cattle.pdf?MOD=AJPERES

van Cleef BA, Graveland H, Haenen APJ et al (2011) Persistence of livestock-associated methicillin-resistant *Staphylococcus aureus* in field workers after short-term occupational exposure to pigs and veal calves. J Clin Microbiol 49(3):1030–1033. doi:10.1128/JCM.00493-10

van der Mee-Marquet N, François P, Domelier-Valentin AS et al (2011) Emergence of unusual bloodstream infections associated with pig-borne-like *Staphylococcus aureus* ST398 in France. Clin Infect Dis 52(1):152–153. doi:10.1093/cid/ciq053

van Duijkeren E, Jansen MD, Flemming SC et al (2007) Methicillin-resistant *Staphylococcus aureus* in pigs with exudative epidermitis. Emerging Infect Dis 13:1408–1410

van Duijkeren E, Ikawaty R, Broekhuizen-Stins M et al (2008) Transmission of methicillin-resistant *Staphylococcus aureus* strains between different kinds of pig farms. Vet Microbiol 126:383–389

van Duijkeren E, Moleman M, Sloet van Oldruitenborgh-Oosterbaan MM et al (2010) Methicillin-resistant *Staphylococcus aureus* in horses and horse personnel: an investigation of several outbreaks. Vet Microbiol 141(1–2):96–102. doi:10.1016/j.vetmic.2009.08.009

van Loo IH, Diederen BM, Savelkoul PH et al (2007) Methicillin-resistant *Staphylococcus aureus* in meat products, the Netherlands. Emerging Infect Dis 13:1753–1755

Vandenesch F, Naimi T, Enright MC et al (2003) Community-acquired methicillin-resistant *Staphylococcus aureus* carrying Panton-Valentine leukocidin genes: worldwide emergence. Emerg Infect Dis 9:978–984

Vanderhaeghen W, Hermans K, Haesebrouck F, Butaye P (2010) Methicillin-resistant *Staphylococcus aureus* (MRSA) in food production animals. Epidemiol Infect 138(5):606–625. doi:10.1017/S0950268809991567

Verasa JF, do Carmob LS, Tongc LC et al (2008) A study of the enterotoxigenicity of coagulase-negative and coagulase-positive staphylococcal isolates from food poisoning outbreaks in Minas Gerais, Brazil. Int J Infect Dis 12(4):410–415

Viana D, Selva L, Segura P et al (2007) Genotypic characterization of *Staphylococcus aureus* strains isolated from rabbit lesions. Vet Microbiol 121:288–298. doi:10.1016/j.vetmic.2006.12.003

Viertel TM, Ritter K, Horz H-P (2014) Viruses versus bacteria—novel approaches to phage therapy as a tool against multidrug-resistant pathogens. J Antimicrob Chemother 69(9):2326–2336. doi:10.1093/jac/dku173

Vitale CB, Gross TL, Weese JS (2006) Methicillin-resistant *Staphylococcus aureus* in cat and owner. Emerging Infect Dis 12(12):1998–2000. doi:10.3201/eid1212.060725

Von Eiff C, Arciola CR, Montanaro L et al (2006) Emerging Staphylococcus species as new pathogens in implant infections. Int J Artific Organs 29:360–367

Voss A, Loeffen F, Bakker J et al (2005) Methicillin-resistant *Staphylococcus aureus* in pig farming. Emerg Infect Dis 11:1965–1966

Wagenaar JA, Yue H, Pritchard J et al (2009) Unexpected sequence types in livestock associated methicillin-resistant *Staphylococcus aureus* (MRSA): MRSA ST9 and a single locus variant of ST9 in pig farming in China. Vet Microbiol 139:405–409

Walker A, Jassim SAA, Holah JT et al (1992) Bioluminescent *Listeria monocytogenes* provide a rapid assay for measuring biocide efficacy. FEMS Microbiol Lett 91(Issue 3):251–255

Wang R, Braughton KR, Kretschmer D et al (2007) Identification of novel cytolytic peptides as key virulence determinants for community-associated MRSA. Nat Med 13(12):1510–1514

Wang L, Li Y, Mustapha A (2009) Detection of viable *Escherichia coli* O157:H7 by ethidium monoazide real-time PCR. J Appl Microbiol 107(5):1719–1728. doi:10.1111/j.1365-2672.2009.04358.x

Weber DJ, Rutala WA, Miller MB et al (2010) Role of hospital surfaces in the transmission of emerging health care-associated pathogens: norovirus, Clostridium difficile, and Acinetobacter species. Am J Infect Control 38(Suppl 1):S25–S33

Weese JS (2010) Methicillin-resistant *Staphylococcus aureus* in animals. ILAR J 51:233–244. doi:10.1093/ilar.51.3.233

Weese JS, Avery BP, Gow S (2009a) Methicillin-resistant *Staphylococcus aureus* (MRSA) surveillance in slaughter-age pigs and feedlot cattle. In: American Society for Microbiology (ASM)-ESCMID conference on methicillin-resistant staphylococci in animals, London

Weese JS, Avery B, Rousseau J, Reid-Smith R (2009b) Methicillin-resistant *Staphylococcus aureus* contamination of retail meat: Canada. In: American Society for Microbiology (ASM)-ESCMID conference on methicillin resistant staphylococci in animals, London

Weese JS, Avery BP, Reid-Smith RJ (2010) Detection and quantification of methicillin-resistant *Staphylococcus aureus* (MRSA) clones in retail meat products. Lett Appl Microbiol 51:338–342. doi:10.1111/j.1472-765X.2010.02901.x

Weese JS, Dick H, Willey BM et al (2006) Suspected transmission of methicillin-resistant *Staphylococcus aureus* between domestic pets and humans in veterinary clinics and in the household. Vet Microbiol 115(1–3):148–155

Weese JS, Sweetman K, Edson H, Rousseau J (2013) Evaluation of minocycline susceptibility of methicillin-resistant *Staphylococcus pseudintermedius*. Vet Microbiol 162(2–4):968–971. doi:10.1016/j.vetmic.2012.10.002

Wendlandt S, Schwarz S, Silley P (2013) Methicillin-resistant *Staphylococcus aureus*: a food-borne pathogen? Annu Rev Food Sci Technol 4:117–139. doi:10.1146/annurev-food-030212-182653

Witte W, Strommenger B, Stanek C, Cuny C (2007) Methicillin-resistant *Staphylococcus aureus* ST398 in humans and animals, central Europe. Emerging Infect Dis 13(12):255–258

Whiteley GS, Derry C, Glasbey T (2012) The comparative performance of three brands of portable ATP-bioluminometer intended for use in hospital infection control. Healthcare Infect 17(3):91–97. doi:10.1071/HI12021

WHO (2009) Guidelines on hand hygiene in health care. First global patient safety challenge clean care is safer care. World Health Organization, Geneva. ISBN-13 978-92-4-159790-6

Yeruham I, Elad D, Avidar Y, Goshen T (2006) A herd level analysis of urinary tract infection in dairy cattle. Vet J 171(1):172–176

Yokoe DS, Mermel LA, Anderson DJ et al (2008) A compendium of strategies to prevent healthcare-associated infections in acute care hospitals. Infect Control Hosp Epidemiol 29 (Suppl 1):S12–S21. doi:10.1086/591060

Zell C, Resch M, Rosenstein R et al (2008) Characterization of toxin production of coagulase-negative staphylococci isolated from foods and starter cultures. Int J Food Microbiol 127:246–251

Zollfrank C, Gutbrod K, Wechsler P et al (2012) Antimicrobial activity of transition metal acid MoO3 prevents microbial growth on material surfaces. Mater Sci Eng, C 32(1):47–54. doi:10.1016/j.msec.2011.09.010

Chapter 5
Reducing Greenhouse Gas Emissions from Livestock Farms

Abstract Greenhouse gas emissions from the agricultural sector are a key contributor to climate change. Bacteriophage 'Phage' biocontrol or harnessing nature's bacterial control mechanisms is becoming increasingly important as a means of eradicating or controlling microbial populations and also has the potential to be used to reduce methane emissions from ruminants. Phages can be programmed for increased effectiveness and integrated into farming practices by harnessing phage biocontrol techniques to enable ruminants' digestive systems and manure degradation to adapt and thereby reduce emissions. Several techniques are proposed to eradicate enteric methane emissions from animal digestion and manure degradation. Understanding the basics about phages' interaction with methanogenic archaea is pivotal in the mitigation strategies to curtail methane production. Strategies for implementing phage biocontrol to reduce the emissions from livestock manure are also discussed.

Keywords Animal waste • Bacteriophage • Climate change • Greenhouse gas • Livestock manure • Methane • Methanogenesis • Nitrous oxide • Phage biocontrol

Contents

S.A.A. Jassim and R.G. Limoges, *Bacteriophages: Practical Applications for Nature's Biocontrol*, DOI 10.1007/978-3-319-54051-1_5

5.1 Introduction

5.1.1 Animal Waste the Environment, and Human Health

The greenhouse gas (GHG) emissions from agriculture accounted for between 6 and 9% of total emissions from the U.S. (https://www3.epa.gov/climatechange/ghgemissions/sources/agriculture.html; Paustian et al. 2006). GHG emissions from agriculture have increased by approximately 11% since 1990. Between 1990 and 2005 in the U.S., methane (CH_4) emissions from dairy cow and pig manure rose by 50 and 37%, respectively (Koneswaran et al. 2008). One driver for this increase has been the 54% growth in combined CH_4 and nitrous oxide (N_2O) emissions from livestock manure management systems, reflecting the increased use of emission-intensive liquid systems over this time period (https://www3.epa.gov/climatechange/ghgemissions/sources/agriculture.html). Livestock farms with hundreds of thousands of cows, chickens or pigs produce massive amounts of animal waste that is dumped on relatively small plots of land. Globally, emissions from pig manure alone account for almost half of all GHG emissions from farm animal manure (Steinfeld et al. 2006). This is considered one of the main environmental issues of animal waste management contributing specially to water pollution (http://www3.epa.gov/region9/animalwaste/problem.html#overview). In the U.S. alone, a large quantity of animal waste is produced, for example one dairy farm with 2500 cows produces as much bio-waste as a city with around 411,000 residents (Wang and Wang 2014). In other words, livestock operations produce about 100 times more manure than the amount of sewage sludge processed in U.S. municipal wastewater plants (Gerba and Smith 2005). However, despite this magnitude, animal waste is not processed like human sewage and there is very little regulation of animal waste disposal (Kilborn 1999; Bravender et al. 2009; http://www3.epa.gov/epawaste/nonhaz/define/rulemaking.htm). Thus farm wastewater discharged without treatment into fresh waters and coastal seawater, are major sources of fecal microorganisms, including pathogens which certainly have a direct impact on the environment and human health (USDA 2002; Gerba and Smith 2005). This practice is not only harmful to the soil, but can also result in contamination of human drinking water and lead to serious public health concerns. A problem of this nature and scale is hard to control and can seriously threaten both humans and aquatic ecosystems, also leading to harmful algal blooms (Jassim and Limoges 2013).

As part of the necessity to control bacterial pathogens in animal waste, research has been conducted to help control the potential problems associated with animal waste. It was found that adding common sodium carbonate to manure can dramatically decrease the amount of the harmful bacterium *Escherichia coli* O157:H7 present (Becker 2000; Jarvis et al. 2001). The downside of this procedure includes side effects and poisons that are emitted (https://www.nlm.nih.gov/medlineplus/ency/article/002486.htm). Other researchers have suggested the use of seaweed extract as a feed additive for cattle that can significantly reduce the amount in of *E. coli* in cattle manure (Bach et al. 2008; Wang et al. 2009). While feed additives are a creative way to address some problems, ultimately they do nothing to address the fact that too much waste is being produced in areas that are too concentrated. Furthermore, eliminating only *E. coli* bacteria does nothing to address other environmental problems and impacts to human health. There are many other manure-borne pathogens aside from *E. coli* on animal farms and in animal manures that urgently require control and prevention, including but not limited to: *Actinomyces pyogenes*, *Bacillus anthracis*, *Bacillus thuringiensis*, *Burkholderia pseudomallei*, *Campylobacter jejuni/coli*, *Clostridium botulinum* neurotoxin (BoNT), *Coxiella burnetii*, *Enterococcus faecium*, *Francisella tularensis*, *Klebsiella pneumoniae*, *Listeria monocytogenes*, *Mycobacterium paratuberculosis*, *Pasteurella haemolytica*, *Pasteurella multocida*, *Pseudomonas aeruginosa*, *Shigella* spp., *Staphylococcus aureus*, *Streptococcus bovis*, *Vibrio cholera*, *Yersinia enterocolitica* and *Yersinia pestis*. It should also be noted that some of these bacteria can be used as potential agents of bioterrorism (see Chap. 7).

5.1.2 Natural Balance of Greenhouse Gases

The atmosphere has a natural supply of GHGs that capture heat and keep the surface of the Earth warm. Before large-scale industrialization began around 150 years ago, the GHGs that released into the atmosphere were in balance with those stored on Earth. Natural emissions of heat trapping gases matched what could be absorbed by nature's plants taking in carbon dioxide (CO_2) during the photosynthesis process and releasing it back into the atmosphere when they die and decompose. Over the last 50 years as many more countries particularly in Asia and the Far East became industrialized, more anthropogenic gases (i.e. emissions produced by human activities) were added to the natural levels in the atmosphere. These are originating from agriculture, animal farms, animal waste, and combustion of carbonaceous fuels, principally wood, coal, oil, and natural gas (Jassim and Limoges 2013), contributing to the increase in CO_2 in the atmosphere from 280 to 390 ppm by 2005 and projected to increase to almost 700 ppm by the end of the twenty-first century (Jassim and Limoges 2013). These GHGs are building up beyond the Earth's capacity to remove them and creating what has been termed 'global warming' or 'climate change'. This has contributed directly to rising temperatures and the resultant depletion of the ozone layer (Jassim and Limoges 2013). In general, the anthropogenic emissions of CO_2 and other gases such CH_4, tropospheric ozone (O_3), and N_2O involved in this process are the fundamental cause of the 'global warming' effect and have driven and will continue to have significant effects on ozone depletion (Jassim and Limoges 2013). For this reason, almost 200 nations participated in the Paris UN Climate Conference 2015, setting the course for a historic agreement in a bid to arrest global warming. The Paris agreement, which will come into effect in 2020, will require all countries to tackle climate change. The deal says the world will aim to stabilise global warming well below two degrees above pre-industrial levels, and less if possible.

Livestock, especially cattle, produce CH_4 as part of their digestion. This process is called enteric fermentation and it represents almost one third of the emissions from the agriculture sector (https://www3.epa.gov/climatechange/ghgemissions/sources/agriculture.html). Over the last 20 years a great deal of research has been conducted in Canada, Australia, New Zealand, Europe, and the U.S. on strategies to reduce methane emissions from animal farm operations (Boadi et al. 2004). The main focus of this research has been on nutritional strategies, especially cows' grazing pastures. Some dietary practices that have been shown to reduce CH_4 include the addition of ionophores and fats, the use of high quality forages and the increased use of grains. These nutritional strategies reduce CH_4 through the manipulation of ruminal fermentation, direct inhibition of the methanogens and protozoa, or by a redirection of hydrogen (H_2) ions away from the methanogens. Relatively new mitigation options have been investigated and include the addition of probiotics, acetogens, bacteriocins, archaeal phage, organic acids and plant extracts (e.g., essential oils) to the diet, as well as immunization and genetic selection of cows (Boadi et al. 2004; Kumar et al. 2009). In general, methane mitigation strategies for dairy cows from all the above methods, save for phage, were not significant and reductions have not exceeded more than 40% in vivo while most of them increased the cost of implementation (Boadi et al. 2004). The conclusion is that a more basic understanding of the natural differences in digestion efficiencies among animals as well as a better knowledge of methanogens and their interaction with other organisms in the rumen would enable researchers to exploit the potential of some of the new CH_4 mitigation strategies for dairy cattle production. The aim of this chapter is to review current bacteriophage biocontrol practices on livestock farms and to suggest new phage strategies to mitigate enteric CH_4 emissions from ruminants and manures.

5.2 Livestock Manure Biodegradation and Production of GHGs

Around 180 million dry tons of livestock and poultry waste are generated every year on animal farms in the U.S. (Dolliver and Gupta 2008). It is well known that various livestock manures are a valuable resource for nutrients such as nitrogen (N), phosphorous (P), potassium (K), calcium (Ca), magnesium (Mg) and sulphur (S). They also contain trace elements of zinc (Zn), iron (Fe), manganese (Mn) and copper (Cu). Obviously manure is very rich in important factors for land reclamation and therefore, livestock manures have been used for thousands of years as a fertilizer for soil to enhance soil quality, tilth and productivity; and traditionally used to fertilize fish ponds (Laçi and Bizhga 2013).

The use of manure rather than chemical fertilizers, as well as the use of untreated sewage or irrigation water containing pathogens undoubtedly contributes to the increased risk associated with fresh fruit and vegetables, especially in countries where an increase in consumption of such products occurred in the last two decades (Beuchat and Ryu 1997). These pathogens are mixed with aerobic and facultative anaerobic bacteria. Such microorganisms thrive and grow rapidly in particular in the warm conditions where the animals live or in manure lagoons with these rich and essential nutrients. The bacterial growth causes GHG emissions through microbial activities during manure decomposition. The main GHGs emitted from this process are CH_4, which is emitted during the anaerobic (without oxygen) decomposition of organic matter during storage and N_2O, which is emitted during storage and soil application. Additional gases emitted from manure include: ammonia (NH_3) and nitrogen oxides (NO_X), which contribute to odour and are indirect sources of N_2O. These factors are affected, in turn, by manure type (livestock type), diet, storage and handling of manure (pile, anaerobic lagoon, etc.), as well as manure application (injected, incorporated, etc.). It is important to note that proper manure management is essential for any agricultural operation because improper use of manure can lead to negative impacts on the environment and productivity.

5.3 The Implication of Farm Waste in Climate Change

The GHG emissions from the agricultural sector are considered to be a key contributor to climate change, accounting for about 25.5% of total global anthropogenic emission (Sejian et al. 2012). Methane production from ruminants has received global attention in relation to its contribution to the GHG effect and global warming (McAllister and Newbold 2008). Compared with cattle, chickens emit no CH_4 and emit less phosphate and CO_2 than other meat-producing animals; therefore, poultry production may have a lesser impact on the environment and global warming than cattle (FAO 2010). The mean three GHGs of concern in animal farms are: CH_4, N_2O, and CO_2, all of which contribute to climactic changes in temperature and precipitation patterns. Generally, these three gases produced from manure are the result of bacterial action on biodegradable parts of animal waste.

Chemical emanations from manure are responsible for more than half of humanity's total GHG emissions (Goodland and Anhang 2009). In Canada, Boadi et al. (2004) reported that approximately 10% of Canadian GHG emissions are attributed to agricultural activities, not including the use of fossil fuels or the indirect GHG emissions from fertilizer production. Nitrous oxide emissions associated with agricultural nitrogen sources (fertilizer and animal manure) represents 61% of agricultural GHG emissions. Methane represents 38%, while net CO_2 accounts for less than 1% of GHG emissions. Livestock production systems contribute about 42% of total GHG production from agriculture where 28% is associated with direct emissions of enteric fermentation and 14% from indirect emissions from livestock operations, related to manure handling, storage, and land application. Canadian cattle account for 97% of total livestock enteric CH_4 emissions; 25% is attributed to dairy and beef cattle account for 72%. Canadian livestock enteric CH_4 emissions were 18 Mt CO_2 equivalents in 1996 and have increased by 14% since 1990, due to an increased number of cattle. This includes 37% of anthropogenic CH_4, and methane gas is 23 times more potent a climate change agent than CO_2 from enteric fermentation by ruminants (LEAD 2006). In Alberta alone, the agricultural industry emits about 9% of the total provincial GHG emissions (Alberta Agriculture and Forestry 2004). Primarily, agricultural sources of GHGs originate from livestock and fertilizer sectors. According to the 2001 GHG Emissions Inventory for Alberta Agriculture, methane emissions from manure accounts for 9% and livestock methane emissions are approximately 28% of the total GHG emissions in Alberta (Alberta Agriculture and Forestry 2004). Yet, despite factory farming's leading role in the climate change crisis, the Environmental Protection Agency (EPA) does not currently have the authority to regulate the U.S. livestock industry's GHG emissions (Bravender et al. 2009). Even though agricultural fertilizer emissions are a leading cause of nitric acid rain, after motor vehicles and coal plants, they remain virtually unregulated in the U.S. (Michael 2010). In general, global methane emission from enteric fermentation and manure management attributable to domesticated livestock are predicted to increase by 21% from 2005 to 2020 (Kumar et al. 2009). As the numbers of farm animals reared for meat, egg, and dairy production increase, so do emissions from their production. By 2050, global farm animal production is expected to double from present levels (Koneswaran et al. 2008). These GHGs from animal waste are all major sources of factory farm air pollution (http://www.sustainabletable.org/266/air-quality), with particulate matter and bacterial toxins found in high concentrations at and around industrialized animal facilities causing serious respiratory (http://www.sciencedaily.com/releases/2007/02/070227105447.htm) and cardiac disorders (Brigham, and Meyrick 1986). Therefore, environmental impacts of animal agriculture require that governments, international organizations, producers and consumers focus more attention on the role played by meat, egg, and dairy production. Mitigating and preventing the environmental harms caused by this sector require immediate and substantial changes in regulation, production practices, and consumption patterns (Koneswaran et al. 2008).

5.4 Methanogenesis and Bacteriophage Interaction

The way in which manure from livestock is managed contributes to CH_4 and N_2O emissions. Manure management accounts for about 14% of the total GHG emissions from the Agriculture sector in the U.S. Recommendations include controlling the way in which manure decomposes to reduce N_2O and CH_4 emissions and capturing CH_4 from manure decomposition to produce renewable energy. (https://www3.epa.gov/climatechange/ghgemissions/sources/agriculture.html). A better understanding of the decomposition of animal waste and natural digestion where methanogens interact with other organisms in the rumen would enable us to exploit the potential of utilizing archaeal phage for CH_4 mitigation strategies in animal farm production.

5.4.1 Methanogenesis

Methanogenesis or biomethanation is the formation of methane by microbes known as methanogens. Methanogenesis is the formation process of organic compound degradation and plays a major role in the global carbon cycle occurring in a variety of natural and artificial environments, such as gastrointestinal tracts of animals, rice paddy soils, deep subsurface, marine or freshwater sediments and anaerobic bioreactors (Garrity and Holt 2001; Liu and Whitman 2008). The methanogens are a large and highly diverse group of archaea comprising 5 genera and 23 species (Baker 1999; Boone et al. 2001). They are nutritionally fastidious anaerobes requiring a very low redox potential i.e. below −300 mV and grow most rapidly in the pH range of 6.0–8.0 (Kumar et al. 2009). These microorganisms are present in diverse natural anaerobic ecosystems for example in the rumen, marshes, wetlands, activated sludge, soil, rhizosphere etc. They obtain their energy from the reduction of CO_2, formate, the methyl moiety of methanol, acetate and some methyl amines (Boadi et al. 2004). In the rumen, methanogens primarily use H_2, CO_2 and formate as a substrate.

Among livestock, methane production is greatest in ruminants, as methanogens are able to produce methane freely through the normal process of feed digestion. *Methanomicrobiales* were reported in cattle-manure and they significantly enhanced the production of CH_4 emissions from decomposition of the manure as compared to swine manure (Kim et al. 2014). Swine manure contains diverse groups of aerobic and anaerobic bacteria. An anaerobic bacterial consortium containing sulfate-reducing bacteria and acetate-utilizing methanogenic bacteria was isolated from swine manure. This consortium used phenol as its sole source of carbon and converted it to CH_4 and CO_2 (Boopathy 1997). Many microorganisms are involved in the decomposition of animal manure, including bacteria. Anaerobic decomposition that leads to methane production involves hydrolytic, acid-forming, and methanogenic organisms. The presence and population of methanogenic bacteria are important for methane production. In the rumen they consist of high anaerobic microbial populations including 10^{10}–10^{11} bacteria and 10^8–10^9 methanogens (Kumar et al. 2009).

5.4.2 Biochemistry of Methanogenesis

Methanogens are physiologically united as methane producers in anaerobic digestion. Although acetate and H_2/CO_2 are the main substrates available in the natural environment, formate, methanol, methylamines, and CO are also converted to CH_4 as described below.

Enteric CH_4 emission is produced as a result of microbial fermentation of feed components. The conversion of feed material to CH_4 in the rumen involves the integrated activities of different microbial species, with the final step carried out by methanogenic bacteria (Boadi et al. 2004). Primary digestive microorganisms (bacteria, protozoa and fungi) hydrolyze proteins, starch and plant cell wall polymers into amino acids and sugars. These simple products are then fermented to volatile fatty acids (VFA), H_2, and CO_2 by both primary and secondary digestive microorganisms. Acetate (C_2), propionate (C_3) and butyrate (C_4), which are the major VFA, are then absorbed and utilized by the host animal to establish CH_4 production (Boadi et al. 2004).

$$\mathbf{C_6H_{12}O_6 + 2H_2O \rightarrow 2C_2H_4O_2(acetate) + 2CO_2 + 8H}$$
$$\mathbf{C_6H_{12}O_6 + 4H \rightarrow 2C_3H_6O_2(propionate) + 2H_2O}$$
$$\mathbf{C_6H_{12}O_6 \rightarrow C_4H_8O_2(butyrate) + 2CO_2 + 4H}$$
$$\mathbf{CO_2 + 8H \rightarrow CH_4 + 2H_2O}$$

The methanogenic archaea are only directly involved in the very final stages of fermentation, yet they are very important because they are capable of effectively utilizing electrons in the form of H_2 to reduce CO_2 to CH_4, thereby maintaining low H_2 pressure in the rumen (Boadi et al. 2004). In absence of methanogenic archaea, organic matter could not be degraded as effectively in the gut (McAllister et al. 1996). However, since CH_4 has no nutritional value to the animal, its production represents a loss of dietary energy to the animal, thus reduction in CH_4 production can result from decreased fermentation in the rumen or from a shift in the VFA pattern towards more propionate and less acetate (Boadi et al. 2004). It was concluded that if decreased feed ruminal degradation is compensated for by an increased digestion in the small intestine instead of in the hindgut, it could be considered an advantage for the animal (Boadi et al. 2004). Controlling ruminant CH_4 emissions in the rumen is an important factor to be considered when developing strategies to reduce global warming. It should be possible to reduce CH_4 production by inhibiting methanogenic archaea using virulent phages. The strategies to reduce ruminal methanogenesis by using phages to target archaea are discussed hereafter.

5.4.3 Bacteriophages (Phages)

Phages are viruses that infect and kill bacteria. Unlike traditional broad-spectrum antibiotics, these viruses target specific bacteria without harming the body's normal microflora. They are the most abundant living entities on earth, (10^{30}–10^{32}) and play a key role in regulating the microbial balance in every ecosystem (Kutter and

Sulakvelidze 2005). They evolve and adapt various host interaction mechanisms which has made them the most numerous category of "organisms" on our planet (Jassim and Limoges 2014). They are widely distributed in locations populated by bacterial hosts, such as hospitals, animal farms, slaughterhouses, wastewater, soil or rivers, seas and oceans. They are found in very high numbers [10^2–10^5 plaque-forming unit (pfu) ml^{-1}] in all types of wastewater (Havelaar 1987), and one of the densest natural sources for phages is sea water, where up to 9×10^8 virions per milliliter have been found in microbial mats at the surface (Wommack and Colwell 2000), and up to 70% of marine bacteria may be infected by phages (Jassim and Limoges 2013). Animal waste of different livestock such as cattle, camel, buffalo, goat, sheep, pig, poultry, quails and pigeons were found to be rich in phages for various bacterial species (Mc Laughlin et al. 2006; Jassim et al. 2010; Shukla et al. 2014; Aldoori et al. 2015).

Little has been known about phages occurring in anaerobic ecosystems, including those in methanogenic digesters for waste treatment (Boadi et al. 2004). Isolation and characterization of some phages of methanogenic archaea were reported in earlier research (Knox and Harris 1986; Roustan et al. 1986). Park et al. (2007) have investigated phages produced in an up flow anaerobic sludge blanket methanogenic digester treating brewery wastes. Phage-like particles (PLPs) in the influent and effluent of the digester were concentrated and purified by sequential filtration then quantified and characterized by transmission electron microscopy, fluorescence assay and field inversion gel electrophoresis (FIGE). Results indicate that numbers of PLPs in the effluent of the digester exceeded 1×10^9 l^{-1} and at least 10 times greater than those in the influent, suggesting that substantial amounts of PLPs were produced in the digester. A production rate of the PLPs was estimated at least 5.2×10^7 PLPs day^{-1} l^{-1}. Transmission electron microscopy and FIGE showed that a variety of phages were produced in the digester, including those affiliated with Siphoviridae, Myoviridae, and Cystoviridae. Three methanogenic archaea virulent phages psi ΨM1, ΨM2, and ΨM100 were reported to infect *Methanobacterium thermoautotrophicum strain Marburg, Methanothermobacter marburgensis* and *Methanothermobacter wolfeii*, respectively, (Jordan et al. 1989; Pfister et al. 1998; Luo et al. 2001).

5.5 Phage Biocontrol to Reduce the Emission of GHGs in Animal Waste

Over the last two decades, research programs in Europe, Oceania, Canada, Australia, New Zealand, and the U.S. have explored a variety of approaches to using other reductive substrates as a means of decreasing methane production in ruminants (Boadi et al. 2004; McAllister and Newbold 2008). Phages were proposed as one of the inhibitors directly targeting methanogens (McAllister and Newbold 2008). Recently phage biocontrol is becoming increasingly important as a means of eradicating or controlling microbial populations and has the potential to be used to reduce methane emissions from ruminants and livestock waste. Phages could be used to directly reduce methanogens from livestock manures, or to help establish rumen phage-biotic feed supplement to inhibit the growth of methanogen microorganisms.

The strategy is to develop a new phage-based biocontrol to abolish methanogen populations thereby preventing the formation of CH_4 using a two-pronged approach (Fig. 5.1):

(1) Use phage biocontrol with intact lytic phages that specifically target a broad spectrum of methanogen genera; and
(2) Phage biocontrol based on phage-encoded products, for example formulation of lytic enzymes or tails.

Both of these approaches would need to be developed with the premise of targeting a range of methanogen genera in order to address the natural complexity and diversity of rumen methanogen community in manure. This could be achieved by utilizing a phage programming technology (this is discussed in greater detail in Chap. 1) to produce smart host-virulent phages capable of infecting a range of methanogen species or having a narrow host range, infecting a single methanogen species. The resultant smart phage could be mixed with manure to target and cause infection across the diversity of methanogen species present in manure (Fig. 5.1a). Methane emissions can also be reduced through supplementing diets with smart phages or 'phage-biotic' (Fig. 5.1b) that can reduce enteric fermentation in rumen. In this case the phages shedding from the animal intestine with manure will prevent methanogen species from producing methane.

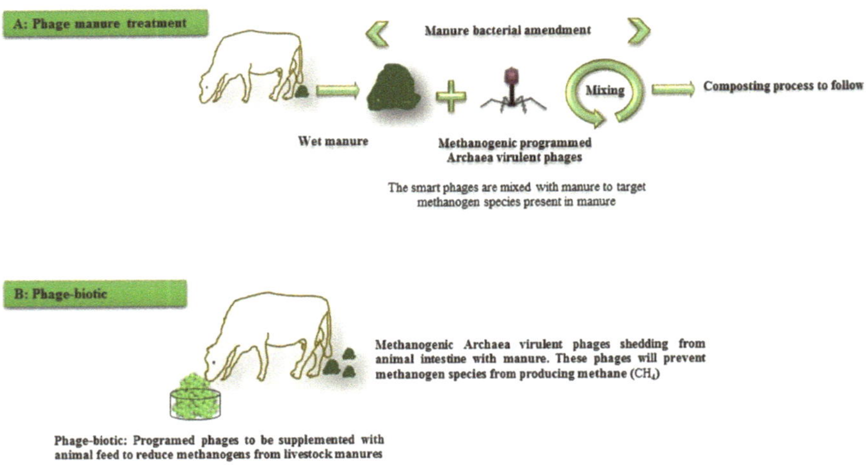

Fig. 5.1 Programmed phage treatment for manure

The phages can also be used to establish alternative microbial populations that will out-compete methanogens for H_2, such as reductive acetogens. Such an approach could be implemented at an early stage of weaning and would potentially overcome any problem that may result from over accumulation of H_2 within the manure. Phage-biocontrol thus represents a novel approach for reducing CH_4 and CO_2 emissions building on the naturally occurring biocidal attributes of these unique microbial predators.

5.6 Conclusion

Limited information is currently available on phage relevant to methanogenic digestion. More research is needed on phage biocontrol for use in animal waste to reduce the emission of GHGs and thereby helping to prevent global warming. Archaeal phage for CH_4 mitigation are abundant in the industrial agricultural waste, yet the potential mitigation strategies using these phages have not been fully researched at the farm level (Park et al. 2007). Strategies using phages with the potential positive effects on mitigation of CH_4 emissions can be effectively achieved by using phage programming technology (Jassim and Limoges 2014) to improve phage virulence and reduce GHGs in livestock waste, while redirecting ruminal fermentation to reduce methanogenesis. Such strategies hold a greater prospect of being adopted by farmers as practical on-farm options for significantly reducing emissions from livestock while simultaneously increasing productivity.

References

Alberta Agriculture and Forestry (2004) Manure management and greenhouse gases—things you need to know. Alberta Environmentally Sustainable Agriculture Council, Number 11, June 2004. http://www1.agric.gov.ab.ca/$department/deptdocs.nsf/all/cl10038

Aldoori AA, Mahdii EF, Abbas AK, Jassim SAA (2015) Bacteriophage biocontrol rescues mice bacteremia of clinically isolated mastitis from dairy cows associated with methicillin-resistant *Staphylococcus aureus*. Adv Microbiol 5:383–403. http://www.scirp.org/journal/PaperInformation.aspx?PaperID=56931

Bach SJ, Wang Y, McAllister TA (2008) Effect of feeding sun-dried seaweed (*Ascophyllum nodosum*) on fecal shedding of *Escherichia coli* O157:H7 by feedlot cattle and on growth performance of lambs. Anim Feed Sci Techn 142:17–32

Baker SK (1999) Rumen methanogens and inhibition of methanogenesis. Aust J Agric Res 50:1293–1298. doi:10.1071/AR99005

Becker H (2000) Carbonating cow manure, the latest strategy in fighting *E. coli* and other microbes. Feb 9, 2000, US Department of Agriculture website. http://www.ars.usda.gov/is/pr/2000/000209b.htm?pf=1

Beuchat LR, Ryu J-H (1997) Produce handling and processing practices. Emerg Infect Dis 3:459–465

Boadi D, Benchaar C, Chiquette J, Massé D (2004) Mitigation strategies to reduce enteric methane emissions from dairy cows: update review. Can J Anim Sci 84:319–335

Boone DR, Whitman WB, Koga Y (2001) Order II. *Methanobacteriacea*. In: Boone DR, Castenholz RW (eds) Bergey's manual of systematic bacteriology, 2nd edn. Springer, New York, pp 246–247

Boopathy R (1997) Anaerobic phenol degradation by microorganisms of swine manure. Curr Microbiol 35(1):64–67

Bravender R, Geman B, Luntz T, Greenwire (2009) Farm interests use EPA spending bill to fight climate regs. The New York Times. June 19, 2009 http://www.nytimes.com/gwire/2009/06/19/19greenwire-farm-interests-use-epa-spending-bill-to-fight-85048.html

Brigham K, Meyrick D (1986) Endotoxin and lung injury. Am Rev Respir Dis 133(5):913–927

Dolliver H, Gupta S (2008) Antibiotic losses in leaching and surface runoff from manure-amended agricultural land. J Environ Qual 37:1227–1237

FAO (2010) Agribusiness handbook: poultry meat and eggs. Rome, Italy. http://www.eastagri.org/publications/pub_docs/6_Poultry_web.pdf

Garrity GM, Holt JG (2001) Phylum AII. Euryarchaeota phy. nov. In Boone DR, Castenholz RW (eds) The archaea and the deeply branching and phototrophic bacteria. In: Garrity GM (ed) Bergey's manual of systematic bacteriology, 2nd ed, vol 1. Springer, New York, pp 211–355

Gerba CP, Smith JE Jr (2005) Sources of pathogenic microorganisms and their fate during land application of wastes. J Environ Qual 34(1):42–48

Goodland R, Anhang J (2009) Livestock and climate change. World Watch Institute. http://www.worldwatch.org/files/pdf/Livestock%20and%20Climate%20Change.pdf

Havelaar AH (1987) Virus, bacteriophages and water purification. Vet Q 9(4):356–360

Jassim SAA, Limoges RG (2013) Impact of external forces on cyanophage–host interactions in aquatic ecosystems. World J Microbiol Biotechnol 29(10):1751–1762. doi:10.1007/s11274-013-1358-5

Jassim SAA, Limoges RG (2014) Natural solution to antibiotic resistance: bacteriophages 'The Living Drugs'. World J Microbiol Biotechnol 30(8):2153–2170. doi:10.1007/s11274-014-1655-7

Jassim SAA; Abdulamir AS, Abu Bakar F (2010) Methods for bacteriophage design. WIPO Patent Application WO2010/064044. http://www.sumobrain.com/patents/wipo/Methods-bacteriophage-design/WO2010064044.html

Jarvis GN, Fields MW, Adamovich DA et al (2001) The mechanism of carbonate killing of *Escherichia coli*. Lett Appl Microbiol 33(3):196–200

Jordan M, Meile L, Leisinger T (1989) Organization of *Methanobacterium thermoautotrophicum* bacteriophage ΨM1 DNA. Mol Gen Genet 220:161–164

Kilborn PT (1999) Hurricane reveals flaws in farm law as animal waste threatens N. Carolina water. The New York Times. October 17, 1999. Sunday, Late Edition-Final. Section 1, p 32

Kim SY, Pramanik P, Bodelier PLE, Kim PJ (2014) Cattle manure enhances methanogens diversity and methane emissions compared to swine manure under rice paddy. PLoS ONE 9 (12):e113593

Knox MR, Harris JE (1986) Isolation and characterization of a bacteriophage of *Methanobrevibacter smithii*. In: Abstract of the XIV International Congress on Microbiology, International Union of Microbiology Societies. Manchester, England

Koneswaran G, Nierenberg D. Global (2008) Farm animal production and global warming: impacting and mitigating climate change. Environ Health Perspect 116(5):578–582. doi:10.1289/ehp.11034

Kumar S, Puniya AK, Puniya M et al (2009) Factors affecting rumen methanogens and methane mitigation strategies. World J Microbiol Biotechnol 25(9):1557–1566

Kutter E, Sulakvelidze A (2005) Bacteriophages biology and applications. CRC Press, Boca Raton, London, New York

Laçi D, Bizhga B (2013) The antibiotic resistance of bacterial pathogens isolated from poultry manure used for fish ponds. Anglisticum J (IJLLIS) 2(4):199–206. doi:10.0001/(aj).v2i4.547

LEAD (Livestock, Environment and Development) (2006) Livestock's long shadow: environmental issues and options. The United Nations: Food and Agriculture Organization. Rome. ftp://fao.org/docrep/fao/010/a0701e/a0701e00.pdf

Liu Y, Whitman WB (2008) Metabolic, phylogenetic, and ecological diversity of the methanogenic Archaea. Ann NY Acad Sci 1125:171–189

Luo Y, Pfister P, Leisinger T, Wasserfallen A (2001) The genome of archaeal prophage ΨM100 encodes the lytic enzyme responsible for autolysis of *Methanothermobacter wolfeii*. J Bacteriol 183:5788–5792

McAllister TA, Newbold CJ (2008) Redirecting rumen fermentation to reduce methanogenesis. Aust J Exp Agric 48:(2)7–13

McAllister TA, Okine EK, Mathison GW, Cheng KJ (1996) Dietary, environmental and microbiological aspects of methane production in ruminants. Can J Anim Sci 76:231–243

McLaughlin MR, Balaa MF, Sims J, King R (2006) Isolation of *Salmonella* bacteriophages from swine effluent lagoons. J Environ Qual 35(522):528

Michael T (2010) Sour showers: acid rain returns—This time it is caused by nitrogen emissions. Scientific American. June 21, 2010. http://www.scientificamerican.com/article/acid-rain-caused-by-nitrogen-emissions/

Park MO, Ikenaga H, Watanabe K (2007) Phage diversity in a methanogenic digester. Microb Ecol 53(1):98–103

Paustian K, Antle M, Sheehan J, Eldor P (2006) Agriculture's role in greenhouse gas mitigation. Pew Center on Global Climate Change, Washington, DC

Pfister P, Wasserfallen A, Stettler R, Leisinger T (1998) Molecular analysis of Methanobacterium phage ΨM2. Mol Microbiol 30:233–244

Roustan JL, Touzel JP, Prensier G, et al (1986) A bacteriophage for Methanothrix sp. In: Duine JA, Van Verseveld HW (eds) Abstract of the 5th international symposium on microbial growth on C1 compounds. Haren, The Netherlands, pp 35

Sejian V, Shekhawat I, Ujor V et al (2012) Global climate change: enteric methane reduction strategies in livestock. In: Sejian V, Naqvi SMK, Ezeji T et al (eds) Environmental stress and amelioration in livestock production. Springer, Berlin Heidelberg, Chap 16, pp 469–499

Shukla SK, Hirpurkar SD, Singh SK, Rajoria R (2014) Isolation of phage from animal waste of different LSF and their utility in phage therapy. IJCMAS 3(8):205–210

Steinfeld H, Gerber P, Wassenaar T et al (2006) Livestock's long shadow: environmental issues and options. Food and Agriculture Organization of the United Nations, Rome

USDA (2002) Agricultural research service. National program 206: Manure and by product Utilization: Research accomplishments http://www.ars.usda.gov/SP2UserFiles/Program/206/WholeNP206AssessmentDocument.pdf

Wang Y, Xu Z, Bach SJ, McAllister TA (2009) Sensitivity of *Escherichia coli* to seaweed (*Ascophyllum nodosum*) phlorotannins and terrestrial tannins. Asian-Austral J Anim Sci 22:238–245

Wang H, Wang L (2014) Animal waste as an energy feedstock availability and sustainability. In: Wang L (ed) Sustainable bioenergy production. CRC Press, NW. Chap 12, pp 245–257

Wommack KE, Colwell RR (2000) Virioplankton: Viruses in aquatic ecosystems. Microbiol Mol Biol Rev 64(1):69–114. doi:10.1128/MMBR.64.1.69-114.2000. PMC 98987. PMID 10704475

Chapter 6
Bacteriophage Biocontrol: Deployment in Aquatic Ecosystems

Abstract This chapter includes descriptions of innovative components for deployment of cyanophages as a biocontrol for cyanobacterial harmful algal blooms (CyanoHABs). The components, comprising a protective cap and the conical shape of the container serve to protect cyanophages from the predominant daytime sunlight, heat and direct UV rays. The depth of the cone coincides with the microbial food chain trophic levels occupied by cyanobacteria and the corresponding cyanophages. The methodology includes introduction of the apparatus to the affected or susceptible waters using the deployment methods outlined in the invention. It should be noted that successful biocontrol, particularly in open aquatic ecosystems, is dependent upon the use of properly programmed phages that are highly lytic, properly targeted and then deployed in a manner which allows them to overcome external forces along with the natural tendency for dispersion and thereby dilution in moving water. As previously stated, some of the many external forces that may impede successful lytic activity include environmental conditions such as increased greenhouse gases, ozone depletion and the resultant increased solar ultraviolet radiation (SUR). Pollution and chemicals affect nutrient levels and phage-host interactions or can cause or impede lysogeny while both cyanophages and cyanobacteria constantly undergo mutation and adaptation in response to their environment and each other. It is also noted that virophages, which coexist with and in turn are the natural predator of phages can also affect in vivo results.

Keywords Aquatic ecosystems · Bacteriophage · Conical cyanophage net · Cyanobacteria · Cyanophage · Harmful algae blooms · Solar ultraviolet radiations

Contents

S.A.A. Jassim and R.G. Limoges, *Bacteriophages: Practical Applications for Nature's Biocontrol*, DOI 10.1007/978-3-319-54051-1_6

6.1 Introduction

6.1.1 Cyanobacteria

Cyanobacteria, (known as blue-green algae, 'BGA') are prokaryotes, or bacteria that obtain their energy through photosynthesis. Based on their morphology (form and structure), they have been divided into five sections (Castenholz 2001). As in other prokaryotes, cyanobacteria contain the polymer peptidoglycan or murein in their cell walls. Their cell diameter ranges from 0.2 to 2 μm and they produce internal gas vesicles, which they use to regulate their buoyancy, allowing them to move up or down into water levels with different intensities of light and nutrient levels (Fristachi and Sinclair 2008).

6.1.2 Cyanobacterial Geographic Distribution

The BGA are resilient predecessors of all higher oxygenic phototrophs, which use light as their main source of energy and can be found in self-sustaining, nitrogen-fixing communities the world over, from Antarctic glaciers to the Sahara Desert (Whitton and Potts 2000). Throughout the terrestrial north and south Polar Regions, they form benthic mats and films at the bottom of lakes, ponds and streams (Schirrmeister et al. 2011). It has been determined that multiple interacting physical, chemical and biological factors lead to the formation of cyanobacterial harmful algal blooms (CyanoHABs) (Fristachi and Sinclair 2008; Jassim and Limoges 2013).

More recent glacial events may have also enabled a cyanobacterial genetic exchange between the Polar Regions, as suggested for cold-water foraminifers. The recent study of Jungblut et al. (2010) used targeted 16S rRNA gene surveys to evaluate the global affinities of cold-dwelling cyanobacteria from lake, stream and ice communities living at the northern limit of high arctic Canada. It revealed that more than 68% of all identified ribotypes at each remote site matched only cyanobacterial sequences from perennially cold terrestrial ecosystems and were <97.5% similar to sequences from warmer environments. These results imply the global distribution of low-temperature cyanobacterial ecotypes throughout the cold terrestrial biosphere. This indicates that cold-adapted cyanobacteria are well equipped to withstand potential nutrient limitations, temperature fluctuation, dehydration and elevated solar ultraviolet radiations (SUR) during long distance aerial transport (Jassim and Limoges 2013). Until recently, only a few regions of the U.S. and Canada were affected by cyanobacterial blooms but now virtually every coastal state has reported major blooms (Kalluri 2008; Winter et al. 2011).

6.1.3 *Cyanobacterial Blooms*

Cyanobacterial blooms occur when cyanobacteria replicate faster than they are consumed by natural processes. The bloom, often many inches thick, usually floats to the surface, especially near the shoreline and in warm, slow-moving waters that are rich in nutrients from fertilizer runoff or septic tank overflows (Jassim and Limoges 2013).

6.1.4 *Economic and Environmental Impacts of CyanoHABs*

CyanoHABs have been documented to impact water quality through production of compounds affecting taste and odor (Carmichael 2008). Selected species within about 40 genera can produce potent Cyanotoxins including hepatotoxins (liver toxins) cause death by liver haemorrhage within a few hours of the acute doses to wild or domestic animals as well as humans (Sivonen and Jones 1999; Carmichael 2008). On the other hand, Cyanotoxins can have significant economic impacts as the cause of mass mortalities of fish and birds and have also been tied to the death of pets and livestock which may be exposed through drinking contaminated water or licking themselves after bodily exposure. Furthermore, unglazed cyanobacterial biomass that accumulates as clotted mats poses a particular threat to dogs, which may lick or eat the toxic mats. For this reason, dog deaths have emerged as an unfortunate early warning that a toxic cyanobacterial bloom is occurring (https://www.whoi.edu/redtide/impacts/freshwater/cyanobacteria). The economic losses in the USA and Canada associated with harmful algal blooms (HABs) are estimated to exceed a few billions of dollars over the next several decades (Bushaw-Newton and Sellner 1999; Kalluri 2008; Bingham et al. 2015). In Australia, the cost of environmental protection schemes attributable to cyanobacteria was estimated \$180–240 million per year (Atech 2000). This estimate included urban sewage and storm water (\$43 million), agriculture and industrial wastewater (\$33 million) and rehabilitation of land and water resources (\$45 million). A large bloom of cyanobacteria spread across Guatemala's Lake Atitlan where the cost estimate of at least US \$32 million to clean up the lake was reported (Duda 2009). These are a few examples of a huge economic problem associated with CyanoHABs. On the other hand, these estimates are conservative, due in part to a lack of quantitative information on the environmental impacts of HAB events and a lack of documentation of socio-cultural impacts.

6.1.5 Cyanophages

Bacteriophages (phages) are described as viruses that infect bacteria. The term 'cyanophage' describes the virus that infects 'cyanobacteria' (Jassim and Limoges 2013). Most phages, at least in the surface ocean, belong to three groups of the tailed phages. The morphological diversity of the phages is high; they vary in capsid size as well as tail length and structure and can have appendices. In surface waters, very large phages can be found, often with capsid diameters of 200–400 nm (Weinbauer and Wilhelm 2011). These phages infect single-celled photosynthetic organisms (phytoplankton). Sequence analysis and isolation of phages showed that some phages have a cosmopolitan distribution, whereas others seem to have a more restricted geographical range (Weinbauer and Wilhelm 2011). The effect of cyanophages on the mortality of cyanobacterial communities is likely to be variable, current estimates suggest that cyanophages are responsible for the removal of approximately 3% of marine *Synechococcus* on a daily basis (Whitton and Potts 2000) and about 1–3% of marine *Synechococcus* spp. contained mature phages (Proctor and Fuhrman 1990). Others have estimated that cyanophages may be responsible for ca. 5–14% of cyanobacterial mortality on a daily basis (Suttle and Chan 1994). We also find estimates that up to 20% of heterotrophic bacteria and up to 51% of cyanobacteria are destroyed daily through phage cell lysis (Fuhrman and Suttle 1993; Suttle 1994). While it has been difficult to obtain accurate in situ estimates of phage-induced host mortality (Suttle 2005), some studies report that phages may be responsible for up to 70% of the daily mortality of marine bacteria and phytoplankton (Proctor and Fuhrman 1990; Suttle et al. 1990; Wommack and Colwell 2000).

6.2 Cyanophage Biocontrol

The single-celled algae and cyanobacteria which can be found residing in oceans, lakes, rivers, ponds and even in snow have contributed significant economic impacts and economic losses (Bushaw-Newton and Sellner 1999; Atech 2000; Kalluri 2008; Duda 2009; Bingham et al. 2015). Many governments are pressed to find suitable solutions to control the tide of HABs. Researchers throughout the world have taken the initiative to develop various biological methods and control strategies to treat HABs in aquatic ecosystems (Sigee et al. 1999; Foflonker 2009; Jassim et al. 2010), but it is not known the extent to which these can be applied to particular HAB species in specific environments or habitats (Drabkova and Marsalek 2007).

Over the last two decades there is a growing body of evidence pointing toward an important role for cyanophages in the dynamics of cyanobacterial populations. The studies of Phlips et al. (1990), Castberg et al. (2001) and Yoshida et al. (2006) involved using cyanophages to infect and kill four common bloom-forming species of cyanobacteria *Lyngbya birgei*, *Anabaena circinalis*, *Anabaena flosaquae*, and *Microcystis aeruginosa*. Nevertheless, the concept of using cyanophages in the control of cyanobacteria had not received widespread attention up to that point. The lack of positive results in using phages as biocontrol agents in these applications is due to the rapid appearance of phage-resistant mutants (Cannon et al. 1976; Waterbury and Valois 1993; Suttle and Chan 1994; Sigee et al. 1999; Jassim and Limoges 2013; Chaps. 1 and 2). This is a natural phenomenon that occurs in phage biology (Stewart et al. 1995), including with cyanophages (Bisen et al. 1986). Moreover, cyanophages are sometimes so obligate (host-specific) that they are often unable to infect different genetic strains of the same host species (Waterbury and Valois 1993). Even after the decline of particular cyanobacteria caused by cyanophages, they can be easily replaced by other cyanobacterial species (Van Hannen et al. 1999).

Resistance to co-occurring cyanophages has been reported for natural communities of *Synechococcus* spp.; however, little was known about the nature of this resistance (Stoddard et al. 2007) or the role of cyanophages to control cyanobacteria (Weinbauer and Rassoulzadegan 2004) especially in lakes (Tucker and Pollard 2005). Remarkably, these cyanophages behave differently under nutrient rich conditions than they do when starved of nutrients. Their ecological behaviour can be altered through the presence of chemical and other environmental factors which makes them difficult to use as a biocontrol for CyanoHABs in aquatic ecosystems (Jassim and Limoges 2013). There is a need to consider factors that influence cyanophage attachment, infection, and lysis of their host, alongside the physical and chemical parameters that drive cyanobacterial growth and production (Tucker and Pollard 2005). Essentially, CyanoHABs form when external forces often related to climate change or pollution, cause the balance of nature to be out of whack. Strategies are therefore required to overcome the imbalance and allow nature's biocontrol to do the work that it was intended for.

There are several environmental physical forces such as SUR that have a direct impact on cyanophage-host interactions in aquatic ecosystems (Jassim and Limoges, 2013). Therefore, in this chapter we have illustrated some novel models developed to provide optimal conditions for cyanophages to be employed to curb HABs.

6.2.1 Cyanophages Isolation and Enhancement

There is no doubt that cyanophages are abundant in aquatic ecosystems. Many wild cyanophages can be isolated from seawater, rivers, ponds, lakes, etc. These phages can be converted as described in Chap. 1 to become highly specific and lytic to selectively reduce or eliminate cyanobacteria from aquatic environments, e.g., the municipal water supply, waste water treatment facilities, marinas, rivers, ponds, lakes, etc.

6.2.2 Phage Encapsulation

To ensure a continuous supply of cyanophages in an aquatic ecosystem, we need to encapsulate them. Encapsulation techniques can be used to enclose in phosphate-buffered saline (PBS) 10^9 plaque-forming units (pfu) ml^{-1} of a cocktail of enhanced cyanophages in a relatively stable shell known as a capsule, allowing us to control their release into local aquatic environments. The capsules, (size 5) length 11.1 mm and external diameter 4.91 mm may be used. Cocktails of 130 μl phage or cyanophages in PBS are made from aqueous solutions of plant polysaccharides or their derivatives such as carrageenans or carrageenins, a family of linear sulphated polysaccharides that are extracted from red edible seaweeds. The phages are released upon physical or chemical disruption in the aquatic environments. Once released, the phages will follow their natural instincts to infect cyanobacteria in the vicinity and upon lysing the cells, release progeny cyanobacteria into the local environment. The advantages of this method include the persistence of the virus in the environment and the transmission of the virus within a given generation and from generation to generation (auto-dissemination). The encapsulation and gradual release of the virus provides an efficacious method to augment the strategy of classical biological auto-dissemination.

6.2.3 Auto-dissemination of Phages or Cyanophages

The auto-dissemination strategy is being used for management of tobacco budworms by baculovirus (Jackson et al. 1992) and in Australia with a fungus in order to control the diamondback moth, *Plutella xylostella* (L.) (Vickers et al. 2004). Auto-dissemination is a technique where infected bacteria or cyanobacteria with appropriate phages or cyanophages, respectively, are purposefully released into the environment to spread and/or transmit the reciprocal viruses to other healthy host cells as a method of biocontrol for HABs. This technique for controlling CyanoHABs has not been previously reported in scientific literature. In this invention, the cyanobacteria become infected, are released and naturally disseminate the pathogen. Cyanophages are released from infected cyanobacteria through the lysing cycle into the aquatic environment. The evolved lytic system weakens the cyanobacterial cell wall resulting in lysis and progeny cyanophages are released to repeat the cycle, infecting more cyanobacteria to spread and transmit the cyanophages into the aquatic environments resulting in the control of CyanoHABs.

6.2.4 Conical Cyanophage Net (CCN) Biocontrol System

There have been many attempts (1960s to present time) to eliminate CyanoHABs using cyanophages in marine and freshwater environments. These attempts have not been a complete success for various reasons, including factors such as dilution, long hours of direct sunlight, UV radiation, and high temperatures (>30 °C) (Jassim and Limoges 2013). Before describing the CCN and its use to protect cyanophage while they work to control CyanoHABs, we need to understand the impact of SUR on cyanophage-host interactions. It is well known that phages or cyanophages are very vulnerable to several stressors occurring in the water column (Wilson et al. 1996; Frederickson et al. 2003; Sommaruga 2003; Clokie and Mann 2006; Jassim and Limoges 2013), and sun light in particular plays a key role in all phases of the cyanophage life cycle by reducing their infectivity (Garza and Suttle 1998). It was found that virus inactivation rates are about 10 times higher in sunlight than in the dark (Sinton et al. 2002; Sabah AA Jassim unpublished data). Thus sunlight is well established as an agent of viral mortality (Clokie and Mann 2006).

It is interesting that cyanophage communities tolerate damage by solar radiation better in summer than in winter (Garza and Suttle 1998). This is due to the cyanobacterial biomass increase during the summer months providing some shade to cyanophages, allowing them to avoid direct solar radiation in lakes (De Haan 1993). Detectable cyanophage production does not occur during daytime (Garza and Suttle 1998), and this was attributed to the significant amount of UV-B on the water surface (Newman et al. 2009). On the contrary, more phages were found during dawn and around dusk and cyanophages were found to reach a maximum abundance after midnight (Clokie and Mann 2006). UV radiation has been demonstrated to damage cyanophages (Asato 1976), and plays a big role in cyanophage decay. To explain this phenomenon, it was found that the virus infectivity is very sensitive to solar radiation and in full sunlight, decay rates were 0.4–0.8 h^{-1} and even when UV-B radiation was blocked, the decay rates were still as high as 0.17 h^{-1} (Suttle and Chen 1992). Their calculations suggested that in clear oceanic waters exposed to full sunlight, most of the virus decay, averaged over a depth of 200 m, would be attributable to solar radiation. When decay rates were averaged over 24 h for a 10-m coastal water column, loss rates of infectivity attributable to sunlight were similar to those resulting from all other processes combined. They concluded that since sunlight destroys infectivity more quickly than virus particles, a large proportion of the viruses in seawater are probably not infective, namely, in the lysogenic state. It was found that UV-B (290–320 nm) and UV-A (320–400 nm) radiation that directly and indirectly damages viral DNA, penetrate to a considerable depth in seawater (Suttle and Chen 1992). This has the potential to cause wide-ranging effects, including mutagenesis in aquatic microorganisms (Vincent and Roy 1993). It also plays an important role in the destruction of viral particles and the decrease in viral infectivity and viral attachment to host cells (Suttle and Chen 1992; Wommack and Colwell 2000; Liao et al. 2010). This will eventually lessen the survival and reproductive rates of viruses and thus reduce their influence on the host cell (Suttle and Chen 1992; Vincent and Roy 1993). The UV-B component of solar radiation can be directly absorbed by viral DNA, resulting in the formation of cyclobutane pyrimidine dimers (CPDs) (Hader and Sinha 2005), which prevent the replication of viral DNA, affecting virus infectivity. Loss of viral infectivity after exposure to SUR seems to be mainly caused by damage to the viral genome, although indirect damage to the capsid has also been suggested to result in inactivation (Wommack and Colwell 2000; Liao et al. 2010).

Like in many other planktonic groups, different viruses appear to have different tolerance towards SUR (Kitamura et al. 2004). The difference in sensitivity between viral nucleic acid types occurs because the most common lethal photoproducts of UV are pyrimidine dimers, particularly thymine dimers (Friedberg et al. 1995), resulting in transcription errors when the DNA replicates (Wommack and Colwell 2000; Clancy 2008). Since DNA but not RNA contains thymine, DNA-containing viruses are generally more sensitive to damage by UV than RNA-containing viruses (Murphy and Gordon 1981). In one common damage event, adjacent thymine bases bond with each other, instead of across the "ladder". This "thymine dimer" makes a bulge and the distorted DNA molecule does not function properly (Svobodova et al. 2012).

In general, sunlight or more specifically, SUR acts as the principal natural virucide in the environment by chemically modifying viral genetic material, whereas, viral nucleic acid plays a crucial role in the absorption of UV radiation and in virus inactivation (Lytle and Sagripanti 2005). In this regard, a mathematical technique was used to quantify how the biological impacts of SUR exposure will affect life, finding a 7% increase in UV yielded a 4.8% increase in damage to DNA (Jassim and Limoges 2013). This may predict that mutation linked to the increase of SUR can also occur in cyanophages DNA (Jassim and Limoges 2013), which prevents cyanophage adsorption in marine cyanobacteria (Sigee et al. 1999; Stoddard et al. 2007).

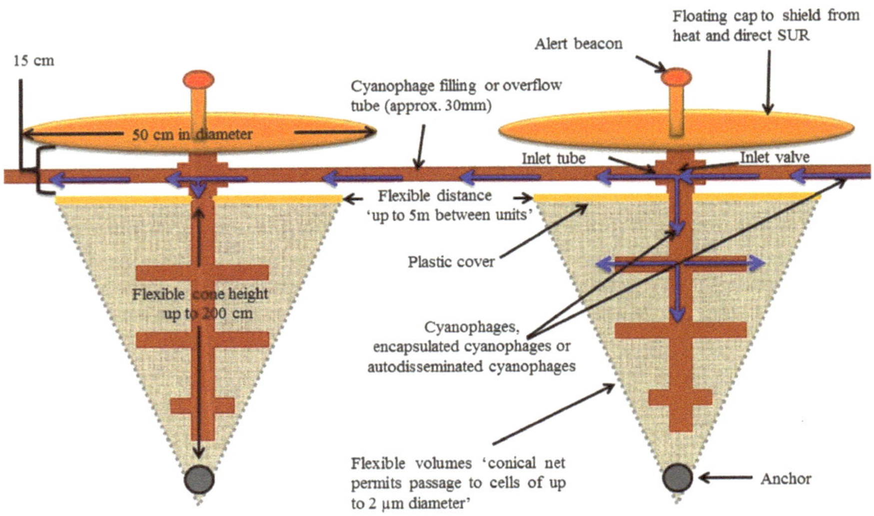

Fig. 6.1 Conical cyanophage net (CCN) biocontrol system

To resolve this environmental problem, we describe a conical CCN biocontrol system (Fig. 6.1) which is able to locate and concentrate the cyanophages directly and precisely among cyanobacteria cells as they are found proliferating either as solitary cells or in long filaments. The filamentous forms can produce large mats which may also attach to the net containing cyanophages.

The containment net in the CCN system will only allow particles less than 2 μm diameter, including cyanobacteria to swim through the net. This pore size will also keep the encapsulated cyanophages or auto-disseminated cyanophages contained within the CCN for longer periods to prolong the direct contact between cyanophages and their host cyanobacterial cells. The containment netting permits motivated cyanophage passage thereby enabling the infection and lysing of any cyanobacterial filaments attached to the rough exterior, again releasing progeny cyanophages to infect other CyanoHABs in the vicinity. The CCN is also equipped with a beacon to alert boats or people in the area.

The floating cap along with the net which penetrates up to 200 cm under water (Fig. 6.1) will prevent direct sunlight, heat and UV radiation, providing improved conditions for the cyanophages to control CyanoHABs. The CCN can be refilled with progeny cyanophages or encapsulated cyanophages or auto-disseminated cyanophages automatically or manually and can be deployed in series (Fig. 6.2) or as a single unit (Fig. 6.3) dependent upon the application (marina, creek, open water).

Nature provides cyanophages with the remarkable ability to recognize and target their corresponding cyanobacterial host while ignoring all other organisms.

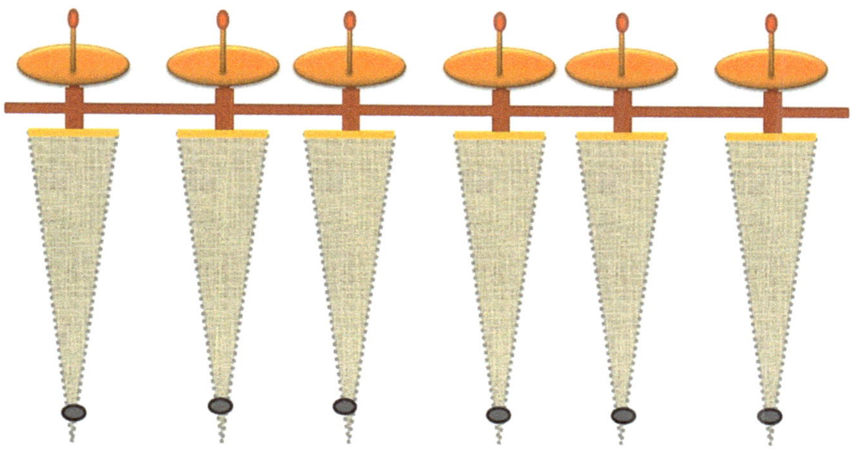

Cyanobacteria can be found as solitary cells or in long filaments. The filamentous forms can produce large mats which can attach to the floating CCN. Cyanophages are a significant parasite of unicellular marine cyanobacteria. When they infect and lyse cells they release progeny phages into the water

Fig. 6.2 Field applications to control cyanobacteria using bio-selective naturally occurring, lytic cyanophages and CCN biocontrol system in series linked through overflow tubes to share the cyanophage population

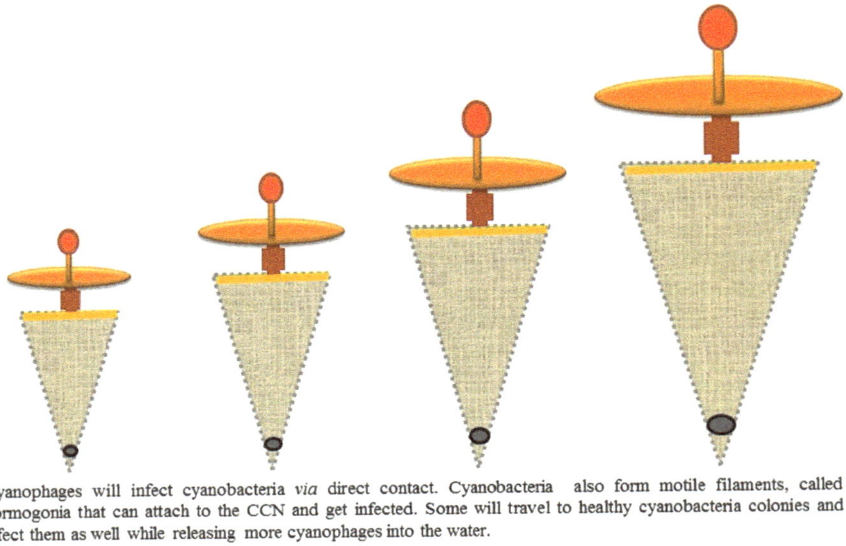

Fig. 6.3 Field applications to control cyanobacteria using various sizes of CCN in standalone applications

The CCN deployment system greatly enhances the likelihood of that infectious encounter thereby empowering this natural control methodology. The current invention is a safe replacement for environmentally harmful chemical compounds or other treatment methods.

6.3 Discussion

The deployment of cyanophages as a biocontrol for CyanoHABs has been problematic since the obligate cyanophage will target cyanobacteria in their immediate environment; however, they are quickly diluted and dispersed into the aquatic environment once released in the water. A solution is required in order to safely harbour the cyanophages awaiting their prey while also attracting the target cyanobacteria.

The encapsulated cyanophages or auto-disseminated cyanophages will concentrate the cyanophages undiluted in aquatic environments and release them sparingly and slowly to ensure safe and efficient cyanophage-cyanobacteria interaction. The PBS in the encapsulated cyanophage suspension will influence cyanophage infection activity longer and also provide suitable conditions for cyanophage-cyanobacteria interaction and auto-dissemination. The design concept illustrated in Fig. 6.1 includes a protective cap made of any non-submersible opaque material that will float on the surface of the water to shield the cyanophages from the sun's heat and UV rays that would otherwise harm the phages or cyanophages. The conical shape of the container below also serves this purpose through the changing angles of the predominant daytime sunlight and direct UV rays. The depth of the cone is made to coincide with the microbial food chain trophic levels occupied by cyanobacteria and the corresponding cyanophage.

The container is made of a resilient fabric or netting that will both contain the cyanophages, encapsulated cyanophages or auto-disseminated cyanophages and attract the cyanobacteria with its rough external texture. The plastic cover serves to preserve the conical shape, contain the cyanophages and also to attach the overflow tubing and valves for deployment in a series. The anchor at the bottom of the cone may be either inside or outside of the enclosure and helps to keep the conical shape while keeping the apparatus in place. Depending on the placement, the apparatus may require further anchoring to account for water currents and weather. The flashing beacon serves to warn boaters, as well as recreational and commercial users of the waterways of the location of the apparatus. Deployment will most often be in shallow coastal or shoreline locations where cyanobacteria are known to propagate which may coincide with other uses of the waterways. The flashing beacon may also serve as a notification activated by microbial test results, allowing the users to know when cyanobacteria are present in excessive amounts or if the cyanophages have successfully controlled the target bacteria.

Once cyanobacteria cells enter the CCN or filaments attach to the surface, the cyanophages naturally penetrate the enclosure and the cells in order to attach to and lyse the cyanobacteria, concluding the process by replicating into progeny cyanophages further stocking our deployment apparatus and surrounding waters with fresh cyanophages. Infected CyanoHABs in the immediate area may also move on to share the infection with other nearby cyanobacteria. The cones may be set up in single (Fig. 6.3), standalone applications or in series (Fig. 6.2) with various biocontrol cones linked through overflow tubes to share the cyanophage population among the series of apparatuses.

6.4 Conclusions 'the Benefits'

1. Encapsulated phages or encapsulated cyanophages can be used directly in freshwater and marine environments to aid in reducing the shedding of target bacteria.
2. The phages or cyanophages can be used as "caplet" or "capsule-shaped tablet" or tablet can be used for treating or preventing harmful algal blooms.
3. Auto-disseminated phages or cyanophage techniques can be used directly in freshwater and marine environments to aid in reducing the shedding of target bacteria with aid of CCN in Fig. 6.1.
4. Conceptual design for phage or cyanophage deployment in Fig. 6.1 can be used for treating or preventing harmful algal blooms such as cyanobacteria, the harmful filamentous cyanobacterium or *Gloeobacter or Synechococcus or*

Prochlorococcus or plastids or all other filamentous and unicellular cyanobacteria, red tides, the harmful dinoflagellate blooms, or *Cochlodinium* sp. or *Cochlodinium polykrikoides* or any other microscopic aquatic harmful algae.

References

Asato Y (1976) Ultraviolet light inactivation and photoreactivation of AS-1 cyanophage in *Anacystis nidulans*. J Bacteriol 126:550–552

Atech (2000) The cost of algal blooms. Report to land and water resources research and development corporation. Canberra, ACT 2601 ISBN 0 64276014 4

Bingham M, Sinha SK, Lupi F (2015) Economic benefits of reducing harmful algal blooms in Lake Erie. Submitted to the International Joint Commission. Environmental Consulting and Technology. Report, 66 pp, Oct 2015. http://ijc.org/files/tinymce/uploaded/Publications/Economic-Benefits-Due-to-Reduction-in-HABs-October-2015.pdf

Bisen PS, Audholia S, Bhatnagar AK, Bagchi SN (1986) Evidence for lysogeny and viral resistance in the cyanobacterium *Phormidium uncinatum*. Curr Microbiol 13:1–5

Bushaw-Newton KL, Sellner KG (1999) Harmful algal blooms. In: NOAA's state of the coast report. Silver Spring, MD: National oceanic and atmospheric administration. NOAA Web http://oceanservice.noaa.gov/websites/retiredsites/sotc_pdf/hab.pdf. Accessed 1999

Cannon RE, Shane MS, Whitaker JM (1976) Interaction of *Plectonema boryanum* (Cyanophyceae) and the LPP-cyanophages in continuous culture. J Phycol 12:418–421

Carmichael W (2008) A world overview—one-hundred-twenty-seven years of research on toxic cyanobacteria—where do we go from here? Adv Exp Med Biol 619:105–125

Castberg T, Larsen A, Sandaa RA et al (2001) Population dynamics and diversity during a bloom of the marine coccolithophorid *Emiliana huxleyi* (Haptophyta). Mar Ecol Prog Ser 221:39–46

Castenholz RW (2001) Phylum BX. Cyanobacteria. Oxygenic photosynthetic bacteria. In: Boone DR, Castenholz RW, Garrity GM (eds) Bergery's manual of systematic bacteriology (The archaea and the deeply branching and phototrophic bacteria), vol 1. Springer, NY, USA, pp 473–599

Clancy S (2008) DNA damage and repair: mechanisms for maintaining DNA integrity. Nature Edu 1(1)

Clokie MRJ, Mann NH (2006) Marine cyanophages and light. Environ Microbiol 8:2074–2082

De Haan H (1993) Solar UV-light penetration and photodegradation of humic substances in peaty lake water. Limnol Oceanogr 38:1072–1076

Drabkova M, Marsalek B (2007) A review of in-lake methods of cyanobacterial blooms control and management. CyanoData Web. http://www.cyanodata.net/review.php. Accessed Apr 2007

Duda K (2009) Harmful bloom in lake Atitlán, Guatemala. EROS Web. http://earthobservatory.nasa.gov/IOTD/view.php?id=41385. Accessed 28 Nov 2009

Foflonker F (2009) Biological methods to control common algal bloom-forming species. MMG 445 basic. Biotech J 5(1):9–24

Frederickson CM, Short SM, Suttle CA (2003) The physical environment affects cyanophage communities in British Columbia inlets. Microb Ecol 46:348–357

Friedberg EC, Walker GC, Siede W (1995) DNA repair and mutagenesis. ASM Press, Washington, D.C., pp 24–31

Fristachi A, Sinclair JL (2008) Occurrence of cyanobacterial harmful algal blooms: Workgroup report. In: Hudnell HK, Back N, Cohen IR, et al. (eds) Cyanobacterial harmful algal blooms: State of the science and research needs, vol 619, chap.3. Springer Science, Ny, pp 45–103. ISBN: 978-0387-75864-0

Fuhrman JA, Suttle CA (1993) Viruses systems in marine planktonic. Oceanogr 6:51–63

Garza DR, Suttle CA (1998) The effect of cyanophages on the mortality of Synechococcus spp. and selection for UV resistant viral communities. Microb Ecol 36:281–292
Hader DP, Sinha RP (2005) Solar ultraviolet radiation-induced DNA damage in aquatic organisms: potential environmental impact. Mutat Res 571:221–233
Jackson DM, Brown GC, Nordin GL, Jackson DW (1992) Autodissemination of a baculovirus for management of tobacco budworms. J Econ Entomol 85:710–719
Jassim SAA, Limoges RG (2013) The impact of changing environmental forces on cyanophage-host interactions in aquatic ecosystems. World J Microbiol Biotechnol 29 (10):1751–1762. doi:10.1007/s11274-013-1358-5
Jassim SAA, AbouFoul KS, Alnajjar AA (2010) Methods and compositions for the control of harmful blooms. WIPO Patent Application WO/2010/109224
Jungblut AD, Lovejoy C, Vincent WF (2010) Global distribution of cyanobacterial ecotypes in the cold biosphere. ISME J 4:191–202
Kalluri SB (2008) A web-based modeling approach for tracking algal blooms in lower Great Lakes. M.Sc. Thesis. State University of New York at Buffalo
Kitamura S-I, Kamata S-I, Nakano S-I, Suzuki S (2004) Solar UV radiation does not inactivate marine birnavirus in coastal seawater. Dis Aquat Org 58:251–254
Liao M-J, Cheng K, Yang J-Y et al (2010) Assessment of UV-B damage in cyanophage PP. Aquat Microb Ecol 58:323–328
Lytle CD, Sagripanti JL (2005) Predicted inactivation of viruses of relevance to biodefense by solar radiation. J Virol 79:14244–14252
Murphy TM, Gordon MP (1981) Photobiology of RNA viruses. In: Fraenkel-Conrat H, Wagner RR (eds) Comprehensive virology. Plenum Press, NY, USA, pp 285–351
Newman PA, Oman LD, Douglass AR et al (2009) What would have happened to the ozone layer if chlorofluorocarbons (CFCs) had not been regulated? Atmos Chem Phys 9(6):2113–2128
Phlips EJ, Monegue RL, Aldridge FJ (1990) Cynophages which impact bloom-forming cyanobacteria. J Aquat Plant Manage 28:92–97
Proctor LM, Fuhrman JA (1990) Viral mortality of marine bacteria and cyanobacteria. Nature 343:60–62
Schirrmeister BE, Antonelli A, Bagheri HC (2011) The origin of multicellularity in cyanobacteria. BMC Evol Bio 11:45. http://www.biomedcentral.com/1471-2148/11/45
Sigee DC, Glenn R, Andrews MJ et al (1999) Biological control of cyanobacteria: principles and possibilities. Hydrobiologia 395(396):161–172
Sinton LW, Hall CH, Lynch PA, Davies-Colley RJ (2002) Sunlight inactivation of fecal indicator bacteria and bacteriophages from waste stabilization pond effluent in fresh and saline waters. Appl Environ Microbiol 68:1122–1131
Sivonen K, Jones G (1999) Cyanobacterial toxins (Chapter 3). In: Chorus I, Bartram J (eds) Toxic cyanobacteria in water: a guide to their public health consequences, monitoring and management. WHO. ISBN 0-419-23930-8
Sommaruga R (2003) UVR and its effects on species interactions. In: Helbling EW, Zagarese HE (eds) UV effects in aquatic organisms and ecosystems. RSC, Cambridge, UK, pp 485–507
Stewart GSAB, Denyer SP, Jassim SAA (1995) Selective virus culture. PCT patent WO95/023848
Stoddard LI, Martiny JBH, Marston MF (2007) Selection and characterization of cyanophage resistance in marine Synechococcus strains. Appl Environ Microbiol 73:5516–5522
Suttle CA (1994) The significance of viruses to mortality in aquatic microbial communities. Microb Ecol 28:237–243
Suttle CA (2005) Viruses in the sea. Nature 437:356–361
Suttle CA, Chen F (1992) Mechanisms and rates of decay of marine viruses in seawater. Appl Environ Microbiol 58:3721–3729
Suttle CA, Chan AM (1994) Dynamics and distribution of cyanophages and their elect on marine *Synechococcus* spp. Appl Environ Microbiol 60:3167–3174
Suttle CA, Chan AM, Cottrell MT (1990) Infection of phytoplankton by viruses and reduction of primary productivity. Nature 347:467–469

Svobodova AR, Galandakova A, Sianska J et al (2012) DNA damage after acute exposure of mice skin to physiological doses of UVB and UVA light. Arch Dermatol Res 304:407–412
Tucker S, Pollard P (2005) Identification of cyanophage Ma-LBP and infection of the cyanobacterium Microcystis aeruginosa from an Australian subtropical lake by the virus. Appl Environ Microbiol 71:629–635
Van Hannen EJ, Zwart G, Van Agterveld MP et al (1999) Changes in bacterial and eukaryotic community structure after mass lysis of filamentous cyanobacteria associated with viruses. Appl Environ Microbiol 65795–801
Vickers RA, Furlong MJ, White A, Pell JK (2004) Initiation of fungal epizootics in diamondback moth populations within a large field cage: proof of concept for auto-dissemination. Entomol Exp Applic 111:7–17
Vincent WF, Roy S (1993) Solar ultraviolet-B radiation and aquatic primary production: damage, protection, and recovery. Environ Rev 1:1–12
Waterbury JB, Valois FW (1993) Resistance to co-occurring phages enables marine Synechococcus communities to coexist with cyanophages abundant in seawater. Appl Environ Microbiol 59:3393–3399
Weinbauer MG, Rassoulzadegan F (2004) Are viruses driving microbial diversification and diversity. Environ Microbiol 6:1–11
Weinbauer MG, Wilhelm SW (2011) Marine viruses. In: Gattuso J-P (ed) The encyclopaedia of earth http://www.eoearth.org/article/Marine_viruses. Accessed 5 Nov 2011
Whitton BA, Potts M (2000) Introduction to cyanobacteria. In: Whitton BA, Potts M (eds) The ecology of cyanobacteria: their diversity in time and space. Kluwer Academic Publishers, Dordrecht, London, Boston, pp 1–11
Wilson WH, Carr NG, Mann NH (1996) The effect of phosphate status on the kinetics of cyanophage infection in the oceanic *Cyanobacterium synechococcus* spp. WH7803. J Phycol 32:506–516
Winter JG, DeSellas AM, Fletcher R et al (2011) Algal blooms in Ontario, Canada: increases in reports since 1994. Lake Reserv Manage 27:107–114. doi:10.1080/07438141.2011.557765
Wommack KE, Colwell RR (2000) Virioplankton: viruses in aquatic ecosystems. Microbiol Mol Biol Rev 64:69–114
Yoshida T, Takashima Y, Tomaru Y et al (2006) Isolation and characterization of a cyanophage infecting the toxic cyanobacterium *Microcystis aeruginosa*. Appl Environ Microbiol 72 (2):1239–1247

Chapter 7
Bacteriophage Biodefense

Abstract Bacteriophages are a double-edged sword in nature and in the hands of a skilled scientist. On the one hand they are able to spread invasive exotoxins, virulence factors of antimicrobial resistance (AMR), to other bacteria via lysogenic bacteriophages which could be used as a possible biothreat. On the other hand, they can be a safeguard or a watchdog to provide enormous benefit via lytic bacteriophages to combat bioterrorism and by developing many novel environmentally friendly technologies. For example, we may develop rapid diagnostics for biothreats or bioalarms and biocontrol to prevent and treat attacks in situ. We constructed a model that can contain the impact of classic agents of biological warfare such as *Bacillus anthracis* spores' using aerosols to protect people and animals in crowded areas, malls, schools, stadia, airports, on farms, etc. Other models are used to illustrate the rapid implementation of programmed bacteriophages to engage or dismantle threats post attack with bacterial pathogens. Please note that although our phage programming technology (Chap. 1) can be used to induce the development of a lytic phage from the prophage state, this approach is not practical when dealing with a biothreat agent since the procedure requires a few days or even weeks to complete on samples that are collected. Thus we use highly virulent, lytic programmed phages that can effectively compete and swiftly target the bacterial biothreat agents and to ensure the bacteria are free of prophage contaminants. We also present some important means of reducing bacterial transmission and infection in this chapter.

Keywords Antimicrobial resistance · Anthrax spores · *Bacillus anthracis* · Bacterial toxins · Bacteriophage · Biodefense · Bioterrorism · Biothreats · Bioalarms · Biocontrol · Biosensors · Bioluminescence · Lysogenic phages · Phage biocontrol · Phage fogger · Rapid bacterial detection

S.A.A. Jassim and R.G. Limoges, *Bacteriophages: Practical Applications for Nature's Biocontrol*, DOI 10.1007/978-3-319-54051-1_7

Contents

7.1 Introduction

Bioterrorism and its potential for mass destruction have been subjects of increasing international concern (Kaufmann et al. 1997). Terrorism is defined as use of unlawful violence or threat of unlawful violence to incite fear; intended to coerce or to intimidate governments or societies in the pursuit of goals that are generally political, social or religious (Binczycka-Anholcer and Imiołek 2011; Daniszewski 2013). Bioterrorism is terrorism by threat or intentional release or dissemination of biological agents, mainly bacteria or viruses (Binczycka-Anholcer and Imiołek 2011). Terrorists groups and individuals with grievances against the government or society have been known to use, threaten or plan to use biological weapons to further personal causes. The bioterrorist attacks in the USA in 2001 resulted in the first 10 confirmed cases of inhalational anthrax caused by intentional release of *Bacillus anthracis*. Epidemiologic investigation indicated that the outbreak, in the District of Columbia, Florida, New Jersey, and New York, resulted from intentional delivery of *B. anthracis* spores through mailed letters or packages (Jernigan et al. 2001).

To produce and effectively use biologic weapons, a terrorist only needs modest microbiologic skills along with access to a virulent strain of the desired agent. Production costs are low, and aerosol dispersal equipment from commercial sources can be adapted for biologic weapon dissemination (Daniszewski 2013). The impact of a bioterrorist attack depends on the specific biological agent (viruses, bacteria and fungus) or biological toxins used, the method and efficiency of dispersal, the population exposed, the level of immunity in the population, the availability of effective post-exposure and/or therapeutic regimens, and the potential for secondary transmission (Kaufmann et al. 1997). Because of our focus on antibiotics resistance, this chapter will cover only the critical bacterial agents (see below). It is also important to recognize that no priority was assigned within the categories and that the list did not rank the probability of deliberate use of an agent.

7.2 Antibiotic-Resistant Bacteria as Potential Agents of Bioterrorism

Antibiotics (antimicrobials) have been used for the last 80 years to treat patients with infectious bacterial diseases. Since the 1940s, humans have abused antibiotics to the point that we threaten their availability for future generations. There is also a lack of new drug development by the pharmaceutical industry due to reduced economic incentives and challenging regulatory requirements (Jassim and Limoges 2014; Ventola 2015). Antibiotics have been used so widely in animal agriculture and by humans for so long that the infectious organisms the antibiotics are designed to kill have adapted, making the antibiotics less effective, this is discussed in greater detail in Chaps. 3 and 4. As a consequence, antimicrobial resistance (AMR) in bacteria is occurring worldwide, endangering the efficacy of antibiotics, which have previously transformed medicine and saved millions of lives (Sengupta et al. 2013; Shute 2013; Gould and Bal 2013; Golkar et al. 2014; Jassim and Limoges 2014; O'Neill 2014; Wright 2014; CRSR-US 2005; Ventola 2015).

Recent reports have revealed that 70% of the world's bacteria have now developed a resistance to antibiotics, including those drugs regarded as our last line of defence which were carelessly sprayed over fields and livestock and have now polluted waterways (Shute 2013; Jassim et al. 2016). As a consequence of this, many common infections will no longer have a cure and hospital surgeries are more dangerous due to the spread of hospital acquired bacteria which are resistant to antibiotics. Therefore, people or farm animals exposed to those agents would become sick with mostly fatal disease. The research effort is aimed at population protection against bacteria affecting skin, blood, lungs, muscles, etc. As a result, AMR has evolved and is increasingly claiming lives. Currently there are more than 25,000 deaths each year in Europe (Shute 2013), and in the US, at least 2 million people become infected with bacteria that are resistant to antibiotics resulting in 23,000 deaths as a direct result of these infections (CDC 2016). By 2050 a continued rise in resistance would result in 10 million people dying worldwide every year costing up to 100 trillion USD (O'Neill 2014). The latter studies estimate that about 300 million people are expected to die prematurely because of drug resistance over the next 35 years. Obviously AMR has developed into a bacterial infections threat (Jassim and Limoges 2014; O'Neill 2014; Spellberg and Gilbert 2014). This unpleasant and grim forecast of new waves of bacterial super-infection challenges global public health and many governments around the world are starting to take an active role in research to address this issue (Shute 2013).

The CDC (2013) has classified a number of bacteria as presenting urgent threats [*Clostridium difficile*, carbapenem-resistant *Enterobacteriaceae* (CRE) and drug-resistant *Neisseria gonorrhoeae*]; serious threats [multidrug-resistant *Acinetobacter*, drug-resistant *Campylobacter*, extended spectrum beta-lactamase-producing *Enterobacteriaceae* (ESBLs), vancomycin-resistant enterococci (VRE), multidrug-resistant *Pseudomonas aeruginosa*, drug-resistant nontyphoidal

Salmonella, drug-resistant *Salmonella typhimurium*, drug-resistant *Shigella*, methicillin-resistant *Staphylococcus aureus* (MRSA), drug-resistant *Streptococcus pneumoniae* and drug-resistant tuberculosis]; and concerning threats [vancomycin-resistant *Staphylococcus aureus* (VRSA), erythromycin-resistant group A *Streptococcus* and clindamycin-resistant group B *Streptococcus*]. Many of these are already responsible for placing a substantial clinical and financial burden on the US health care system, patients, and their families. Coordinated efforts to implement new policies, renew research efforts, and pursue steps to manage the crisis are greatly needed (Ventola 2015). Certainly the impact of three classic agents of biological warfare, *B. anthracis* spores, *Brucella melitensis*, and *Francisella tularensis* with their potential use for bioterrorism, coupled with antibiotic resistance puts humanity in the face of a terrifying threat (Kaufmann et al. 1997; Grilló et al. 2006; WHO 2007; Weigel and Morse 2009; Daniszewski 2013). Antibiotic resistance in bacteria from environmental samples may cause serious problems in treating anthrax (Stroud et al. 2012; Jończyk-Matysiak et al. 2014). Other researchers also have considered *Yersinia pestis* (plague), *Coxiella burnetii*, *Burkholderia mallei*, *Burkholderia pseudomallei*, *Escherichia coli* O157:H7 and *Vibrio cholerae* as potential agents of bioterrorism (Khan et al. 2000; Weigel and Morse 2009).

The biggest nightmare that haunts homeland security in many countries around the world is when, where, and how governments can protect their nations if terrorists deliberately disseminate one or a combination of these bacteria via a number of different routes, including air, water, food, and infected vectors, to affect the health of humans. Thus bioterrorism is an important theoretical threat to be taken seriously in order to protect humanity. In this regard, over the last decade the US Congress began to address this challenge by providing funding to the Centers for Disease Control and Prevention (CDC) to enhance the ability of the nation's epidemiology and laboratory systems to respond to the deliberate release of a biological agent (Khan et al. 2000). The review of Weigel and Morse (2009) addressed this and many more issues such as Strategic National Stockpile (SNS) which was established to provide large quantities of essential medical material to states and communities during such an emergency, including intravenous administration kits, airway maintenance supplies, and medical/surgical items. The antibiotics in the SNS were selected, in part, for their effectiveness on the basis of the current data for antimicrobial susceptibility of each bacterial species. The review has also elaborated upon the revelations during the last decade suggesting that in the former Soviet Union, a priority of the offensive biological weapons program was the development of recombinant organisms that were resistant to common therapies. Since the aim of this book is to enable the use of phages as novel biocontrol agents to treat drug-resistant bacteria, we can address health security and public health needs against such biothreats as well.

7.3 Bacteriophages

Bacteriophages (phages for short) are bacterial viruses and represent one of the simplest life forms that are potentially the most numerous organisms on earth. They are found in the guts of many animals and are also found in great numbers in marine settings. Oceans and seas contain an unusually high number of phages, and some estimates put the number of bacteria infected with phages in sea waters to a staggering 70%. They are natural bacterial killers and most bacterial species have sensitivity to one or more phages. Phages that infect bacteria can be generally classified as virulent 'lytic' or temperate, this is discussed in greater detail in Chap. 2.

Different kinds of phages contain different material; however, all phages contain nucleic acid and protein. The nucleic acid of the phage can be either deoxyribose nucleic acid (DNA, in which most of them have double-stranded DNA) or ribonucleic acid (RNA, most have single-stranded). Some phages have single-stranded or double-stranded genomes, spanning between 4 and 200 genes. Some phages kill (known as lytic or virulent) every cell they infect; others are temperate, which means that they are able to integrate themselves into the bacterial genome as dormant 'prophages', also called lysogens (this is discussed in greater detail in Chap. 2). Phages also contain a head structure, collar, neck, tall, fibers, and pins (Fig. 7.1). Within the head is the nucleic acid, and the structure itself is composed of several copies of one or more proteins. Many phages also have a tail through which they pass the nucleic acid during infection. The phage infective cycle consists of either the lytic stage or the lysogenic stage. This is discussed in greater detail in Chaps. 1, 2 and 4.

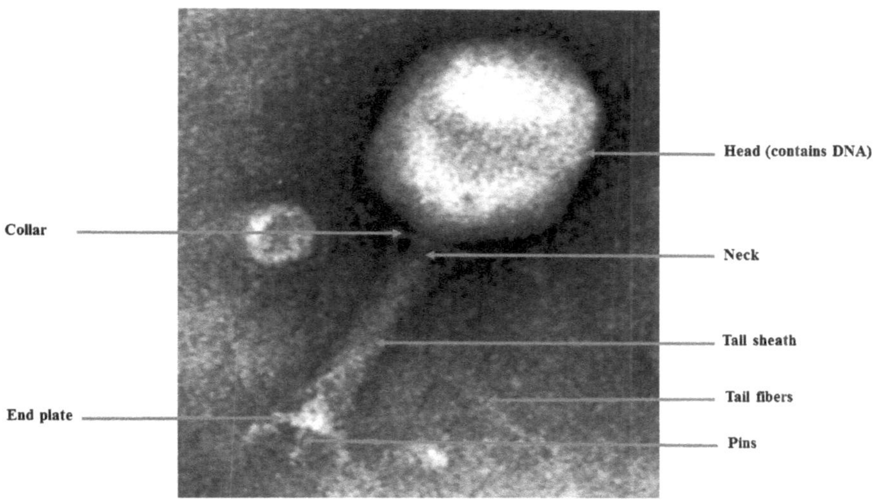

Fig. 7.1 Structure phage *E. coli* O157:H7

7.3.1 Lytic Cycle

When a lytic phage infects a bacterial cell, it undergoes replication 'multiplies vegetatively' followed by the release of viral progeny, which results in host lysis killing the host cell at the end of the cycle. The lytic cycle is composed of the following stages:

1. Searching for a susceptible host cell.
2. Phage adherence/adsorption/binding: Phage attaches itself to the target bacterial cell. Adsorption of specific receptors on the bacterial surface takes place, which can involve lipopolysaccharides, proteins or even flagella.
3. Injection of genetic material: The phages via a syringe-like tail, inject their genetic material. After finding the right receptor, the end (base) plate is brought closer to the cell surface. Tail fibers help accomplish this, and after attachment the tail fibers contract, releasing genetic material into the host membrane.
4. Latent or eclipse phase: This is similar to the bacterial lag phase. During the latent phase, the genetic material of the phage takes over the cell machinery of the host, and specified phage m-RNA's, as well as proteins, are produced. The production of the host's genetic material is halted, and the host cell becomes a "factory" for viral material. Phage DNA is also produced, followed by the production of late m-RNA's and late proteins. The purpose of the late proteins is the lysis of the bacterial cell.
5. Progeny virion assembly: Helper proteins assemble new virions, which infect other host cells upon release. New phages are assembled as follows: First, the base plates, followed by the tails. The head capsids are separately assembled and join the tails. Genetic DNA is placed within the head capsids. While all this is happening, the cell is drawing raw material from its environment. Also, the genes of the phage allow only viral components to be built. The host's DNA becomes completely inactive or is destroyed.
6. Lysis phase: In this final phase, two specific enzymes break down the cell wall peptidoglycan, resulting in the host cell bursting and new phages being released to search for new host cells, so the cycle can continue. In this process there are two enzymes:

   (i) Holins are a diverse group of small proteins produced by dsDNA phages in order to trigger and control the degradation of the host's cell wall at the end of the lytic cycle. Holins form pores in the host's cell membrane, allowing lysins to reach and degrade peptidoglycan, a component of bacterial cell walls.

   (ii) Lysins, also known as endolysins or murein hydrolases, are hydrolytic enzymes produced by phages in order to cleave the host's cell wall during the final stage of the lytic cycle. Lysins are highly evolved enzymes that are able to target one of the five bonds in peptidoglycan (murein), the main component of bacterial cell walls, which allows the release of progeny virions from the lysed cell. These enzymes are being used as antibacterial agents due to their high effectiveness and specificity in comparison with antibiotics, which are susceptible to bacterial resistance (see Chap. 2).

7.3.2 Temperate Phages

Temperate phages can have two ecological roles after infection of a bacterium. Either there is production of viral progeny and host cell lysis similar to what occurs for lytic phages 'or predators', or lysogenic 'parasite' cycles occur, whereby the phage genome can exist indefinitely by being inserted in the bacterial chromosome or as an episome (known as the prophage state).

The lysogenic life cycle of phages lambda and P1, for example, ensures the replication of the integrated prophage along with the bacterial genome for many generations. In the lysogenic cycle, lysis is delayed, and the phage becomes part of the host for some period of time. The phage remains dormant and can become active at any time, directing the synthesis of viral material. The following stages are part of the lysogenic cycle:

1. Adsorption: This stage is similar to the lytic cycle phase of adsorption.
2. Injection of genetic material: Similar to the lytic cycle phase.
3. Replication: In this phase, the genetic material of the phage is neither manufactured nor transcribed significantly.
4. Repression: In this phase, a protein, called a repressor, is made to bind to a specific site on the DNA of the phage. This site is called the operator. The purpose of the repressor is to shut off the transcription process of nearly all the phage genes, except the repressor gene. This results in a genome that is repressed and becomes assimilated into the host chromosome.
5. End of Lysogeny: When stressful conditions (such as exposure of the cell to UV radiation, or ionizing radiation or thymine starvation) occurs through damage of the DNA, which subserve the imminent death of the host, the cell will produce proteases, which in turn cause the destruction of the repressor protein. The phage switches to the lytic cycle, which results in the host cell lysis and in the release phage genes and lytic multiplication, bringing the lysogenic cycle to an end (Jassim and Limoges 2013; Chap. 2).

However, many lysogenic phages express gene products that have subtle effects on the phenotype of the host cell. This process is called lysogenic conversion. Recently, it became clear that phages establish conditional symbiotic relationships with their bacterial hosts through lysogeny which increases host fitness or virulence factors allowing the bacteria to successfully infect hosts and expand their distribution (Alexa et al. 2001; Hayashi et al. 2001; Krylov 2001; Brüssow et al. 2004; Skurnik and Strauch 2006; Sousa 2006; Schuch and Fischetti 2006; Skurnik et al. 2007; Merabishvili et al. 2009; Mokrousov 2009; Gill and Hyman 2010;

Kutter et al. 2010; Vondruskova et al. 2010; Casas and Maloy 2011; Charles and Ryan 2011; Keen 2012; Barra et al. 2013; Jassim and Limoges 2014; Chap. 2). Interestingly, it was found that lysogenic strains have higher metabolic rates than non-lysogenic (Gama et al. 2013). Therefore, such lysogens reproduce faster than non-lysogenic strains (Edlin et al. 1975, 1977; Lin et al. 1977), outcompeting them in environments where nutrients are scarce (Gama et al. 2013). This reduction of nutrients concentration would reduce Q protein expression to a level that provides more opportunity for lysogenic response (Jassim and Limoges 2013). In general, it has been suggested that in nutrient-rich waters, lytic phages dominate, whereas in nutrient-poor waters, lysogenic viruses dominate (Jassim and Limoges 2013).

7.4 Phages as Potential Bioterrorism Agents

Phages are realized to be numerous and important components of therapy, principally because of their lytic capabilities. On the other hand, it has been proposed that pathogens and parasites in general may be useful to their hosts as biological weapons (Dionisio 2007). In particular, phages may be useful to bacteria as antagonistic allelopathic entities (Brown et al. 2006). Allelopathy is a biological process by which an organism produces one or more biochemicals that influence the germination, growth, survival, and reproduction of other organisms. However, this allelopathic agent is not a biochemical agent, but a "replicating agent" antagonistically affecting the growth, reproduction, or survival of other organisms (Brown et al. 2009a, b). Thus lysogenic cells may use viral particles in a way similar to the use of toxins to kill non-lysogenic susceptible strains, thus possessing a fitness advantage over them (Gama et al. 2013). Whereas, the phages may replicate inside bacteria, which die in the process of phage release (Gama et al. 2013). Therefore, in contrast to typical toxins, the ability of phages to replicate inside their victims would give an additional advantage to the lysogenic bacterial population (Brown et al. 2006; Joo et al. 2006; Dionisio 2007). The amplification of phages inside their victims is particularly significant when only a small number of carriers of phages invade a susceptible bacterial population: within a few hours, the victims can substantially amplify the number of viral particles (Brown et al. 2006). Alternatively, as mentioned in Chap. 2, temperate phages can establish a symbiotic relationship with the host. In this case, the cell is not killed and becomes immune to similar viruses, thereby bacterial cells may use temperate viruses as biological shields, after a few days only the classical view of temperate bacterial viruses as parasitic agents prevails (Gama et al. 2013).

Phages would be useful as biological weapons for some bacteria to displace other bacteria in habitats such as mammalian guts, precisely because of the capacity of phages to replicate themselves inside their victims (Brown et al. 2006). The subtle changes that temperate phages impart to their hosts in the environment reduce the effectiveness of phage therapy (Jassim and Limoges 2014). It has been hypothesized that there is a 'mutualism' in the host's body, where bacteria were less virulent until the prophages were induced by environmental conditions such as exposure to antibiotics dismutase (Wanger and Waldor 2002). Phage-encoded virulence factors could then be up-regulated, which could contribute to the pathogenesis of the bacteria. Subsequently, the lytic release of virions could also amplify the number of infected bacteria carrying the phage-encoded virulence factor. Phages can contribute to bacteria virulence through transduction and regulation of phage-encoded virulence factors, alteration of bacterial virulence factors and function in the control of bacterial virulence through in situ prophage induction (Wanger and Waldor 2002). Some phages carry DNA that changes the phenotype of the bacterial host (lysogenic conversion genes) and this lysogeny is a form of short-term bacterial evolution (Canchaya et al. 2003). The lysogenic phage or 'prophage' will drive the adaptive evolution of bacteria to achieve more powerful virulence factors inherited from previously infected bacteria via transduction, *i.e.*, the transfer of genetic material to a bacterial cell via phage infection (Jassim and Limoges 2014).

Transduction is one of the main driving forces of horizontal gene transfer in microbial evolution, and it is also an important mechanism for altering bacterial virulence, because lysogenic phages often carry virulence genes (this is discussed in greater detail in Chap. 2). The expression of any virulence factor can benefit the bacterium host and the bacterial population, as the encoding phage/prophage allows more efficient adaptation to changing host challenges, potentially leading to fitter and more virulent clones (Abedon and Lejeune 2007; Vojtek et al. 2008; Casas and Maloy 2011). This may represent a potentially serious hazard to humans, animals and plants (Saunders et al. 2001; Verheust et al. 2010). It is noteworthy that lateral gene transfer virulence factors can also be accomplished through the lysogenic phages, which harbour a multitude of prophages and each phage-encoded virulence factor makes an incremental contribution to the fitness of the lysogenic phage (Biscay and Murphy 1988; Brüssow et al. 2004; Abedon and Lejeune 2007; Vojtek et al. 2008; Verheust et al. 2010; Casas and Maloy 2011). A significant concern with using phages is their considerable propensity to transfer genetic material among their host cells via the process referred to as transduction. With transduction estimated to occur at the phenomenal rate of 2×10^{16} times per second in aquatic environments, phages are quite possibly one of the primary drivers of bacterial evolution (Canchaya et al. 2003). The role of phages as natural vectors for some of the most potent bacterial toxins is well recognized and includes classical type I membrane-acting super antigens, type II pore-forming lysins, and type III exotoxins, such as diphtheriae and botulinum toxins 'BoNT' and among Gram-negative pathogens. A novel class of bacterial virulence factors called effector proteins (EPs) are phage encoded among distinct pathovars such as enterohemorrhagic *E. coli* (EHEC) serotype O157:H7, a human pathogen responsible for outbreaks of bloody diarrhea and hemolytic uremic syndrome worldwide,

Table 7.1 Examples of some phage encoded exotoxins (virulence factors) that will lead to the evolution of pathogenic bacteria (Boyd and Brüssow 2002; Boyd 2012)

Host bacteria	Phage 'vector'	Virulence factor	Disease
Streptococcus pyogenes	T12 CS112	Erythrogenic toxin A Erythrogenic toxin C	Scarlet fever
Staphylococcus aureus	N315 Mu50A Φ13 ΦETA ΦPVL	Enterotoxin P Enterotoxin A Enterotoxin A Exfoliative toxin A Panton-valentine Leucocidin S, F	Vomiting Food poisoning syndrome Gastroenteritis Scalded-skin syndrome Leukocytes destruction
Vibrio cholerae	CTXΦ	Cholerae toxin A, B	Cholerae
Corynebacterium diphtheriae	Phage-β	Diphtheriae toxin	Diphtheriae
Escherichia coli	H-19B, 933 W ΦC3208	Shigalike toxin 1A, 1B, 2A, 2B Hemolysin	Hemorrhagic diarrhea, dysentery Hemolysis
Pseudomonas aeruginosa	ΦCTX	Cytotoxin	Cytotoxicity
Clostridium botulinum	Phage C1	Neurotoxin	Botulism (food poisoning)

Shigella spp., and *Salmonella enterica* (Biscay and Murphy 1988; Boyd and Brüssow 2002; Boyd 2012; Table 7.1), these natural vectors have led to the evolution of pathogenic bacteria (Verheust et al. 2010).

The phage exotoxin-encoded toxins (e.g. botulism toxin, diphtheriae toxin, cholerae toxin, and Shiga toxin) are secreted polypeptides that modulate the virulence of the host bacteria. Exotoxin genes carried by phages can be integrated into the bacterial chromosome and are responsible for many of the symptoms associated with human diseases (Mokrousov 2009; Charles and Ryan 2011; Keen 2012). However, phage infection of bacteria is increasingly associated with additional effects on bacterial virulence. Phages can alter host bacterial properties relevant to all stages of the infectious process including bacterial adhesion, colonization, invasion, and spread through human tissues; resistance to immune defenses; exotoxin production; sensitivity to antibiotics; and transmissibility among humans (Wagner and Waldor 2002). Some examples of virulence factors carried by phage are shown in the data of Vojtek et al. (2008) indicating that horizontal transfer of lysogenic phages among group A *Streptococcus* can occur across the M-type barrier; these data also provide further support for the hypothesis that toxigenic conversion can occur via lysogeny between species. The conversion of harmless strains of *Corynebacterium diphtheriae* or *Vibrio cholerae* by phages to highly virulent ones, cause diphtheriae or cholera, respectively (Mokrousov 2009; Charles and Ryan 2011).

Strategies to combat certain bacterial infections by targeting these toxin-encoding prophages have been proposed (Keen 2012). The effect of lysogenic conversion can be seen clearly in the disease cholerae. *Vibrio cholerae* toxin gene ctxABT is caused by a Gram negative, curved rod called *V. cholerae*. The bacterium is transmitted through contaminated water and results in severe diarrhea and rapid dehydration of the infected person. Infection with the lysogenic phage CTXΦ gives the bacterium its toxinogenicity. The phage recognizes a pilus on the surface of the bacterium and uses it to enter the cell. Once inside the cell, the CTXΦ integrates into the chromosome and the lysogen expresses cholerae toxin (Thomas 1966). The CTXΦ has received special attention because it is the first filamentous phage that encodes cholerae toxin and is given as the sole example to date of a single-stranded DNA phage that encodes a bacterial toxin. The important lesson from this discovery is that many different types of phages may carry virulence factors, and transfer of virulence genes by phage may be a major mechanism in the evolution of new bacterial diseases. It was also found that diphtheriae toxin is produced when bacillus is infected by the β-coryneform phage (Waldor and Mekalanos 1996). On the other hand, lysogenic phage mediation has also been shown to induce the production of Shiga toxins (Stx-I, Stx-II) by the EHEC (Hamelin et al. 2007). The shiga toxin-producing *E. coli* epidemic was reported in Germany in 2010 (Bielaszewska et al. 2011).

Lysogenic phages can readily integrate deadly virulence genes into a host bacterial chromosome, thus turning a harmless bacterium into a deadly killer. In some cases, the expression of these genes may be driven in part by induction of the prophages that encode them. Since prophage induction contributes to virulence gene expression and since the human body is replete with prophage-inducing chemicals, prophage induction in the human body may be an important, under recognized mechanism of bacterial pathogenesis (Wagner and Waldor 2002). Based on this principle, lysogenic phages containing virulence genes and/or broad-spectrum drug-resistance genes could be used as tools to turn non-pathogenic bacteria into pandrug resistant bacteria which are resistant to all known antimicrobial agents (Fan and Tong 2012). The latter authors have concluded that such lysogenic phages could be directly used as biological weapons that could be dispersed in an appropriate environment to generate highly pathogenic, drug-resistant biological killers. Bioterrorist attacks of this nature would be well concealed and it would be extremely difficult to predict, detect or prevent dispersal of the biological agents derived from lysogenic phages. Moreover, it would be very difficult to distinguish bacterial agents transformed by lysogenic phages from naturally occurring bacteria, and tracing them would also be problematic. Therefore, lysogenic phages could be exploited by terrorists in the future. Because terrorists constantly upgrade their methods of attack, an unpredicted threat of this nature would pose an unexpected safety risk (Fan and Tong 2012).

7.5 Smart Lytic Phages as Effective Anti-bioterrorism Agents

Phages have made important contributions to the field of molecular biology because they have simple structures and a small number of genes. In this regard, phages are introduced in genetic engineering and in exploring molecular biological processes. However, in terms of bioterrorism and anti-bioterrorism, few reports have considered the use of phages as a potential incapacitative agent against bioterrorism, in particular for *B. anthracis* and antimicrobial resistance (AMR) in bacteria (Inal 2003; Fan and Tong 2012; Jończyk-Matysiak et al. 2014). Currently 70% of bacterial strains/species have developed AMR (Shute 2013; Jassim et al. 2016). Coupled with the decline in research for new classes of antibiotics these pose a real threat to mankind as well as the animal and plant kingdoms. This strongly suggests that phages can be effective biological tools in treating bacterial diseases and should also be considered a potential tool against bioterrorism.

Phages have been used in therapy for many decades in the former Soviet Union and more recently there have been many phage products developed in the western world with FDA approval for use in therapy and as a food biocontrol (see Chap. 3). Nevertheless, they have been described as a double-edged sword (Brown et al. 2006; Gama et al. 2013). The core problem with phages is that lysogenic phages appear to play an important role in transferring virulence genes and/or broad-spectrum drug-resistance genes which could be used as tools in new biological weapons. In this regard, if the lysogenic phage infects its host bacteria, it will transfer new aggression and aggressive pathogenic components. It has been shown that such a pathogen may be able to cause infection to both humans and animals in parallel. This compatibility relationship of the lysogenic phages to host pathogen has unfortunately been largely ignored by pathologists. One good reason for this may be that compatibility relationships are relatively difficult to investigate empirically. In addition, it requires rapid diagnostic tools to uncover empirical evidence to describe a convincing genetic mechanism of the gene-to-gene interaction for each phage lysogenesis with a bacterial host. Human epidemiology has repeatedly proven the dramatic importance of these compatibility relationships between lysogenic phage and bacteria for developing a new wave of diseases. There are many examples of bacteria benefiting from phage lysogenesis by inheriting new aggression and aggressive pathogenic components and antibiotic resistance. Previously harmless to humans, they became lethal as illustrated in Table 7.1. There is now convincing logic that the novel weapons hypothesis can be expanded to include phage lysogenesis, this includes threats from naturally occurring phages as well as man-made, able to colonise different bacterial strains and species. The bacterial hosts work as vectors carrying new waves of bacterial diseases which can act as biological weapons if they infect and kill humans and animals. Governments around the world have struggled with various responses to bio events including biosurveillance systems to enhance early detection as well as training and other initiatives. However, the reliability of these approaches has not been absolute, therefore presenting the dilemma of how to prepare and respond to potential biothreats.

Let's hypothesize two theories as potential biothreats these are the AMR bacteria and lysogenic phages. In order to prepare for such an attack, we need to formulate the basic strategies to prevent them, and be prepared to use biocontrol agents that are inexpensive and environmentally friendly to rapidly neutralize and kill the bacterial pathogens in situ, as discussed below:

7.5.1 AMR as a Biothreat

7.5.1.1 Rapid Bacterial Detection

In the fight against terrorist attacks with pandrug resistant bacteria, we obviously need rapid detection and diagnostics, qualitative and quantitative (Q&Q) of bacterial pathogens. Current food and drug administration methods for detecting contamination of foodstuffs with foodborne pathogens consist of conventional culture techniques, immunoassays, and biochemical analyses (Andrews and Hammack 2001). Conventional culture methods used for the detection and isolation of bacterial pathogens include, non-selective pre-enrichment followed by selective enrichment and plating on selective and differential agars. Suspect colonies are then confirmed biochemically and serologically. The complete test is laborious and time consuming, generally requiring 5 days to obtain a negative result and up to 7 days to confirm a positive result. However, despite all these drawbacks, fundamental microbiological assays remain the benchmark of most pathogen detection schemes. With significant demand to detect pathogens in minutes rather than hours or days, alternatives to culture methods have been introduced. The enzyme-linked immunosorbent assay (ELISA) depends on the binding of a specific antibody to an antigen on the bacterium for rapid bacterial detection in food products and is commercially available in different formats (dePaula et al. 2002; Fukuda et al. 2005). Some of these immunoassays, which also require a pre-enrichment cycle that may take up to 24 h, therefore requiring up to 36 h before results are known, have been validated as rapid detection methods (Andrews and Hammack 2001; Taitt et al. 2004), with the detection limits determined in the range of 10^3 colony forming unit (cfu) ml^{-1} (Taitt et al. 2004). The nucleic acid hybridization and polymerase chain reaction (PCR) techniques which rely on the detection of a target organism's nucleic acid, have also been developed and have proven successful in detecting *Salmonella*, EHEC and *Campylobacter* in foods and poultry-related samples. PCR is an exquisitely sensitive technique and is able to detect 1–100 *Salmonella* in pure culture (Malorny et al. 2003; Taitt et al. 2004). A large number of papers describing PCR-based detection of *Salmonella* from spiked or naturally contaminated foods have claimed bacterial limits of detections (LODs) ranging from 0.1 to 100 cfu g^{-1}

(Taitt et al. 2004). Noteworthy, many of these LODs represent the concentrations of inoculums used, rather than actual bacterial counts after enrichment; post-enrichment concentrations can reach 10^6–10^9 cfu ml^{-1} (Taitt et al. 2004). A PCR-enzyme-linked immunosorbent assay system able to detect 10^2–10^4 cfu of *Campylobacter coli*, *C. jejuni*, and *Salmonella enterica* in a cm^2 poultry carcasse sample without pre-enrichment has also been described (Hong et al. 2003). Current, PCR technology allows for rapid detection of pathogens in real time and can provide Q&Q information of infectious diseases for biodefence (Saunders and Lee 2013).

In general, nucleic acid-based analysis PCR does have its limitations, for example some foods and fecal matter have proven challenging due to the presence of inhibitors of PCR amplification, such as proteases and phenolic compounds (Dickinson et al. 1995). This gives false-negative and false-positive results that may be encountered with the daily running of PCR assays by a diagnostic laboratory (Maurer 2011). A pre-enrichment (8–24 h) step is typically required to increase the number of viable cells while effectively diluting inhibitory substances present in the sample for foodborne pathogens detection by PCR at a limit of 10^2 cfu ml^{-1} of sample (Taitt et al. 2004; Myint et al. 2006; Josefsen et al. 2007). Furthermore, PCR testing requires highly trained, certified technicians and laboratories because the procedures are complex and varied and other technical issues have made neither a routine diagnostic tool. Thus, the need for a rapid, simple, accurate Q&Q, portable and cost-effective device for detection of AMR is of major interest against bioterrorism.

Phage-based diagnostics can be developed according to the need of each bacterial infection case. The principal intent is to enable the detection and identification in situ of the bacterial pathogens Q&Q within a few minutes. This information is vital if biothreats occur in heavily crowded areas such as malls, sport events, schools, airplanes, etc., in order for rapid response to control the pathogens from spreading and also to have enough time to warn the public and to prepare an appropriate response. In comparison, the massive number of phages from which one can choose represents a powerful set of tools for reporting, sensing and detecting target bacteria and offers several advantages that can operate with sufficient sensitivity and speed to complement and/or replace conventional detection methods, PCR and ELISA. Another advantage is that phages unlike PCR and ELISA, are able to differentiate between living and dead cells in assays where host infectivity is exploited. In addition, the natural ability of a phage to eclipse adsorb to, infect, and propagate within only a unique set of host cells offers another mode of capture that can be exploited to detect, monitor, and/or enumerate pathogenic bacteria. They can be produced within only a few hours using basic microbiology lab techniques and they are sufficiently robust and stable to maintain long shelf lives. Obviously phage mediated detection methods hold promise for a wide selection of novel assay formats to meet user-defined needs.

One of the first examples of reporter phage sensing bacteria using *luxAB*-based detection and phage amplification assays were reported by Ulitzur and Kuhn (1987); Jassim and co-inventors (Ress et al. 1992; Stewart et al. 1998; Denyer et al. 1998).

More recently, phage mediated detection protocols have been comprehensively investigated by many researchers using several novel diagnostic approaches such as phage typing, labeling of phage DNA, *luxAB*-based bioluminescent reporter phages, adenosine triphosphate (ATP), green fluorescent protein (GFP) reporter phages, lacZ-based colorimetric reporter phages, OmniLog™ system, reporter phages ice nucleation, phage amplification, quantum dots, thermal change, changes in conductance, phage components, phage-based limulus amoebocyte lysate assay (Jassim et al. 1996; Goodridge and Griffiths 2002; Jassim and Griffiths 2007; Jassim et al. 2010a, b; Smartt and Ripp 2011; Henry et al. 2012; Smartt et al. 2012; Schofielda et al. 2012; Klumpp and Loessner 2014; Mortari et al. 2015; Peltomaa et al. 2016; Sharp et al. 2016). Phages lytic for *Y. pestis*, *B. anthracis* and *Brucella* species are routinely used for the detection, identification and typing of the host bacteria (Filippov et al. 2013). Recently, *B. anthracis*, the causative agent of anthrax, is considered a high-priority agent that may be used in a food-related terrorist attack because it can be contracted by ingestion and it also forms spores with heat and chemical resistance. The phage display technique with a magnetoelastic wireless detection platform was developed using the affinity-based biosensor and using a phage-derived diagnostic probe as the biomolecular recognition element to capture target agents multivalently. The sensitivity of the magnetoelastic acoustic sensor was tested to be 130 Hz per order of magnitude of *B. anthracis* spores' concentration with a detection limit of 10^3 ml^{-1} (Wan et al. 2007a). A phage-based bioluminescence assay for the detection of *B. anthracis* on deliberately contaminated foods was also developed (Sharp et al. 2015). The engineered *B. anthracis* reporter phage (Wβ: *luxAB*-2) which transduces bioluminescence to infected cells was able to detect 1 cfu of *B. anthracis* spores in 8 h from pure cultures and as low as 10 cfu g^{-1} in sterile soil (Sharp et al. 2016). Filamentous phages were shown to specifically bind to *B. anthracis* spores (Brigati et al. 2004; Fu et al. 2011a). The real-time detection of down to 10^3 ml^{-1} of *B. anthracis* spores in water suspension was conducted using a biosensor based on filamentous phage as bioprobe and magnetostricitive milli/microcantilever as a sensor platform (Wan et al. 2007b; Shen et al. 2009; Fu et al. 2011a). Another spore-binding phage, 8a, was recently isolated and studied in terms of mechanisms of DNA injection (Fu et al. 2011b). Recent identification in the gamma phage lysin PlyG a domain for recognizing *B. anthracis* spores (Yang et al. 2012) can also help to improve phage-based spore detection. Despite the diversity of experimental phage diagnostics that demonstrate valid proof-of-concept and effectiveness, only a few phage-based detection assays are commercially available. The reluctance to use phages as biodetection agents can be attributed to several key perceived or factual disadvantages (Smartt and Ripp 2011). Primary among these is the concern over host resistance to phage infection. A subpopulation of phage resistant mutants within a sample can lead to false-negative results. As well, the emergence over time of phage-resistant bacterial populations can render a phage-based detection assay obsolete. In general, all new highly efficient indirect approaches with the use of genetically engineered or wild type phages have advanced diagnosis of the biothreat infections but require acceptance to be moved into practical use (Filippov et al. 2013).

Phage cocktails can be successfully used for the specific decontamination of chemically sensitive surfaces and expensive equipment (Filippov et al. 2013). Jassim et al. (1995, 2010a, b) developed an unprecedented countermeasure using a non-genetic manipulation phage biotechnology technique called phage breeding and phage design technology, respectively. The more recent phage programming technology (Chap. 1) have overcome these resistances and enhanced phages to recognize host cell receptors. This has been elaborated in greater detail in Chaps. 1–4. Indeed, the ideal diagnostic method to prevent biothreats needs detection limits of less than 10 bacterial cell per 25 gm or 25 ml of food, water, blood, feces, etc., with results within working hours. In this regard, several novel methods using smart phages to diagnose bacterial pathogens or as a bioalarm were discussed in greater detail in Chaps. 3 and 4. In general, novel rapid diagnostic biothreats and biosensor prototypes need more research to demonstrate the power of smart phage assays as marketable devices applicable to the biotechnological, food, medical, and agricultural industries and to prevent any potential biothreats.

7.5.1.2 Phage Biocontrol

The emergence of multidrug-resistant bacteria of *Brucella* and *Y. pestis* strains are serious public health concerns for biothreat infections (Kinsara et al. 1999; Memish et al. 2000; Galimand and Courvalin 2012; Kiefer et al. 2012). Other concerns are the potential for the induction of antibiotic resistance in *B. anthracis* using spores induced to transform into their vegetative form by streaking them onto a BHI agar plate and incubating for 18 h at 37 °C (Athamna et al. 2004), alongside with the potential application of genetically engineered multidrug resistant strains of biothreat bacteria in biological attacks (Gorelov et al. 1991; Pomerantsev and Staritsyn 1996; Block 1999).

The severity of these potential biothreat bacterial infections requires efficient biocontrol methods. Phages active against *Y. pestis*, *B. anthracis* and *Brucella* species have been described (Filippov et al. 2013), which can be used as biosurveillance, biodefense, biocontrol and disinfectant in animal farms, barns, feed, water, and food processing environments. Methods of using phages as biocontrol are presented in greater detail in Chaps. 3 and 4.

B. anthracis is a Gram-positive spore forming bacterium that exists in two morphologically distinct states, the spore and the vegetative cell. Anthrax commonly affects hoofed animals such as sheep, cattle, and goats. Humans who come into contact with infected animals can get sick with anthrax as well. There are three main routes of anthrax infection: cutaneous, inhalation, and gastrointestinal. Infections of humans usually happen through direct contact with infected animals or contaminated animal products. Their spores, not the vegetative cells, are the etiologic agent of the disease anthrax. Anthrax infections result from their inert spores entering a suitable host, recognizing that the environment is suitable for growth, and then transitioning to rapid growth and toxin production. It is well known that bacterial spores are not metabolically active, and they may be inactivated by physical methods (gamma irradiation, ultraviolet light, and high pressure) that are also not safe for humans (Koper et al. 2002; Walter 2003). Phages need viable vegetative forms of bacteria for their reproductive cycle, therefore using phages against the classic agents of biologic warfare *B. anthracis* spores is problematic. This form of *B. anthracis* is the most dangerous one as a potential terrorist bio-weapon. Therefore, there is a need for highly effective and safe methods that can rapidly disinfect the spores. Despite the great promise of phage-based nano-tools technology, their potential as anti-microbial therapeutics and detection tools has been limited to vegetative bacteria, since most phage are typically ineffective against bacterial spores. There are however many filamentous phages that specifically bind to *B. anthracis* spores (Brigati et al. 2004; Wan et al. 2007b; Shen et al. 2009; Fu et al. 2011a). The recent discovery of spore-binding dsDNA phage 8a, a *B. anthracis*-specific Myoviridae phage that not only destroys vegetative *B. anthracis* cells, but also at phage titre 10^8 pfu ml^{-1} can recognize and adhere to anthrax spores and eject its DNA inside the spore, suggests that phage-based therapeutics or detection/decontamination tools effective against sporulating bacteria can be developed (Fu et al. 2011b). One year after this unprecedented spore-binding phage discovery Henry et al. (2012) reported application of phages (isolated from giraffe feces from a zoo on Long Island, New York) in aerosol form to germinated spores of the *B. anthracis* Sterne strain caused effective destruction of spores, but the effect was mainly observed when high titer lysate was also applied at 10^8 pfu ml^{-1}.

Novel phages able to recognize and adhere to anthracis spores' surface and eject their DNA inside the spore would be important, especially for the disinfection of spores. More importantly the spores' anthracic phages arsenal should be resistant to harsh environmental conditions, for example, dryness, ultraviolet radiation, extreme temperatures, and bodily fluids, to maintain the ability to kill bacteria (Walter 2003) since solar UV radiation is a key factor in phage mutation (Jassim and Limoges 2013). This could result in lysogenic conversion inside the anthracic spores which will be a nightmare scenario, see also Chap. 2.

Gamma phage lysin PlyG can recognize *B. anthracis* spores (Yang et al. 2012). Interestingly, gamma phage is found to contain a functional fosfomycin resistance gene, which is evidence of lysogenic conversion and could be the cause of the spread of fosfomycin resistance among *B. anthracis* strains that is, in turn, important for survival in soil in the presence of *Streptomyces* producing this antibiotic (Schuch and Fischetti 2006). It was observed that lysogenic strains of *B. anthracis* serve as a reservoir for phages that attack certain sensitive strains of anthrax bacillus (Inal and Karunakaran 1996). As a caveat, many phages specific to *Bacillus* species mediate generalized transduction (Inal et al. 1996), which during phage therapy could result in the transfer of anthrax toxin genes (so-called phage conversion) (Inal 2003; see Chap. 2), and this occurs via temperate phage-mediated gene transfer between species (Brabban et al. 2005). However, Inal (2003) concluded this would be the major flaw with using phage CP-5131 to "destroy stockpiles, of *B. anthracis*, or even as a therapy for the disease". Moreover, the infection with temperate phages and lysogenization of *B. anthracis* can cause multiple phenotypic alterations: blockade or promotion of sporulation, induction of exopolysaccharide production and biofilm formation associated with the ability for long-term colonization of soil and the earthworm gut in a non-sporulating form (Schuch and Fischetti 2009). Consider this a 'call to action' for researchers interested in corrective action to prevent evolving anthracic spores from becoming more virulent.

7.5.2 *Lysogenic Phages as Surrogates for Bacterial Pathogens in Biothreats*

Lysogenic phages carrying virulence factors are a potential threat since they can rapidly invade different bacterial species, increase host fitness or virulence factors and cause resistance to antibiotics. Here are some ways that we can prepare against such threats:

7.5.2.1 Fight Phage with Phage

Lysogenic phages have several criteria as a potential biothreat as summarized in Table 7.2. In this regard, the most important goal is to prevent the lysogenic phage, the vector of virulence factors and AMR genes, from infecting their target bacteria. If the best defense is a good offense, we should seek to rapidly disinfect their target bacteria using highly lytic phages.

Highly lytic phages and their lysins are a promising antibacterial for the treatment of drug-resistant plague, anthrax and brucellosis (Filippov et al. 2013). Therefore, we need to develop smart highly lytic phages to target bacterial biothreats. This can be achieved by developing a library of smart programmed lytic phages for each anticipated bacterial pathogen that could be used in biothreats. The

Table 7.2 Lysogenic phage characteristics as a potential biothreat

Attribute	Remarks
Safety	Won't cause illness or infection in humans, animals or plants
Ease of hunt	Available in all environmental samples for example hospitals, sewage, river, soil, manure, human and feces etc.
Ease to culture	Able to produce with standard microbiological methods in a few hours and have reproducibility
Ease and speed of transfer virulence factors and AMR	Drives the adaptive evolution in bacterial host to achieve more powerful virulence factors and AMR inherited from previously infected bacteria
Cost	Surrogate production should not be excessive

technology to produce these smart programmed phages can be developed as discussed in greater detail in Chap. 1. These smart programmed lytic phages can rapidly kill the target bacteria before the lysogenic phages can infect them and begin their reproductive cycle. In this way the lysogenic phages will not survive long in an environment without their host cells. Using lytic phage to fight lysogenic phage is an ideal rapid, inexpensive and environmentally friendly way to control and prevent manmade lysogenic phages spread by air or water. To implement safe and rapid phage biocontrol, particularly in heavily crowded areas will require several applications and specialized apparatus for phage biocontrol such as the dry phage fog system, phage electric mist fan, phage showers, phage spray, phage-based hand rubs, phage probiotics, etc., to provide appropriate disinfectant levels (see Chaps. 3 and 4).

7.5.2.2 Fight Spores Using Phage

The nightmare scenario for homeland security is if terrorists deliberately disseminate serious biological warfare such as *B. anthracis* spores in the air, water or food. Experts believe that if such an attack occurred in a large city, hundreds of thousands of people could be at risk of the deadly disease, while the present systems of defense are insufficient. In this regard, understanding and quantifying the impact of a bioterrorist attack are essential to developing an effective response. Therefore, in this chapter we have illustrated models of using smart programmed phages to rapidly lyse and kill bacterial warfare agents when released as aerosols in major cities. In this regard, there is an urgent need for intelligent devices 'automatic biothreat alarm system' (BTA), namely an automatic detector system that can be placed in sensitive areas. The basic principle is to monitor the air 24/7/365 for any biological threats such as *B. anthracis* spores. If the spores are detected in the air even in low numbers, the device can automatically send electronic signals to alert authorities or biological inspectors. They can also be set up to automatically release phages as a biocontrol agent, to kill and inhibit the spores.

In order to develop an intelligent BTA we need phages that are able to discover both *B. anthracis* spores and vegetative cells. As previously noted, some phages able to bind and destroy *B. anthracis* spores have already been found (Brigati et al. 2004; Wan et al. 2007b; Shen et al. 2009; Fu et al. 2011a, b; Henry et al. 2012). Some phages contain specific receptors enabling them to bind to the spores resulting in their effective destruction. The odds of such productive encounters can however be increased by adding excess phages $\geq 10^8$ pfu ml^{-1} to the sample, which, including processing, we can accomplish within only a few hours. These natural anthracis spore-binding phages can be enhanced using phage programming technology (Chap. 1) to equip them with aggressive virulent properties against anthracic spores. These reprogrammed phages can be used to disinfect anthracic spores via phage foggers, phage electric mist fan, phage spray, etc., (see Chaps. 3 and 4). Furthermore, the programmed phages can also be used to develop a rapid anthracic spores automatic detector based on the same principles as the 'Phage

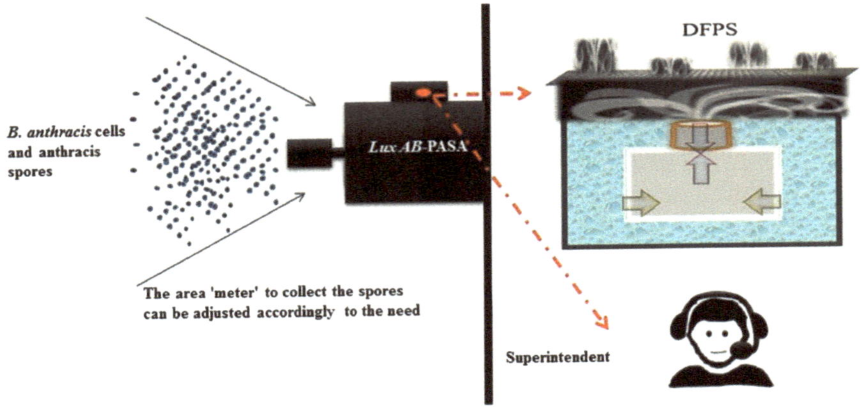

Fig. 7.2 *LuxAB* phage for anthracis vegetative, *B. anthracis* cells and anthracic spores alarm with DFPS system

Alarm and Detector' (PAD)' described in Chap. 3 and can be connected to the dry fog phage system (DFPS) described in Chap. 4 with some alterations as presented in the Fig. 7.2.

7.5.2.3 Phage-Based Spores Biosensors and Biocontrol Strategies

No matter how efficient a phage is in recognizing and infecting its host cell, the phage must by necessity first find its host cell, and this is typically a random event. Under low target bacterial densities, this can occur far too slowly for a detection assay to be practical. For example, a rough approximation predicts that it would take 1000 years for one phage to randomly find one bacterium in 1 ml of liquid (Hagens and Loessner 2010). Add in particulates from environmental samples that would non-specifically bind up free phages and the chances of productive phage-host encounters diminish even further. In order to increase the opportunities for the phage-spores encounter mechanisms, the vertical and horizontal biokinetic interaction can be applied through a programming technology (Chap. 1) to develop smart virulent lytic anthracic spores-binding phages. Once smart programmed phages are able to encounter spores swiftly and efficiently, there will be many applications and protocols to develop rapid diagnostic methods as noted above. In this chapter the *luxAB*-based phage mediated detection protocol was adopted since the *luxAB* genes are more convenient, rapid and reliable tools to detect bacteria (Ulitzur and Kuhn 1987; Denyer et al. 1998). Since the causative biothreat agents are the *B. anthracis* spores, it is likely to be one of the first-choice weapons of bioterrorists. The engineered transduces bioluminescence phage Wβ:*luxAB*-2 was able to detect as low as 1 cfu of *B. anthracis* spores in 8 h and as low as 10 cfu g^{-1} in sterile soil (Sharp et al. 2016). Therefore, the introduction of *luxAB* genes into *B.*

anthracis, using *luxAB*-based bioluminescent reporter phages would be a novel approach to specifically detect low numbers of anthracis spores (Sharp et al. 2016). In this regard, the test principle would be using the bioluminescence phage Wβ: *luxAB*-2 in the *luxAB*-phage anthracic spore alarm (*luxAB*-PASA) (Fig. 7.2) as a real-time detection device for *B. anthracis* spores in air. The technology can also be developed for use in aqueous samples using the phage filters tube technique described in Chap. 4.

The '*luxAB*-PASA' (Fig. 7.2), is an air monitoring system which can be an effective biodefense tool in anti-terrorism for *B. anthracis* spores. In theory, if the air is contaminated with *B. anthracis* spores, the *luxAB*-PASA can be designed to send electronic signals to alert authorities. In order to prevent such a biological hazard and to prevent the spores from spreading, the *luxAB*-PASA should be connected to DFPS (Fig. 7.2). The DFPS opens automatically with electronic signals received from *luxAB*-PASA (Fig. 7.2), this idea is discussed in greater detail in Chaps. 3 and 4). As soon as the *luxAB*-PASA detects the *B. anthracis* spores in the air it will instantly send two electronic signals, one to the biothreats inspectors and the other to open DFPS to automatically release fog containing billions of smart lytic phage particles to target and rapidly kill vegetative *B. anthracis* cells and their spores.

7.5.2.4 Biocontrol for Inhalational Anthrax Spores

Inhalational anthrax develops when anthrax spores enter the lungs through the airways. It is most commonly contracted when people breathe airborne anthrax spores. The bacterial spores must germinate before the actual disease occurs. This process usually takes 1–6 days. Once the spores germinate, they release several toxic substances. These substances cause internal bleeding, swelling, and tissue death. Bioterrorism-related inhalational anthrax was previously reported in the US (Jernigan et al. 2001). The inhalation route of infection leads to death (Friedlander et al. 1993). The life-threatening nature of inhalational anthrax spores requires intravenous therapy with two or more antibiotics at the very first consideration of this disease. Natural resistance of *B. anthracis* to antibiotics has also been documented and in vitro studies have shown that *B. anthracis* can develop resistance to sulfamethoxazole, trimethoprim, cefotaxime, ciprofloxacin, doxycycline and β-lactam antibiotics (Athamna et al. 2004; Weigel and Morse 2009; Habrun et al. 2011; Daniszewski 2013). The close monitoring of patients treated for anthrax is mandatory (Athamna et al. 2004). As an alternative treatment for anthrax infection (besides the obvious development of new classes of antibiotics), phage biocontrol should also be considered. With the current multidrug resistance in bacteria and the ease of preparing *B. anthracis* spores, it is likely to be one of the biothreat agents of choice. To overcome the problem of exact biocontrol for inhalation of anthrax spores, we need to first develop smart programmed phages such as the ones we described earlier and in Chap. 1. The smart programmed phages, that are able to recognize and bind on *B. anthracis* spores and vegetative cells, should be

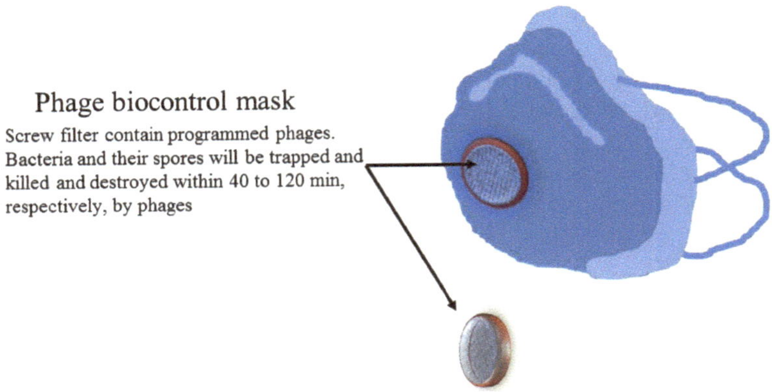

Fig. 7.3 Smart phage biocontrol mask

impregnated on 0.22 μm pore sizes of filter. The filter can be fixed to a plastic pad on a mask (Fig. 7.3). These filters can be replaced as needed depending on the spores load and bacterial contamination and also to reduce infiltration of the biothreats agents. The smart programmed phages as a living biocontrol, will amplify in the process of killing bacteria and destroying spores. This means there will be new phages to continue killing the biothreat agent as long as the host is present in air. Figure 7.3 illustrates a smart phage biocontrol mask as a potential use of smart programmed phages for bacterial lysis and to destroy the spores.

This type of phage biocontrol mask can be worn if needed for example in large religious gatherings, football matches or fairs and festivals, etc. Phage mask biocontrol can cover all possible biothreats agents by using a phage cocktail for different bacteria. This can be an appropriate biocontrol kit in particular in countries with hot and humid weather conditions where bacteria can thrive. Overcrowded and confined areas present many public health challenges. These crowded and tight quarters, combined with very hot and humid air, create a superb breeding ground for germs of all types, thus people face numerous communicable health hazards. Many countries are spending millions of dollars every year just to protect residents and tourists from communicable diseases that are introduced and shared in huge crowds for example at sporting events. A phage biocontrol mask can be used in the event of onset for example of tuberculosis or bacterial pneumonia or in the event of a biological threat to save lives and improve health care services.

7.5.2.5 Biocontrol for Cutaneous Anthrax Spores

Cutaneous anthrax occurs when anthrax spores touch a cut or scrape on the skin. It is the most common type of anthrax infection. The main risk is contact with animal hides or hair, bone products, wool, or with infected animals. People most at risk for cutaneous anthrax include farm workers, butchers, veterinarians, tanners, and wool workers. The phage deployment strategy to treat and prevent cutaneous anthrax spores is by using a mist fan device emitting tiny droplets of water containing millions of smart phage particles. If the anthrax vegetative cells or their spores are present on the human or animal skin, the phage will quickly (5–10 min) bind, then kill the target cells and destroy their spores, within 2–3 h (this is discussed in greater detail in Chaps. 1, 3 and 4). In this regard, the electric misting fan can be used to control anthrax and other bacterial infections in barns, farm environments, farm workers, contaminated animals, slaughterhouses, veterinarians, tanners, wool workers, silos, tools, equipment, feeding systems and with the community at large. This will reduce infectious anthrax spores transmitted by direct contact or if used in biothreats. This safe and natural supplement will not affect beneficial human microflora, nor harm skin or eyes, and is safe if inhaled or digested. These applications will prevent and control dangerous bacteria in crowded places.

7.6 Conclusions

The concept of phage biocontrol and therapy to treat bacterial infections was born with the discovery of the phage 100 years ago. The past several decades have proven that bioterrorism is a real threat which needs to be properly controlled. Recent developments in phage therapy confirm that it may provide a reliable countermeasure, preventing serious consequences of a terrorist attack using deadly bacteria, especially those resistant to antibiotics. Phage-mediated elimination of *B. anthracis*, seems to be an efficient tool against the potential use of such bacteria as a terrorist bioweapon.

AMR in bacterial species are mostly introduced via lysogenic phages, where they become established and expand in hospitals, on farms and in the environment (Chaps. 2–4). They undergo rapid proliferation following the colonization of new habitats, thus having a negative impact on animals and humans and representing a biological hazard. Lysogenic phages appear to play a major role in this context. Sometimes a lysogen also known as episome or prophage may provide benefits and protection to the host bacterium while they are dormant by adding new characteristics to the bacterial genome in a phenomenon called lysogenic conversion (Chap. 2). If lysogenic phages carrying virulence factors are released by terrorists into water, foods, or directly in crowded areas, these phages can rapidly invade different bacterial species. Infected bacteria will be more competitive in newly colonized habitats because the lysogenic phages have supplied them with new virulence skills to become more pathogenic and also more resistant to antibiotics or subsequent infection by virulent phages. There is now convincing evidence that the lysogenic phage weapons hypothesis can be expanded to include pathogens and virulence factors that can act as biological weapons if they infect humans, animals and even plants. Therefore, the potential for phage misuse should be considered by anti-terrorism agencies when preparing for bioterrorist attacks (Fan and Tong 2012).

The horizontally lysogenic biased transmission routes "horizontal gene transfer (HGT)" (biased toward nonrelated bacterial species) (Chap. 2) select hosts to increase virulence genes transmission. In this case lysogenic phages will help in the evolution of bacterial pathogens by enhancing transmission to encode exotoxins and other virulence factors via phage-pathogenicity island interactions. Bacterial hosts infected by lysogenic encoded exotoxins and other aggression and aggressive virulence factors can swiftly transmit diseases to humans, animals and plants. They may also spread AMR to other previously harmless bacteria and render them harmful to humans, animals and plants at large.

There is a considerable need to improve our understanding of the lysogenic epidemiology of contagious pathogens with relation to biological weapons:

1. to argue in theory that terrorists may use lysogenic phages as biological weapons, and
2. to argue that if this is likely to happen, what we should do to prevent it.

However, there are four elements that we need to implement to prevent the use of lysogenic phage via their host cells as biological weapons:

(i) control the therapeutic use as well as research and development on lysogenic phages and genetically engineered lysogenic phages in particular;
(ii) alert law enforcement and anti-terrorism agencies in preparation for such types of attack;
(iii) develop an arsenal or library of smart lytic phages that are able to rapidly storm, invade and kill bacterial pathogens including those that carry man-made lysogenic phages.
(iv) theoretically since a phage exists for any bacterial target, they are special living diagnostic agents which can be used for reporting, sensing and detecting target bacteria and offer several advantages compared to classical bacteriology, PCR and ELISA in terms of speed, detection limit, potential for automation, and cost;
(v) smart lytic phages should be available for all bacterial pathogens that are potential biothreats to prevent infection via rapid bacterial detection and biocontrol.

References

Abedon ST, Lejeune JT (2007) Why bacteriophage encode exotoxins and other virulence factors. Evol Bioinform Online 1:97–110

Alexa P, Stouracova K, Hamrik J, Rychlik I (2001) Gene typing of the colonisation factors K88 (F4) in enterotoxigenic *Escherichia coli* strains isolated from diarrhoeic piglets. Vet Med Czech 46:46–49. http://old.vri.cz/docs/vetmed/46-2-46.pdf

Andrews WH, Hammack TS (2001) Salmonella. In: Bacteriological analytical manual, 8th edn. U. S. Department of Agriculture, Washington, DC

Athamna A, Athamna M, Abu-Rashed N et al (2004) Selection of *Bacillus anthracis* isolates resistant to antibiotics. J Antimicrob Chemother 54:424–428

Barra JJ, Auroa R, Furlana M et al (2013) Bacteriophage adhering to mucus provide a non-host-derived immunity. PNAS 10(26):10771–10776. doi:10.1073/pnas.1305923110

Bielaszewska M, Mellmann A, Zhang W, Köck R et al (2011) Characterisation of the *Escherichia coli* strain associated with an outbreak of haemolytic uraemic syndrome in Germany, 2011: a microbiological study. Lancet Infect Dis 11:671–676

Binczycka-Anholcer M, Imiołek A (2011) Bioterrorism as a form of modern terrorism. Hygeia Public Health 46(3):326–333

Biscay WR, Murphy JR (1988) Bacteriophage gene products that cause human disease. In: Calendar R (ed) The bacteriophages. Plenum Press, New York, pp 683–724

Block SM (1999) Living nightmares: biological threats enabled by molecular biology. In: Drell SD, Sofaer AD, Wilson GD (eds) The new terror: facing the threat of biological and chemical weapons. Hoover Institution Press, Stanford, pp 39–75

Boyd EF (2012) Bacteriophage-encoded bacterial virulence factors and phage-pathogenicity island interactions. Adv Virus Res 82:91–118. doi:10.1016/B978-0-12-394621-8.00014-5

Boyd EF, Brüssow H (2002) Common themes among bacteriophage-encoded virulence factors and diversity among the bacteriophages involved. Trends Microbiol 10(11):521–529. doi:10.1016/S0966-842X(02)02459-9

Brabban AD, Hite E, Callaway TR (2005) Evolution of foodborne pathogens via temperate bacteriophage-mediated gene transfer. Foodborne Pathog Dis 2:287–303. doi:10.1089/fpd.2005.2.287

Brigati J, Williams DD, Sorokulova IB et al (2004) Diagnostic probes for Bacillus anthracis spores selected from a landscape phage library. Clin Chem 50(10):1899–1906

Brown SP, Le Chat L, De Paepe M, Taddei F (2006) Ecology of microbial invasions: amplification allows virus carriers to invade more rapidly when rare. Curr Biol 16:2048–2052. doi:10.1016/j.cub.2006.08.089

Brown SP, Inglis RF, Taddei F (2009a) Evolutionary ecology of microbial wars: within-host competition and (incidental) virulence. Evol Appl 2:32–39. doi:10.1111/j.1752-4571.2008.00059.x

Brown SP, West SA, Diggle SP, Griffin AS (2009b) Social evolution in micro-organisms and a Trojan horse approach to medical intervention strategies. Philos Trans R Soc B Biol Sci 364:3157–3168. doi:10.1098/rstb.2009.0055

Brüssow H, Canchaya C, Hardt WD (2004) Phages and the evolution of bacterial pathogens: from genomic rearrangements to lysogenic conversion. Microbiol Mol Biol Rev 68(3):560–602. doi:10.1128/mmbr.68.3.560-602.2004

Canchaya C, Fournous G, Chibani-Chennoufi S et al (2003) Phage as agents of lateral gene transfer. Curr Opin Microbiol 6(4):417–424

CDC (Centers for Disease Control and Prevention) (2013) Office of infectious disease antibiotic resistance threats in the United States, 2013. Apr 2013. Available at: http://www.cdc.gov/drugresistance/threat-report-2013. Accessed 28 Jan 2015

CDC (Centers for Disease Control and Prevention) (2016) Antibiotic/Antimicrobial resistance Available at: https://www.cdc.gov/drugresistance/. Accessed 14 July 2016

Casas V, Maloy S (2011) Role of bacteriophage-encoded exotoxins in the evolution of bacterial pathogens. Future Microbiol 6:1461–1473. doi:10.2217/fmb.11.124

Charles RC, Ryan ET (2011) Cholera in the 21st century. Curr Opin Infect Dis 24(5):472–477. doi:10.1097/QCO.0b013e32834a88af

CRSR-US-Congressional Research Service Report Life Expectancy in the United States. Mar 2005 (2015) Available at: http://www.cnie.org/nle/crsreports/05mar/RL32792.pdf. Accessed 5 Jan 2015

Daniszewski P (2013) *Bacillus anthracis*—as biological weapons. Int Lett Soc Humanist Sci 9:74–83

Denyer SP, Jassim SAA, Fearon PS et al (1998) Genetically engineered reporter bacteria for the detection of bacteriophage. United States Patent 5723330. http://www.patentgenius.com/patent/5723330.html

DePaula AMR, Gelli DS, Landgraf M et al (2002) Detection of Salmonella in foods using Tecra Salmonella VIA and Tecra Salmonella UNIQUE rapid immunoassays and a cultural procedure. J Food Prot 65:552–555

Dickinson JH, Kroll RG, Grant KA (1995) The direct application of the polymerase chain reaction to DNA extracted from foods. Lett Appl Microbiol 20:212–216

Dionisio F (2007) Selfish and spiteful behaviour through parasites and pathogens. Evol Ecol Res 9:1199–1210

Edlin G, Lin L, Kudrna R (1975) *Lambda lysogens* of *Escherichia coli* reproduce more rapidly than non-lysogens. Nature 255:735–737. doi:10.1038/255735a0

Edlin G, Lin L, Bitner R (1977) Reproductive fitness of P1, P2, and Mu lysogens of *Escherichia coli*. J Virol 21:560–564

Fan H, Tong Y (2012) Potential duel-use of bacteriophage related technologies in bioterrorism and biodefense. J Bioterr Biodef 3:121. doi:10.4172/2157-2526.1000121

Filippov AA, Sergueev KV, Nikolich MP (2013) Bacteriophages against biothreat bacteria: diagnostic, environmental and therapeutic applications. J Bioterr Biodef S 3:010. doi:10.4172/2157-2526.S3-010

Friedlander AM, Welkos SL, Pitt ML et al (1993) Post exposure prophylaxis against experimental inhalation anthrax. J Infect Dis 167:1239–1243

Fu L, Li S, Zhanget K et al (2011a) Detection of *Bacillus anthracis* spores using phage-immobilized magnetostrictive milli/micro cantilevers. J IEEE Sens J 11(8):1684–1691. doi:10.1109/JSEN.2010.2095002

Fu X, Walter MH, Paredes A et al (2011b) The mechanism of DNA ejection in the *Bacillus anthracis* spore-binding phage8a revealed by cryo-electron tomography. Virology 421(2):141–148

Fukuda S, Tatsumi H, Igimi S, Yamamoto S (2005) Improved bioluminescent enzyme immunoassay for the rapid detection of Salmonella in chicken meat samples. Lett Appl Microbiol 41:379–384

Galimand M, Courvalin P (2012) Plague treatment and resistance to antimicrobial agents. In: Carniel E, Hinnebusch BJ (eds) Yersinia: systems biology and control. Caister Academic Press, UK

Gama JA, Reis AM, Domingues I, Mendes-Soares H et al (2013) Temperate bacterial viruses as double-edged swords in bacterial warfare. PLoS ONE 8(3):e59043. doi:10.1371/journal.pone.0059043

Gill JJ, Hyman P (2010) Phage choice, isolation and preparation for phage therapy. Curr Pharm Biotechnol 11:2–14

Golkar Z, Bagazra O, Pace DG (2014) Bacteriophage therapy: a potential solution for the antibiotic resistance crisis. J Infect Dev Ctries 8(2):129–136

Goodridge L, Griffiths M (2002) Reporter bacteriophage assays as a means to detect foodborne pathogenic bacteria. Food Res Int 35(9):863–870

Gorelov VN, Gubina EA, Grekova NA, Skavronskaia AG (1991) The possibility of creating a vaccinal strain of *Brucella abortus* 19-BA with multiple antibiotic resistance. Zh Mikrobiol Epidemiol Immunobiol 9:2–4

Gould IM, Bal AM (2013) New antibiotic agents in the pipeline and how they can overcome microbial resistance. Virulence 4(2):185–191

Grilló MJ, De Miguel MJ, Muñoz PM et al (2006) Efficacy of several antibiotic combinations against *Brucella melitensis* Rev 1 experimental infection in BALB/c mice. J Antimicrob Chemother 58:622–626. doi:10.1093/jac/dkl289

Habrun I, Racic G Kompes et al (2011) The antimicrobial susceptibility and virulence factors of *Bacillus anthracis* strains isolated in Croatia. Vet Med 56(1):22–27

Hagens S, Loessner MJ (2010) Bacteriophage for biocontrol of foodborne pathogens: calculations and considerations. Curr Pharm Biotechnol 11(1):58–68

Hamelin K, Bruant G, El-Shaarawi A, Hill S et al (2007) Occurrence of virulence and antimicrobial resistance genes in *Escherichia coli* isolates from different aquatic ecosystems within the St. Clair River and Detroit River areas. Appl Environ Microbiol 73:477–484

Hayashi T, Makino K, Ohnishi M et al (2001) Complete genome sequence of enterohemorrhagic *Escherichia coli* O157:H7 and genomic comparison with a laboratory strain K-12. DNA Res 8:11–22

Henry M, Biswas B, Vincentetal L et al (2012) Development of a highly throughput assay for indirectly measuring phage growth using the OmniLog system. Bacteriophage 2(3):159–167

Hong Y, Berrang ME, Liu T et al (2003) Rapid detection of *Campylobacter coli*, *C. jejuni*, and *Salmonella enterica* on poultry carcasses by using PCR-enzyme-linked immunosorbent assay. Appl Environ Microbiol 69(6):3492–3499

Inal JM (2003) Phage therapy: a reappraisal of bacteriophages as antibiotics. Archivum Immunologiae et Therapiae Experimentalis 51(4):237–244

Inal JM, Karunakaran KV (1996) φ20 a temperate bacteriophage isolated from *Bacillus anthracis* exists as a plasmidial prophage. Curr Microbiol 32:171–175

Inal JM, Karunakaran KV, Jones DR (1996) *Bacillus thruingiensis* subsp. Aizawai generalized transducing phage φHD248: restriction site map and potential for fine-structure chromosomal mapping. Microbiology 142:1409–1416

Jassim SAA, Denyer SP, Stewart GSAB (1995) Selective virus culture. International Patent Application, No. WO 9523848. http://patentscope.wipo.int/search/en/WO1995023848

Jassim SAA, Abdulamir AS, Abu Bakar F (2010a) Methods for bacteriophage design. WIPO Patent Application WO2010/064044 A1 http://www.sumobrain.com/patents/wipo/Methods-bacteriophage-design/WO2010064044A1.pdf

Jassim SAA, Abdulamir AS, Abu Bakar F (2010b) Phage-based limulus amoebocyte lysate assay for rapid detection of bacteria. WO2011/098820A1. http://www.lens.org/images/patent/WO/2011098820/A1/WO_2011_098820_A1.pdf

Jassim SAA, Akoush S, Griffiths MW (1996) Rapid detection using thermal change to monitor infection by host specific bacteriophage. IUFOST meeting food associated pathogens Uppsala, Sweden, May 1996

Jassim SAA, Griffiths MW (2007) Evaluation of a rapid microbial detection method via phage lytic amplification assay coupled with Live/Dead fluorochromic stains. Lett Appl Microbiol 44 (6):673–678

Jassim SAA, Limoges RG (2013) The impact of changing environmental forces on cyanophage-host interactions in aquatic ecosystems. World J Microbiol Biotechnol 29 (10):1751–1762. doi:10.1007/s11274-013-1358-5

Jassim SAA, Limoges RG (2014) Natural solution to antibiotic resistance: bacteriophages 'The Living Drugs'. World J Microbiol Biotechnol 30(8):2153–2170

Jassim SAA, Limoges RG, El-Cheikh H (2016) Bacteriophage biocontrol in wastewater treatment. World J Microbiol Biotechnol 32(4):70. doi:10.1007/s11274-016-2028-1

Jernigan JA, Stephens DS, Ashford DA et al (2001) Bioterrorism-related inhalational anthrax: the first 10 cases reported in the United States. Emerg Infect Dis 7(6):933–944

Jończyk-Matysiak E, Kłak M, Weber-Dąbrowska B et al (2014) Possible use of bacteriophages active against *Bacillus anthracis* and other *B. cereus* group members in the face of a bioterrorism threat. Biomed Res Int 2014:735413

Joo J, Gunny M, Cases M, Hudson P, Albert R et al (2006) Bacteriophage-mediated competition in Bordetella bacteria. Proc Biol Sci 273:1843–1848. doi:10.1098/rspb.2006.3512

Josefsen MH, Krause M, Hansen F, Hoorfar J (2007) Optimization of a 12-hour TaqMan PCR-based method for detection of Salmonella bacteria in meat. Appl Environ Microbiol 73 (9):3040–3048. doi:10.1128/AEM.02823-06

Kaufmann AF, Meltzer MI, Schmid GP (1997) The economic impact of a bioterroristic attack: are prevention and postattack intervention programs justifable? Emerg Inf Dis 3:83–94

Keen EC (2012) Paradigms of pathogenesis: targeting the mobile genetic elements of disease. Front Cell Infect Microbiol 2:161. doi:10.3389/fcimb.2012.00161

Khan AS, Morse S, Lillibridge S (2000) Public-health preparedness for biological terrorism in the USA. Lancet 356(9236):1179–1182

Kiefer D, Dalantai G, Damdindorj T et al (2012) Phenotypical characterization of Mongolian Yersinia pestis strains. Vector Borne Zoonotic Dis 12:183–188

Kinsara A, Al-Mowallad A, Osoba AO (1999) Increasing resistance of Brucellae to co-trimoxazole. Antimicrob Agents Chemother 43(6):1531

Klumpp J, Loessner MJ (2014) Detection of bacteria with bioluminescent reporter bacteriophage. In: Thouand G, Marks R (eds) Bioluminescence: fundamentals and applications in biotechnology. Springer, Berlin, vol 1:144 of the series Advances in Biochemical Engineering/Biotechnology pp 155–171. doi:10.1007/978-3-662-43385-0_5. ISBN 978-3-662-43384-3

Koper OB, Klabunde JS, Marchin GL et al (2002) Nanoscale powders and formulations with biocidal activity toward spores and vegetative cells of Bacillus species, viruses, and toxins. Curr Microbiol 44(1):49–55

Krylov VN (2001) Phagotherapy in terms of bacteriophage genetics: hopes, perspectives, safety, limitations. Genetika 37:869–887

Kutter E, De Vos D, Gvasalia G et al (2010) Phage therapy in clinical practice: treatment of human infections. Curr Pharm Biotechnol 11:69–86

Lin L, Bitner R, Edlin G (1977) Increased reproductive fitness of *Escherichia coli* lambda lysogens. J Virol 21:554–559

Malorny B, Hoorfar J, Bunge C, Helmuth R (2003) Multicenter validation of the analytical accuracy of Salmonella PCR: towards an international standard. Appl Environ Microbiol 69 (1):290–296

Maurer JJ (2011) Rapid detection and limitations of molecular techniques. Annu Rev Food Sci Technol 2:259–279. doi:10.1146/annurev.food.080708.100730

Memish Z, Mah MW, Al Mahmoud S et al (2000) Brucella bacteraemia: clinical and laboratory observations in 160 patients. J Infect 40(1):59–63

Merabishvili M, Pirnay JP, Verbeken G et al (2009) Quality-controlled small-scale production of a well-defined bacteriophage cocktail for use in human clinical trials. PLoS ONE 4:e4944. doi:10.1371/journal.pone.0004944

Mokrousov I (2009) Corynebacterium diphtheriae: genome diversity, population structure and genotyping perspectives. Infect Genet Evol 9(1):1–15. doi:10.1016/j.meegid.2008.09.011

Mortari A, Adami A, Lorenzelli L (2015) An unconventional approach to impedance microbiology: detection of culture media conductivity variations due to bacteriophage generated lyses of host bacteria. Biosens Bioelectron 67:615–620

Myint MS, Johnson YJ, Tablante NL, Heckert RA (2006) The effect of pre-enrichment protocol on the sensitivity and specificity of PCR for detection of naturally contaminated Salmonella in raw poultry compared to conventional culture. Food Microbiol 23(6):599–604

O'Neill J (2014) Antimicrobial resistance: tackling a crisis for the health and wealth of nations. The Review on Antimicrobial Resistance. Wellcome Trust and the UK Government. http://amr-review.org/sites/default/files/160518_Final%20paper_with%20cover.pdf

Peltomaa R, López-Perolio I, Benito-Peña E et al (2016) Application of bacteriophages in sensor development. Anal Bioanal Chem 408(7):1805–1828. doi:10.1007/s00216-015-9087-2

Pomerantsev AP, Staritsyn NA (1996) Behavior of heterologous recombinant plasmid pCET in cells of *Bacillus anthracis*. Genetika 32:00–509

Rees CE, Rostas-Mulligan K, Park SF, Denyer SP, Stewart GSAB, Jassim SAA (1992) Methods for rapid microbial detection. PCT WO92/02633

Saunders JR, Allison H, James CE, McCarthy AJ, Sharp R (2001) Phage-mediated transfer of virulence genes. J Chem Technol Biotechnol 76(7):662–666

Saunders NA, Lee MA (2013) Real-time PCR: advanced technologies and applications. Caister Academic Press, Norfolk

Schofielda DA, Bullb CT, Rubioc I et al (2012) Development of an engineered bioluminescent reporter phage for detection of bacterial blight of crucifers. Appl Environ Microbiol 78:3592–3598. doi:10.1128/AEM.00252-12

Schuch R, Fischetti VA (2006) Detailed genomic analysis of the Wß and? Phages infecting *Bacillus anthracis*: implications for evolution of environmental fitness and antibiotic resistance. J Bacteriol 188:3037–3051

Schuch R, Fischetti VA (2009) The secret life of the anthrax agent *Bacillus anthracis*: bacteriophage-mediated ecological adaptations. PLoS ONE 4:e6532

Sengupta S, Chattopadhyay MK, Grossart HP (2013) The multifaceted roles of antibiotics and antibiotic resistance in nature. Front Microbiol 4:47

Sharp NJ, Molineux IJ, Page MA, Schofield DA (2016) Rapid detection of viable *Bacillus anthracis* spores in environmental samples by using engineered reporter phages. Appl Environ Microbiol 82(8):2380–2387. doi:10.1128/AEM.03772-15

Sharp NJ, Vandamm JP, Molineux IJ, Schofield DA (2015) Rapid detection of *Bacillus anthracis* in complex food matrices using phage-mediated bioluminescence. J Food Prot 78(5):963–968. doi:10.4315/0362-028X.JFP-14-534

Shen W, Lakshmanan RJ, Mathison LC et al (2009) Phage coated magnetoelastic micro-biosensors for real-time detection of *Bacillus anthracis* spores. Sens Actuators B Chem 137:501–506

Shute J (2013) Too much of a good thing. The telegraph. http://s.telegraph.co.uk/graphics/projects/antibiotic-resistance/

Skurnik M, Strauch E (2006) Phage therapy: facts and fiction. Int J Med Microbiol 296(1):5–14. doi:10.1016/j.ijmm.2005.09.002

Skurnik M, Pajunen M, Kiljunen S (2007) Biotechnological challenges of phage therapy. Biotechnol Lett 29:995–1003. doi:10.1007/s10529-007-9346-1

Smartt AE, Ripp S (2011) Bacteriophage reporter technology for sensing and detecting microbial targets. Anal Bioanal Chem 400:991–1007. doi:10.1007/s00216-010-4561-3

Smartt AE, Xu T, Jegier P et al (2012) Pathogen detection using engineered bacteriophages. Anal Bioanal Chem 402:3127–3146

Sousa CP (2006) The versatile strategies of *Escherichia coli* path types: a mini review. J Venom Anim Toxins incl Trop Dis 12:363–373

Spellberg B, Gilbert DN (2014) The future of antibiotics and resistance: a tribute to a career of leadership by John Bartlett. Clin Infect Dis 59(suppl 2):S71–S75

Stewart GSAB, Jassim SAA, Denyer SP et al (1998) The specific and sensitive detection of bacterial pathogens within 4 h using bacteriophage amplification. J Appl Bacteriol 84(5):777–783. doi:10.1046/j.1365-2672.1998.00408.x

Stroud C, Viswanathan K, Powell T, Bass RR (2012) Prepositioning antibiotics for anthrax. Committee on Prepositioned Medical Countermeasures for the Public Board on Health Sciences Policy, National Academy of Sciences, Washington. ISBN 978-0-309-21808-5

Taitt CR, Shubin YS, Angel R, Ligler FS (2004) Detection of *Salmonella enterica* serovar Typhimurium by using a rapid, array-based immunosensor. Appl Environ Microbiol 70 (1):152–158. doi:10.1128/AEM.70.1.152-158.2004

Thomas R (1966) Control of development in temperate bacteriophages: I. Induction of prophage genes following hetero-immune super-infection. J Mol Biol 22:79–95

Ulitzur S, Kuhn J (1987) Introduction of lux genes into bacteria, a new approach for specific determination of bacteria and their antibiotic susceptibility. In: Scholmerich J, Andreesen R, Kapp A et al (eds) Bioluminescence and chemiluminescence: new perspectives. Wiley, New York, pp 463–472

Ventola CL (2015) The antibiotic resistance crisis: part 1: causes and threats. Pharm Ther 40 (4):277–283

Verheust C, Pauwels K, Mahillon J, Helinski DR, Herman P (2010) Contained use of bacteriophages: risk assessment and biosafety recommendations. Appl Biosaf 15(1):32–44

Vojtek I, Pirzada ZA, Henriques-Normark B, Mastny M et al (2008) Lysogenic transfer of group A Streptococcus superantigen gene among streptococci. J Infect Dis 197(2):225–234

Vondruskova H, Slamova R, Trckova M et al (2010) Alternatives to antibiotic growth promoters in prevention of diarrhoea in weaned piglets: a review. Vet Med Czech 55(5):199–224

Waldor MK, Mekalanos JJ (1996) Lysogenic conversion by a filamentous phage encoding cholerae toxin. Science 272:1910–1914

Wagner PL, Waldor MK (2002) Bacteriophage control of bacterial virulence. Infect Immun 70 (8):3985–3993. doi:10.1128/IAI.70.8.3985-3993.2002

Walter MH (2003) Efficacy and durability of *Bacillus anthracis* bacteriophages used against spores. J Environ Health 66(1):9–15

Wan J, Shu H, Huang S et al (2007a) Phage-based magnetoelastic wireless biosensors for detecting *Bacillus anthracis* spores. IEEE Sens J 7(3):470–477

Wan J, Johnson ML, Guntupalli R et al (2007b) Detection of *Bacillus anthracis* spores in liquid using phage-based magnetoelastic micro-resonators. Sens Actuators B Chem 127:559–566

Weigel LM, Morse SA (2009) Implications of antibiotic resistance in potential agents of bioterrorism. In: Mayers DL (ed) Antimicrobial drug resistance. Publisher Humana Press, a part of Springer Science+Business Media, LL. Ch 90, pp 1315–1338. doi:10.1007/978-1-60327-595-8_44

Wright GD (2014) Something new: revisiting natural products in antibiotic drug discovery. Can J Microbiol 60(3):147–154

WHO (2007) WHO guidelines on tularaemia. ISBN 978 92 4 154737 6

Yang H, Wang DB, Dong Q et al (2012) Existence of separate domains in lysin PlyG for recognizing *Bacillus anthracis* spores and vegetative cells. Antimicrob Agents Chemother 56:5031–5039

Chapter 8
Discussion and General Conclusions

Abstract In this chapter we discuss the vast number of potential applications for bacteriophages (phages) in our daily lives as an effective biocontrol from farm to fork and from house to hospital or vice versa. Phages were discovered over 100 years ago and have been used ever since to treat topical infections in Eastern Bloc Countries. The time is right and the science is available to rekindle their safe use in day to day applications. We prefer to avoid the use of genetic modification in these applications as they may set off a sequence of events that have unpredictable results in the environment. Furthermore, the use of lysogenic phages could just as easily trigger bacterial adaptations and new issues with resistance to treatment as are now found rampant due to the misuse of antibiotics. This book is meant to guide scientists and to present phages with all of their advantages and disadvantages, thereby allowing readers to decide on the best uses for Nature's bacterial control mechanism.

Keywords Antibiotics · Antibiotic resistance · Animal farms · Aquatic ecosystems · *Bacillus anthracis* · Bacteriophage · Biocontrol · Biothreats · Climate change · Cyanobacteria · Cyanophage · Diagnostic · Environment · Food · Foodborne pathogens · Horizontal gene transfer · Infections · Livestock · Lysogenic phages · Phage programming technology · Phage therapy · Poultry · Solar ultraviolet radiations

Contents

S.A.A. Jassim and R.G. Limoges, *Bacteriophages: Practical Applications for Nature's Biocontrol*, DOI 10.1007/978-3-319-54051-1_8

8.1 Remarks and Recapitulation

The use of bacteriophages (phages) as an effective biocontrol, anti-biothreats and rapid bacterial detection strategy faces significant challenges for controlling bacterial diseases in an open environment that has external interactions. A number of factors must be taken into account when considering phage biocontrol for bacterial pathogens. Effective phage titer requires high phage populations to be present nearby to the bacterial pathogen at critical times, however, we also need highly virulent lytic phages to rapidly control and prevent the disease in external surroundings. Challenges include all of the biotic and abiotic factors that surround and affect phage efficacy as a biocontrol, including the emergence of phage-resistant mutants which represent a major concern. Resistance can arise due to environmental factors such inactivation by solar ultraviolet radiation (SUR) with direct sunlight reducing phage persistence on surfaces. Other environmental forces can cause the alteration or loss of the bacterial cell surface receptor or blocking the receptor by the bacterial extracellular matrix, inhibiting phage DNA penetration, production of modified restriction endonucleases degrading phage DNA or inhibition of phage intracellular development (Chap. 2; Jassim and Limoges 2013).

Mutations affecting phage receptors represent the most frequent cause of phage resistance. Therefore, in this book we have illustrated alternative designs for phage biocontrol strategies that minimize this effect. Other strategies employed include non-genetic modification of phages using programming technology, which have been developed to increase phage adherence, adsorption, binding and persistence on open environmental surfaces and in aquatic ecosystems. These techniques rely on establishing highly virulent lytic phages at attenuated titre to the target bacterium. Selecting the correct phages for disease control is critical. This requires careful monitoring of bacterial strains in the field to minimize development of bacterial strains with resistance to the deployed phages. Therefore, we have also described other strategies starting with proper phage selection as a first step to using phage programming technology. Selecting the phages based on phage adherence, adsorption, binding and persistence on open biotic and abiotic environmental surfaces and in aquatic ecosystems relies upon establishing highly virulent lytic phages at attenuated titre to the target bacterium for field application. Although phages have great potential in biological control for bacterial pathogens and disease control, there are major obstacles, which must be considered to avoid similar issues that we have encountered with antibiotics.

8.2 The Need for Smart Lytic Phages

Phages may have a lytic cycle or a lysogenic cycle and some phages are capable of carrying out both. Lytic phages are more suitable for phage biocontrol as they are able to lyse ‘break open’ the bacterial cells and destroy them after mass replication of the virion (Chap. 2). Once the cell is destroyed, the phage progeny can find new hosts to infect. In contrast, the lysogenic cycle does not result in immediate lysing of the host cell. Phages able to undergo lysogeny are known as temperate phages. Their viral genome will integrate with host DNA and replicate along with it, or may even become established as a plasmid. As the lysogenic cycle allows the host cell to continue to survive and reproduce, the phage is also replicated in all of the cell's offspring, effectively protecting the bacteria from lytic phages in the process. The phage remains dormant until host conditions deteriorate, perhaps due to depletion of nutrients (macronutrients and micronutrients), temperature, light, SUR, seasonal variations, *p*H, cation (+ve ion) concentrations, salinity, microbial community and other environmental forces (Chap. 1; Jassim and Limoges 2013) where the prophages can become active (lytic). At this point they initiate the reproductive cycle, resulting in lysis of the host cell.

The prophage may also transfer genes that enhance host virulence and resistance to the immune system via horizontal gene transfer (HGT) (Chap. 2). The development of phage resistance by bacteria is an issue facing scientists investigating phage-bacteria interaction. Phage-mediated transduction of bacterial genes likely reflects an infrequent mistake in the assembly of the phage particle, rather than a bacterial adaptation (Chap. 2). The mechanism that caused the spread of antibiotic resistance genes between bacteria occurs most often by the gene transfer process of plasmid mediated conjugation and sometimes by phage-mediated transduction via HGT (Chap. 2). This process is a natural phenomenon causing the evolution of many human and zoonotic pathogens of current concern (Chaps. 2–4 and 7). As a matter of fact, all issues of resistance must be investigated and well characterized before an effective biocontrol agent can be established and to put to use in the field. Therefore, the success of phage biocontrol to greatly reduce harmful bacteria entering the food chain at farm level requires the production of virulent phages that can survive in extreme environments, having a broad host range for the target genus, while lacking bacterial virulence genes. In this regard, Chap. 1 describes methods that avoid genetic modification of phages to control phage-host interactions and to produce highly lytic phages with no or far less phage-resistant mutants, along with broad host targeting capabilities. These programmed phages can play an important role in phage biocontrol from farm to fork!

8.3 New Wave of Super Bacteria While Antibiotics Are Failing!

Bacterial pathogens are associated with various bacterial diseases in intensive animal farming or industrial livestock production. Factory farming refers to the keeping of livestock, such as cattle, poultry (including in "battery cages") and fish at higher stocking densities than is usually the case with other forms of animal agriculture. The main products of this industry are meat, fish, shrimp, milk and eggs for human consumption in temperate, sub-tropical and tropical environments. These pathogens can account for major economic losses to agricultural production and disease control for many bacterial-incited diseases are challenging. The most challenging pathogens include but are not limited to: *Mycobacterium tuberculosis*, *Staphylococcus aureus*, *Listeria*, *Clostridium difficile*, *Clostridium perfringens*, *Bacillus anthracis* and members of the Proteobacteria (classes Gammaproteobacteria and Epsilonproteobacteria) include diversity of pathogen; for example, *Salmonella*, *Campylobacter*, *Escherichia* (for example *E. coli* O157:H7), *Vibrio cholera*, *Yersinia*, *Acinetobacter*, *Pseudomonas*, *Serratia* and *Aeromonas*. Most antibiotics are not effective against antimicrobial resistant (AMR) bacteria which cause serious diseases that reach populations of human, poultry, livestock and fish in a relatively short period of time, especially when conditions are conducive for disease development. Management strategies in hospital environments, healthcare settings and animal farms are difficult to carry out. In addition to healthcare challenges there will be huge economic losses along with the spread of zoonoses among animals and humans causing serious illnesses and death (Chaps. 3, 4 and 7).

Antibiotics have been a major component of poultry and livestock disease management, especially for diseases caused by bacteria. Unfortunately, bacterial plant pathogens have been recalcitrant to antibiotics treatments. Farm disease management approaches have relied heavily on antibiotics, in addition to antibiotic growth-promoters (AGPs) in use since the 1950s (Chap. 2). As a result of overuse, antibiotic-resistant strains became prevalent in a very short period of time (i.e., within several years), which in turn limited their efficacy for managing bacterial diseases. This has also contributed a considerable amount of pressure on the natural microbial environment, including beneficial bacteria, human and animal nutrition and immunity and resulting in the evolution of dangerous super-bacteria, resistant to most if not all antibiotics (Chaps. 2–4). This has put considerable pressure on governments around the world in particular in Europe and North America to put an end to AGPs on animal farms.

8.4 Need to Establish New Protocols to Control Bacterial Diseases

Recent advances in the development of phage therapy to treat topical infections have offered another promising approach for disease control. Nevertheless, phage control management is not yet integrated into animal farm management strategies because of the rapid occurrence of phage-resistant mutants in bacteria which have challenged the idea of using phage therapy on animal farms to replace antibiotics (Chap. 2). Although there are a few phage biocontrol products in the market that have gained FDA approval (Chaps. 2 and 3), nevertheless, the idea of using phage to improve performance on animal farms has not yet taken hold. To get phages to effectively kill bacteria we need to optimize the phages, application rates and intervals or make applications directly to animals and all the surrounding areas including the air, barns, feed, water, etc., (Chap. 3). Regardless of phage efficacy, phages alone have not provided sufficient control of bacterial diseases without proper mechanical delivery systems to enable integration on farms to effectively control bacterial pathogens without using antibiotics (Chaps. 3 and 4).

In Chap. 1 we described novel methods to enhance phage lytic properties via a phage programming technology. The phage is analogous to a biological computer, it contains intelligent nucleic acids, able to store, retrieve and exchange biological information between bacterial cells, connecting the chromosomes of all cells into a holistic continuum (series of biochemical reactions), a kind of biological internet inside the bacteria keeps evolving, over time with a unique self-contained storage capability. The phage, like a computer, allows the countless complex biochemical events that take place inside their host bacterial cell to be regulated in a coherent way. The information exchange (the phage DNA) in the bacteria by means of HGT via lysogenic phage is able to produce new virulent aggression/aggressive factors or antibiotic resistance as an example.

In the phage-bacteria interaction the phages have the ability to perfectly reprogram their own DNA within their host cells under the influence of external environmental forces, such as nutrients (macronutrients and micronutrients), temperature, light, SUR, *p*H, cation (+ve ion) concentrations, salinity, seasonal variations and microbial community (Chap. 2; Jassim and Limoges 2013). In 1989 Jassim delivered the first proof that bacterial DNA can be reprogrammed under starvation conditions. He placed a fastidious bacterium, *S. aureus* strain Wood 46 which typically need rich media containing 18 amino acids, 3 vitamins, minerals and carbohydrates to grow, into an environment that only contained 5 amino acids, as their only possible source of nutrients. Instead of dying off, these bacteria were able to reprogram their own DNA (evolve), they grew but also produced alpha-haemolysin toxin with strong hydrophobicity which helped them to resist the phagocytosis of polymorphonuclear leucocytes, intracellular killing and serum bactericidins. Additionally, Meyers (2014) has provided more evidence to shed light on reprogramming bacterial DNA in harsh environmental conditions. The latter author reported that bacterial DNA is able to cope with harsh environmental factors to synthesise the required nutrients for their survival in the absence of their essential nutrient. More recently, scientists from the Harvard medical school demonstrated that adaptive changes of the bacteria could take place within 11 days as they used a giant petri dish providing powerful evidence of how the bacterium *E. coli* is able to adapt to doses of the antibiotic trimethoprim, 1000 times higher than the dose that killed their progenitors 11 days earlier (http://hvrd.me/m0Bv3044yNZ).

Consider how this translates to our study of phages. When the phage infects a donor cell and some progeny phages accidentally incorporate a fragment of donor DNA rather than a copy of the phage chromosome, this is known as lysogenic phage transduction (Chap. 2). These phages are able to "mobilize" bacterial genes and carry them from one bacterial cell to another through the process of phage transduction by integrating phage genome with host DNA and replicating along with it, or may even become established as a plasmid, replicating independently. The lysogenic phage remains dormant until host conditions deteriorate, perhaps due to depletion of

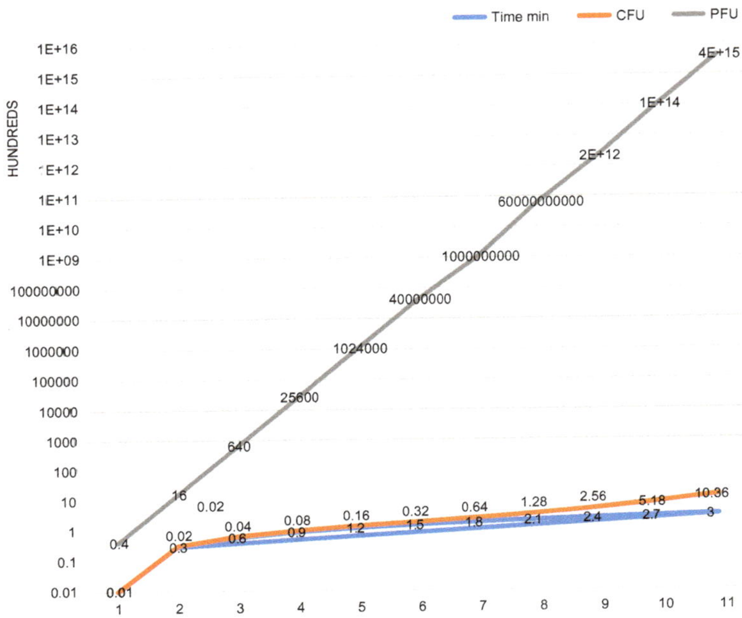

Fig. 8.1 Three hundred minutes (5 h) growth curve of *E. coli* compared to one step growth curve of *E. coli* phage. The doubling generation time for *E. coli* in a rich medium at 37 °C is 30 min. The phage burst size is 40 phage progeny released from an infected cell after 40 min infection

nutrients; then, the endogenous phages (known as prophages) become active. At this point they initiate the reproductive cycle, resulting in lysis of the host cell. As the lysogenic cycle allows the host cell to continue to survive and reproduce, the phage is reproduced in all of the cell's offspring unless and until something induces it to adopt a lytic cycle. The lambda phage of *E. coli* is a typical example of a phage known to follow the lysogenic cycle and the lytic cycle. So we could claim that the opportunistic DNA of lysogenic phages uses their bacterial cell as a vehicle to reprogram their own DNA when the host 'bacterial' DNA experiences external environmental forces (Jassim and Limoges 2013). The basic idea of phage evolution is that gene mutations take place in the live bacterial state and this can occur fairly rapidly since phage replication reaches 10^{17} particles within 5 h compared to bacterial growth in the same period of only 1036 bacterial cells (Fig. 8.1).

Therefore, phage mutations allow for an endless number of possible gene combinations taking place at the same time. Changes in the genome of a new bacterial form can create antibiotic resistance or new toxins in this process, where eventually a new species emerges. This may explain why we have rapid appearance of antibiotic resistant bacteria where 70% of bacterial strains/species have developed AMR (Chap. 7). This could be related to AGP coupled with the effects of global warming and the increase of SUR (Chaps. 2–5; Jassim and Limoges 2013).

The intelligent lysogenic phage pervasively inherited new genes from previous host cells and is also able to serve as a reporter medium between many genetic fields of different bacterial DNA, much like a biological computer. Thus we believe that the lysogenic phage evolution is to blame for the inevitable end of antibiotics therapy somewhere in the future. Thus we recommend using smart lytic (virulent) phages as the first option before the real chaos will occur when antibiotic resistance could cause 10 million deaths every year by 2050, with a cumulative economic cost of around 100 trillion USD (http://amr-review.org/).

Notwithstanding, the fact that we apparently have very limited knowledge of how to make a good lytic phage, scientists nevertheless took the liberty to start modifying the genome of phages and bacteria to produce patentable medicines and foods. This enterprise could be very hazardous to the biosphere and the very survival of life on Earth, since the restructuring of the DNA molecule could be hijacked by lysogenic phages able to destroy the wave properties of many useful bacteria such as probiotics that took nature millions of years to perfect, providing essential health benefits for humans and animals. Therefore, we must reduce the use of lysogenic phages in therapy since these phages can provide harmless bacteria with new virulence and aggression/aggressive factors, along with resistance to antibiotics which could prompt a new wave of pathogenic bacterial diseases. Smart lytic (virulent) phages can replace antibiotics as they become ineffective and can also be used for rapid bacterial diagnostics and phage pathogen detector alarms (Chaps. 3, 4 and 7) to avoid or reduce bacterial infections, which will reduce the incidence and spread of drug resistance. Nevertheless, we need international collaboration to put lytic phages in action on animal farms to reduce AMR bacteria. This action is complementary to promising first steps made by governments, and by the farm and food industries and healthcare sectors.

8.5 Smart Phage Biocontrol

8.5.1 Human and Animal Healthcare

As soon as phage was hypothesized in 1915 by the British scientist Frederick Twort, and discovered in 1917 by French-Canadian microbiologist Felix d'Herelle who named the bacteriophage, both scientists envisioned a strategy for controlling human and animal diseases. Since their discovery 100 years ago and despite the fact that these naturally occurring predators of bacteria are ubiquitous in the environment, having high host specificity and a capacity to evolve to overcome bacterial resistance, they still not have been utilized to their full potential in therapy and biocontrol for humans and animals. With the sharp increase of AMR bacteria, the world needs to understand that the antibiotic golden age has come to an end as they are currently being phased out of animal production in many countries. The inexorable rise in the incidence of antibiotic resistance in bacterial pathogens, coupled with the disappointingly low rate of emergence of new, clinically useful antibiotics, has refocused attention on the potential utility of phages in biocontrol and preventing or treating human and animal diseases.

For the past 30 years' phages were studied in animal biocontrol and agribusiness and have shown a remarkable success presenting solutions that will help to replace, curb, or promote judicious use of antibiotics on farms (Chaps. 3 and 4). Phages have been extremely effective at treating a number of bacterial infections in controlled animal studies, especially as a biocontrol agent in the prevention of food-borne illnesses, due to its target specificity, rapid bacterial killing and ability to self-replicate. Phages have the potential to treat bacterial infections afflicting animals and in particular to prevent fatal *E. coli* respiratory infections in broiler chickens (Chap. 3). Aerosol spraying and intramuscular injection have given the best results over using oral delivery of phages via direct administration or through addition to drinking water and/or feed (Chap. 3). Veterinary biocontrol applications require appropriate administration. They must target specific bacteria with a strategy that includes a comprehensive methodology, detailing the phage-host interactions, dose optimization and accounting for all chemical and physical factors on the farms. It is not an easy task to control diseases since humans interact in the community, at the hospital, on the farm, etc., (Chaps. 2–4). Therefore, a great effort and deep understanding of intrinsic phage properties is critical to designing biocontrol interventions to isolate each part of the farm and outside community (Chaps. 2–4). Accordingly, the reduction of zoonotic diseases and foodborne pathogens requires a comprehensive phage biocontrol program on the farm, where the animals are born, hatched or raised, before shipment to processing plants. Potential pre-harvest sources of foodborne pathogen contamination include breeder herds and flocks, hatcheries, contaminated feed and water, along with environmental sources and vectors, such as litter, animal caretakers, insects, wild birds and wind.

Phage therapy in healthcare is still seen as a cost or a social program rather than an economic driver and although phage therapy has been practiced for several decades in some of the former Soviet Union countries and Poland, there are still many doubts as to its ability to replace antibiotics for treatment. This is probably due to the complexity of a phage therapy management program which also conflicts with our healthcare system based on patented drugs, since we are dealing with a 'living drug', phage (Jassim and Limoges 2014; Fig. 8.2). Furthermore, from the pharmaceutical point of view, the pharmacology (pharmacokinetics and pharmacodynamics) principles of phage therapy don't easily compare to a single chemical compound as in antibiotics. Therefore, it seems the pharmacokinetics and pharmacodynamics aspects of phage therapy need considerable research in order to obtain rigorous pharmacological data concerning both lytic and lysogenic phages, including full-scale toxicological studies, before lytic phages can be used therapeutically in humans (Jassim and Limoges 2014).

Since phages are naturally occurring predators of bacteria, we believe they can play important roles to prevent and control the bacterial pathogens in the environment, as opposed to using them as therapy to treat sick humans, since prevention is always better than treatment. They can be effective antibacterial agents due to their specificity against a particular bacterial species and lack of impact on other

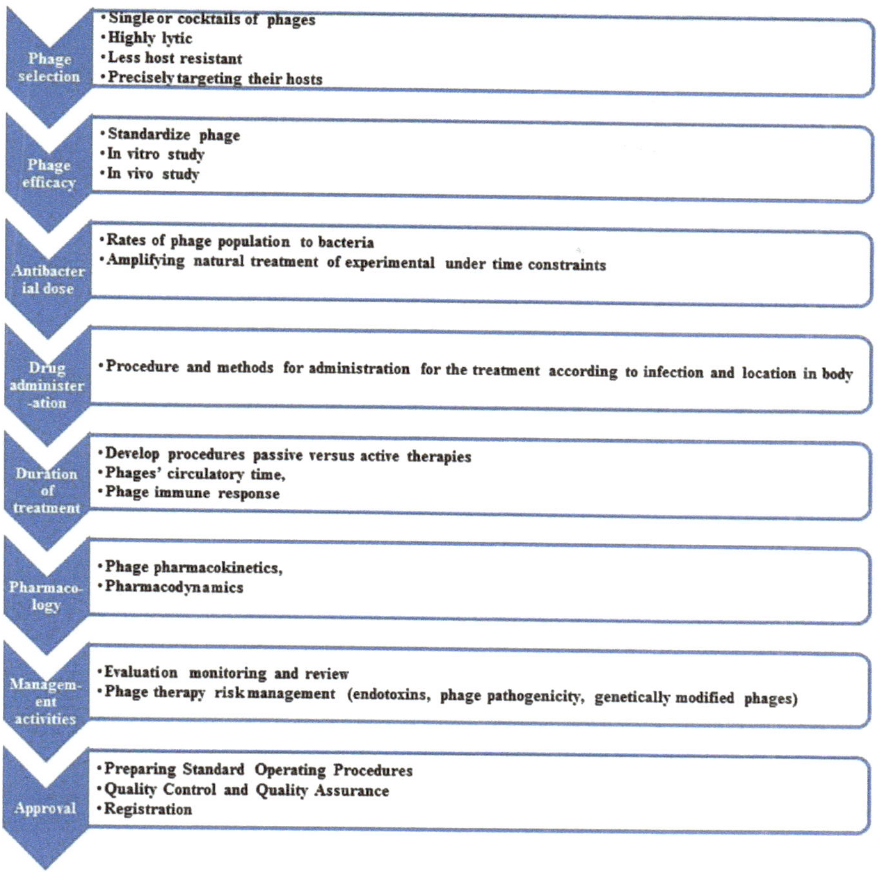

Fig. 8.2 Development of a phage therapy management program (Jassim and Limoges 2014)

microflora. Nevertheless, there are potential problems with their use as environmental biocontrol: (1) the isolation and cultivation of phages from natural sources is time consuming and problematic for producing large amounts of active inoculums, and (2) just as bacteria have become resistant to antibiotics, they may also develop resistance to phages. This problem can be avoided by using phage enhancing methods (Chap. 1) to develop smart lytic phage for use against phage-resistant bacteria mutants. Farm or hospital environments also contain many distinct parts and each part has their own conditions and require management of disinfectants. For this reason, we have presented various innovative suggestions to deliver the phages to the infected/contaminated area with effective dosing 'phage titre plaque-forming units (pfu) and to assure that the phages quickly reach the target bacteria. Phage delivery strategies such as dry phage fog system, phage mist fan, phage showers, phage spray, phage-based hand rubs, phage probiotics and phage mask (Chaps. 3, 4 and 7) can be tailored to achieve effective biocontrol programs on farms and in human healthcare settings to provide reliable control of bacterial pathogens.

8.5.2 Animal Production and Global Warming: Mitigating Greenhouse Gases (GHG)

Each year, farm animals produce billions of tons of manure (solid and liquid waste) around the globe. Storing and disposing of these immense quantities of manure can lead to significant anthropogenic emissions of greenhouse gases (GHG) such as methane (CH_4), nitrous oxide (N_2O), ammonia (NH_3) and nitrogen oxides (NO_X) which trap heat in the atmosphere and contribute to climate change and ozone depletion (Chap. 5; Jassim and Limoges 2013). Besides GHG's, farm animals' manure contributes to harmful algal blooms (CyanoHABs) (Chap. 6; Jassim and Limoges 2013), *Salmonella* spp., coliform, *E. coli*, *Campylobacter* spp., *Clostridium perfringens*, *Shigella* spp., *Staphylococcus* spp., groundwater contamination, and bad smells (Chaps. 3–5). Governments around the world need to reduce greenhouse-gas emissions along with microbial contamination in rivers, ponds, lakes and groundwater from agriculture animals to cope with climate change and to eradicate AMR, respectively. But climate change and AMR are only a few of many pressures on agriculture. Faced with growing global demand and competition for resources, the world's food production and consumption needs to be seen in a broader context, linking agriculture, energy, food security and health security. In this regard we believe that phages can solve many problems facing agricultural industries (Chaps. 2–4). An overview of a phage agriculture management system is proposed (Fig. 8.3). The system can offer great potential in agriculture for climate change mitigation by reducing the GHG, and AMR along with complementary ways to keep animals healthy by increasing probiotic bacteria in animal guts.

Phage biocontrol can play a unique role to lower agriculture GHG emissions because of its ability to 'capture' and kill the methanogenic archaea (*Methanobacterium thermoautotrophicum strain Marburg, Methanothermobacter marburgensis* and *Methanothermobacter wolfeii*) to control ruminant CH_4 emissions (Chap. 5). This is an important factor to be considered when developing agricultural GHG management strategies to reduce global warming (Fig. 8.3). It should also be possible to reduce CH_4 production by inhibiting methanogenic archaea by using virulent phages. The strategies to reduce ruminal methanogenesis by using phages to target archaea are discussed in Chap. 5 where farm field applications were elaborated in greater detail. Phage biocontrol can also be used to target manure-borne pathogens (Fig. 8.3), such as *E. coli*, *Actinomyces pyogenes*, *Bacillus anthracis*, *Bacillus thuringiensis*, *Burkholderia pseudomallei*, *Campylobacter jejuni/coli*, *Clostridium botulinum* neurotoxin (BoNT), *Coxiella burnetii*, *Enterococcus faecium*, *Francisella tularensis*, *Klebsiella pneumoniae*, *Listeria monocytogenes*, *Mycobacterium paratuberculosis*, *Pasteurella haemolytica*, *Pasteurella multocida*, *Pseudomonas*

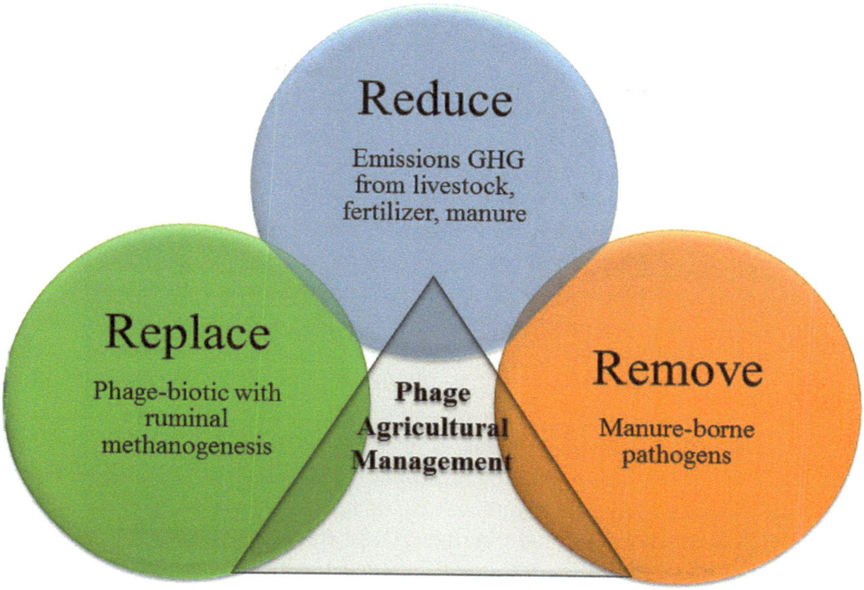

Fig. 8.3 Smart lytic phage for agricultural management

aeruginosa, *Shigella* spp., *S. aureus*, *Streptococcus bovis*, *V. cholera*, *Yersinia enterocolitica* and *Yersinia pestis* (Chap. 5), where some of these bacteria can also be used as potential agents of bioterrorism (Chap. 7).

In general, reducing GHG emissions by simply using smart biocontrol means that livestock are raised more efficiently and even healthier (Chaps. 3 and 4), thus reducing wasteful input losses like nitrogen and energy. Adoption of phage biocontrol practices will not only help to reduce GHG emissions but can also benefit the livestock with phage-biotics (Fig. 8.3), improve air and water quality while increasing the resilience of soils to climate change.

8.5.3 Cyanobacterial Harmful Algal Blooms a Global Problem—A Natural Solution!

Cyanobacteria, through their long history of evolution, have developed unique adaptive capabilities that allow them to take advantage and proliferate in many variable environmental conditions. They are the major harmful algal blooms (HABs) group in freshwater environments and are recognized as a rapidly expanding global problem that threatens human health and ecosystems. Cyanobacterial harmful algal blooms (CyanoHABs) have been documented to impact taste-and-odor problems in drinking water and aquaculture resources, animal deaths, and reduced recreational opportunities and tourism due to Cyanotoxins (Chap. 6). The economic losses worldwide associated with CyanoHABs are estimated to exceed a few billions of dollars over the next several decades (Chap. 6). Many governments are pressed to find suitable solutions to control the tide of HABs. Researchers throughout the world have taken the initiative to develop various biological methods and control strategies to treat HABs in aquatic ecosystems (Jassim and Limoges 2013), but the extent to which these can be applied to particular HABs species in specific environments or habitats is not known (Chap. 6).

Cyanophages have not received widespread attention to control Cyanobacteria due to (1) the rapid appearance of phage-resistant mutants (Jassim and Limoges 2013; Chaps. 1–3 and 6), (2) cyanophages are often host-specific and unable to infect different genetic strains of the same host species, (3) cyanophages are capable of entering into a lysogenic relationship with their hosts under the influence of external forces in aquatic ecosystems such as SUR, direct sunlight, phosphate concentrations, GHG, CO_2, ozone and virophage (Jassim and Limoges 2013; Fig. 8.4). Over the last few decades, there have been increases in GHG (Chap. 5), coupled with the depletion of the ozone layer and the corresponding increase of SUR. These factors have been overlooked for their role in cyanophage–cyanobacterial interaction in aquatic environments (Jassim and Limoges 2013). Research published over the last decade on changing environmental conditions has demonstrated how these factors

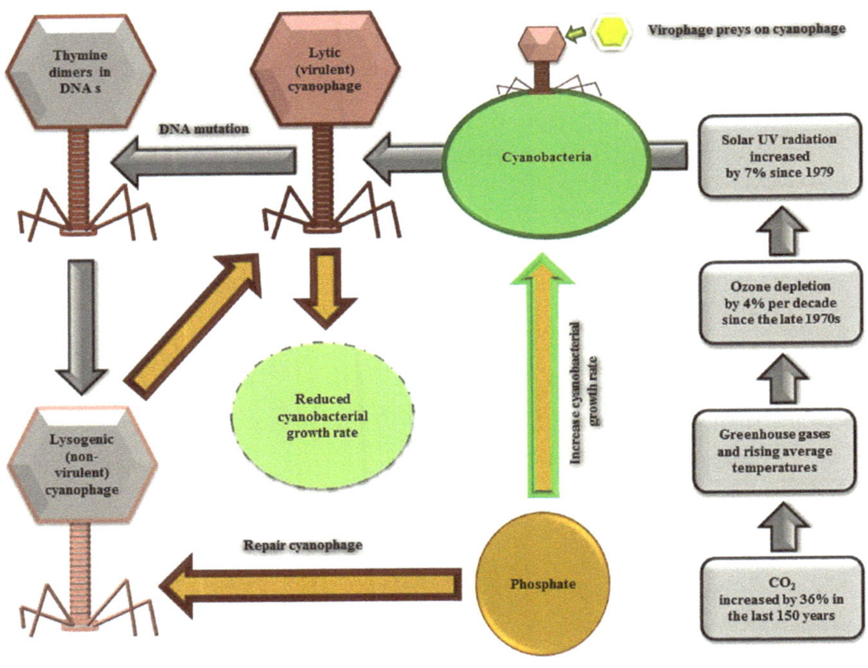

Fig. 8.4 Abiotic and biotic factors that regulate cyanophage–cyanobacteria interactions and their influence on the mechanism of cyanobacterial growth and bloom formation in aquatic ecosystems (Jassim and Limoges 2013)

have induced mutations in the highly adaptive cyanophage DNA. These mutations have led to the increase of lysogenic cyanophages coupled with a corresponding reduction of lytic cyanophages. Thus similarly evolved cyanobacteria can escape the lysis induced by cyanophages and replicating without this natural predator, they contribute to the increasing incidence of HABs worldwide (Jassim and Limoges 2013).

From Fig. 8.4 surely the effects of phosphate concentrations and other external forces, in particular the SUR, are the main factors having a direct impact on the cyanophage–host interactions in aquatic ecosystems. In order to achieve an effective biocontrol strategy in aquatic ecosystem to eradicate Cyano-HABs we need:

1. Smart lytic cyanophages able to infect different genetic strains of the same host species of Cyanobacteria which can be achieved using phage programming protocol (Chap. 1)
2. Conical cyanophage net (CCN, see Chap. 6) to provide protection against SUR and to concentrate cyanophages on site where the Cyano-HABs are present
3. Increasing phage persistence, including the use of protective formulations, application scheduling for sunlight avoidance and co-application of bacterial hosts for in vivo phage propagation

In several studies solar protectants were identified that increased biocontrol efficacy not only for phages, but also for entomopathogenic viruses and proteinaceous biopesticides (Shapiro et al. 1992; Behle et al. 1996; Ignoffo et al. 1997; Balogh et al. 2003; Arthurs et al. 2006). The formulated virus/phage with fluorescent brighteners, skim milk alone or in combination with sucrose were shown to enhance their activity under intense UV irradiation (Shapiro et al. 1992; Balogh et al. 2003; Iriarte et al. 2007).

8.5.4 *Prevention of Biothreats*

Under the current circumstances where AMR affects over 70% of pathogenic bacteria, we can't prevent people who are exposed to pathogenic enteric or aerosolized bacteria from developing infections. Prophylactic antibiotics strategies can no longer be the principle solution to treat the community, therefore, we need other strategies to help alleviate the burden on the public health system. Concerns remain about the nation's ability to respond to a big outbreak of infectious diseases such as, *V. cholera*, enterohemorrhagic *E. coli* (EHEC), methicillin-resistant *S. aureus* (MRSA), or even biothreats such as *Bacillus anthracis* spores if a large scale attack is carried out.

Ironically enough, we use multifunction smart mobile phones and we sent spaceships to Mars to discover water; yet we are unable to detect qualitatively and quantitatively (Q&Q) 1–100 pathogenic bacterial cells on small piece of a bread within a few minutes (Chaps. 3, 4 and 7). Since we are unable to detect pathogenic bacterial diseases rapidly *on site*, therefore, we are unable to curb biothreats swiftly if they occurred in heavily crowded areas (Chaps. 3, 4 and 7). We believe there is an urgent and fundamental need for a smart anti-biothreats device able to detect Q&Q bacterial biothreats and at the same time decontaminate them within 15–30 min (Chaps. 3, 4 and 6). We believe the existence of phages as the natural enemy of bacteria can be used as a smart tool against biothreat agents. This can be established by using centralized phage biocontrol distribution and dispensing stations as have been illustrated in Chaps. 3, 4 and 7. These strategies will provide more flexibility following biothreats if necessary. However, other strategies may be needed if an attack occurs in an unexpected location or uses bacterial strains resistant to the prepositioned antibiotic or manmade genetically engineered lysogenic phage encoded with aggression and aggressive pathogenic components. In such cases mobile phage stations should be prepositioned closer to the intended users.

Concerns have been raised about AMR bacteria since laboratory testing of susceptibility of a bacterial strain to antibiotics is likely to take 2–3 days (Chaps. 4 and 7; Figs. 8.5 and 8.6).

People exposed to EHEC, *V. cholera* or to anthrax spores or *Clostridium botulinum* must receive antibiotics to prevent disease. During such attacks we have a short window of time to identify the biothreat agents (at least 12–36 h if PCR are used; Chaps. 4 and 7; Fig. 8.6), secondly to find the appropriate antibiotics for treatment (using disc diffusion method 24–72 h), and thirdly to fully decontaminate the site. These processes are complicated and critical as we are fighting against time to save the victims. This is so difficult in particular if anthrax spores or *C. botulinum* are used in the attack, whereas, here 'time is life'. In this regard, we believe that using the concept of smart lytic "virulent" phage can fight against time to control biothreats agents (Fig. 8.7; see Chap. 7)

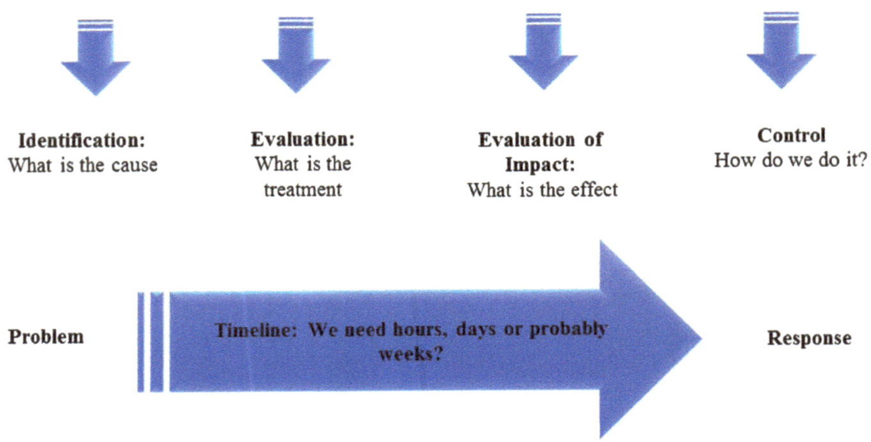

Fig. 8.5 The current public health approach to AMR bacteria

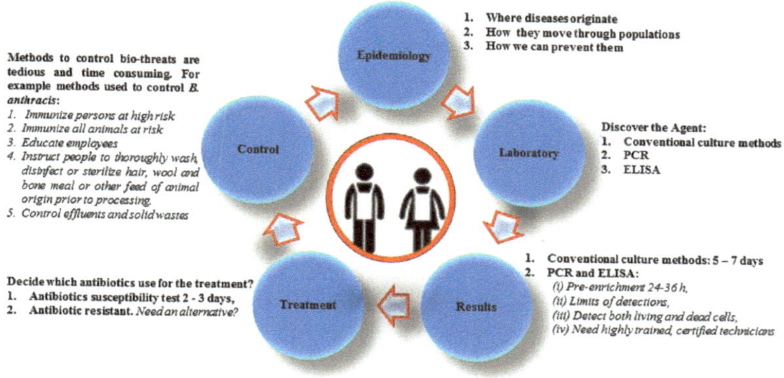

Fig. 8.6 Public health core works are lengthy and time is consumed with many tedious steps

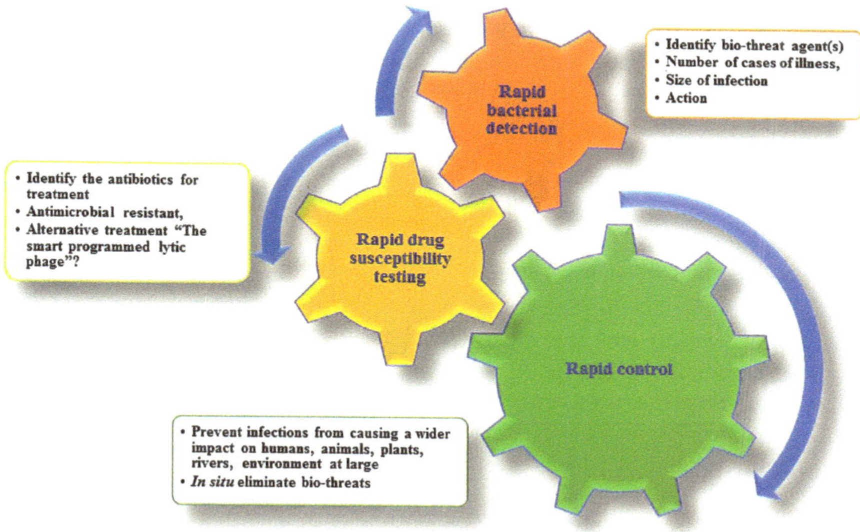

Fig. 8.7 Phage work against time to win the war

8.6 Future Considerations for Phage Biocontrol

As we stated in previous chapters, phage biocontrol requires highly lytic virulent phages with the ability to persist at high levels of killing index to the target bacterium. Furthermore, the phages must reach and attach to their hosts before environmental factors reduce phage populations below levels effective for biological control. Several considerations exist with regard to improving phage-bacterium interactions: population density and accessibility of the target bacterium; timing of phage application to optimize phage efficacy; the ability of the phage to infect and replicate in the target environment; phage concentration at the site of interaction (under i.e., sunlight, SUR, dry surfaces, *p*H, temperature, food, aqueous solutions, etc.); phage binding time, infective ratio, burst time, burst size, eclipse period, latent period, biokinetic interaction, and rates of virion degradation (phage vary in degradative properties).

Another key issue to be considered for using phage as an effective biocontrol is to improve phage efficacy through proper phage selection. The proper assay for phage selection is a critical and often overlooked factor in ensuring the success of phage biocontrol in animal agricultural, hospital environments, healthcare settings, water treatment, etc. Although in vitro assays are frequently used for selecting phages, these may not be adequate predictors of biological control efficacy. These assays provide optimal conditions for phage infections such as exponentially growing susceptible bacterial culture, controlled constant temperature, constant level of free moisture, more or less constant *p*H, the availability of a wide range of nutrients, and protection from sunlight exposure. These conditions do not represent the actual conditions in the environment, where nutrients and water are scarce and the environmental conditions such as solar SUR, dust, sand, heat, frost are crucial factors affecting phage-host interaction. Direct introduction of phage into the environment is also feasible but the phage would be exposed to solar radiation in this scenario. The stability of phages, as well as host pathogens is adversely affected by exposure to environmental factors such as solar radiation, especially the SUR portion of the spectrum (Jassim and Limoges 2013). In summer the hemisphere tends to have more SUR exposure exacerbated by ozone depletion, thus spikes in SUR-B can be expected at the surface (Jassim and Limoges 2013). These factors increase the probability of failure while using phage biocontrol under these conditions, thus we propose using encapsulated phage (Chap. 3) or auto-disseminated phages of the pathogen via contaminated or infected host (Chap. 6). These strategies will increase the persistence of phages on surfaces, ["lithosphere" (land), "hydrosphere" (water), "biosphere" (living things), phyllosphere (plant), and "atmosphere" (air)], since these are the major limiting factors in using phage for bacterial control due to SUR and direct visible sunlight (Jassim and Limoges 2013).

There is considerable need for identifying formulations that can provide superior protection of phages under SUR. This requires careful monitoring of bacterial strains in the field to minimize development of bacterial strains with resistance to the deployed phages or the proliferation of wild-type strains that are not sensitive to the phages that are used for field application. Phages have shown the potential to confer effective disease control as part of an integrated management strategy (Jones et al. 2012). Therefore, we need continuous research efforts to identify comprehensive strategies that are able to improve the phage fitness (strictly lytic which do not carry pathogenicity, antibiotic resistance genes, or transduce host DNA). In addition, we need to develop novel shields to protect the phages from SUR. In this book we attempt to narrow the knowledge gap and set a direction for constructive use of phages as a biocontrol, in diagnostics and as a potential alternative or supplement to antibiotics. These ideas are intended to be useful to many industries and we believe in the potential of phages for biocontrol and detection of human and animal pathogenic bacteria in biotic and abiotic environments.

References

Arthurs SP, Lacey LA, Behle RW (2006) Evaluation of spraydried lignin-based formulations and adjuvants as solar protectants for the granulovirus of the codling moth, Cydia pomonella (L). J Invertebr Pathol 93:88–95

Balogh B, Jones JB, Momol MT et al (2003) Improved efficacy of newly formulated bacteriophages for management of bacterial spot on tomato. Plant Dis 87:949–954

Behle RW, McGuire MR, Shasha BS (1996) Extending the residual toxicity of *Bacillus thuringiensis* with casein based formulations. J Econ Entomol 89:1399–1405

Jassim SAA (1989) Aspects of staphylococcal growth, haemolysis and phagocytosis. Ph.D. thesis. University of Loughborough, UK. ASIN: B001A1RT34

Jassim SAA, Limoges RG (2013) Impact of external forces on cyanophage–host interactions in aquatic ecosystems. World J Microbiol Biotechnol 29(10):1751–1762

Jassim SAA, Limoges RG (2014) Natural solution to antibiotic resistance: bacteriophages 'The living drugs'. World J Microbiol Biotechnol 30(8):2153–2170

Jones JB, Vallad GE, Iriarte FB et al (2012) Considerations for using bacteriophages for plant disease control. Bacteriophage 2(4):208–214

Ignoffo CM, Garcia C, Saathoff SG (1997) Sunlight stability and rain-fastness of formulations of Baculovirus heliothis. Environ Entomol 26:1470–1474

Iriarte FB, Balogh B, Momol MT et al (2007) Factors affecting survival of bacteriophage on tomato leaf surfaces. Appl Environ Microbiol 73:1704–1711

Meyers BA (2014) Pulsed electromagnetic field. The 5th element of health. Balboa Press, USA, pp 66

Shapiro M, Dougherty EM, Hamm JJ (1992) Compositions and methods for biocontrol using fluorescent brighteners. US Patent 5,124,149

Index

S.A.A. Jassim and R.G. Limoges, *Bacteriophages: Practical Applications for Nature's Biocontrol*, DOI 10.1007/978-3-319-54051-1

Kolibri 360 GmbH
Ferdinand-Jühlke-Straße 7
99095 Erfurt, Deutschland
produktsicherheit@kolibri360.de